MW00475369

Directing the ERP Implementation

A Best Practice Guide to Avoiding Program Failure Traps While Tuning System Performance

Series on Resource Management

RECENT TITLES

Directing the ERP Implementation: A Best Practice Guide to Avoiding Program Failure Traps While Tuning System Performance
by Michael W. Pelphrey
ISBN: 978-1-4822-4841-8

Building Network Capabilities in Turbulent Competitive Environments: Business Success Stories from the BRICs
by Paul Hong and YoungWon Park
ISBN: 978-1-4665-1575-8

Principles of Supply Chain Management, Second Edition
by Richard E. Crandall, William R. Crandall, and Charlie C. Chen
ISBN: 978-1-4822-1202-0

Supply Chain Risk Management: An Emerging Discipline
by Gregory L. Schlegel and Robert J. Trent
ISBN: 978-1-4822-0597-8

Supply Chain Optimization through Segmentation and Analytics
by Gerhard J. Plenert
ISBN: 978-1-4665-8476-1

Vanishing Boundaries: How Integrating Manufacturing and Services Creates Customer Value, Second Edition
by Richard E. Crandall and William R. Crandall
ISBN: 978-1-4665-0590-2

Food Safety Regulatory Compliance: Catalyst for a Lean and Sustainable Food Supply Chain
by Preston W. Blevins
ISBN: 978-1-4398-4956-9

Driving Strategy to Execution Using Lean Six Sigma: A Framework for Creating High Performance Organizations
by Gerhard Plenert and Tom Cluley
ISBN: 978-1-4398-6713-6

Building Network Capabilities in Turbulent Competitive Environments: Practices of Global Firms from Korea and Japan
by Young Won Park and Paul Hong
ISBN: 978-1-4398-5068-8

Integral Logistics Management: Operations and Supply Chain Management Within and Across Companies, Fourth Edition
by Paul Schönsleben
ISBN: 978-1-4398-7823-1

Lean Management Principles for Information Technology
by Gerhard J. Plenert
ISBN: 978-1-4200-7860-2

Directing the ERP Implementation

A Best Practice Guide to Avoiding Program Failure Traps While Tuning System Performance

Michael W. Pelphrey

CRC Press
Taylor & Francis Group
Boca Raton London New York

CRC Press is an imprint of the
Taylor & Francis Group, an **informa** business

CRC Press
Taylor & Francis Group
6000 Broken Sound Parkway NW, Suite 300
Boca Raton, FL 33487-2742

Printed on acid-free paper
Version Date: 20150126

International Standard Book Number-13: 978-1-4822-4841-8 (Hardback)

Visit the Taylor & Francis Web site at
http://www.taylorandfrancis.com

and the CRC Press Web site at
http://www.crcpress.com

This book is dedicated to Brock and Isaac, my grandsons,
that they may be inspired and led by wisdom.

Contents

SECTION III WRAP-UP

Preface

As the business climate has yo-yoed up and down over the years, we have seen the enterprise resource management (ERP) marketplace equally oscillate as well. Back in the early days of material requirements planning (MRP), computers were introduced to assist a business, automate their planning, and scheduling using a somewhat integrated model (inventory, purchase orders, work orders, and sales orders) to calculate dependent demand and explode through a bill of materials. During this volatile time, there was a tremendous industry learning curve that was impacted by such things as record accuracy, new roles and responsibilities, new tools, and a variety of other variables. The Kardex inventory ledgers and string scheduling boards were being replaced by electronic file systems. During these early days, I worked for a pharmaceutical firm that wanted to be forefront on the new computerization era. The company decided to "grow its own solution" and created a reasonably large budget to deploy its custom shop tool. The first deployment effort failed miserably. There were the typical issues of inadequate management support, inadequate requirements definition, inadequate training, record accuracy issues, and no locked stockroom, among other issues. The MRP process was laborious whereby once a week MRP was run for a single level of the bill of material, it was run over the weekend, it was flown from the Midwest (Chicago) to West Coast (Glendale) on Monday, the printout was burst on Tuesday, and the planners and schedulers began their analysis of material requirements. Analysis and transaction updates had to be submitted by Friday and the cycle began again with level 2 of the bill of material being exploded. Needless to say, most bill of materials had more than four levels; therefore, a complete top-to-bottom MRP run could not be completed within a month. The company leadership pulled the plug after flushing down over $3 million down the drain. However, they recognized the competitive need for the tool, teamed up with a large consulting firm, and did it right the second time around. Within six months, there was a functioning MRP-based solution working. I was full time on the core project team on the successful implementation, so I had invaluable knowledge of the wrong way to proceed as well as the right way, and with this knowledge, I joined the MRP revolution wave.

During these formative years, there were a plethora of software companies developing MRP solutions, a host of education and training programs was developed, numerous consulting firms jumped on the bandwagon to support MRP deployments, and a movement began. The heydays of MRP were exciting to say the least. Businesses were signing up for this revolutionary new tool and there were ardent projects undertaken with few truly successful implementations. The term MRP transitioned to an integrated solution (MRP II) and has since migrated to ERP taking into account a broad span of control of computerized tools, processes, and capability.

Jump ahead to the current day… there are a variety of ERP solutions, many very sophisticated, on the market with still tepid, if any, ERP implementation return on investment (ROI). In addition, there are frequent ERP implementation derailments and out-and-out failures. Why this poor report card? Computers are faster, software are more comprehensive, and projects are energized with budget and commitment, yet where is the ROI?

The purpose of this book is to discuss why things go wrong and to create a framework that will allow a business venturing on an ERP implementation to do it right the first time. Even if you have completed or are currently on your ERP journey, you may want to look at the best practice opportunities discussed in this book and reimplement with an eye toward creating or increasing the project ROI.

Michael W. Pelphrey
Author

Author

Michael W. Pelphrey obtained his BA in business administration from California State University, Fullerton, Fullerton, California, and his MS in finance from West Coast University, Los Angeles, CA. He has over 30 years of experience with a broad and varied background in the ERP marketplace. Not only has he been a user of ERP solutions, but he was also a director of technical and business consultants at Comserv, a leading ERP software company located at Mendota Heights, Minnesota; an ERP project manager; an ERP business architect; and a partner at BDO Seidman, an accounting and consulting practice. Functionally, he was the president/CEO of a mid-market just-in-time manufacturing firm with nine divisions spread across the United States.

Throughout his career, he has supported, in varying degrees, over 300 ERP projects for Fortune 500 companies as well as mid-market and small businesses. As an executive and business architect, he was highly successful at driving dynamic gains in revenue and profit and market share in small to Fortune 500 arenas. As an expertise in IT Program Management, he focused upon cost/schedule deliverables, customer service, and collaborative team proficiency. In addition, he was awarded IT Employee of the Year for Operational Performance (Fortune 100 Company).

He has been a frequent article contributor and speaker at the American Production and Inventory Control Society, the American Society for Quality Control, and the Society of Manufacturing Engineers. He has authored over 20 ERP implementation guides and methodologies, including the following:

- ERP Software Selection Guide
- Cycle Inventory Turnkey Guide
- Physical Inventory Turnkey Guide
- Process Performance Management Guide
- Master Scheduling Workshop
- Buyers Workshop
- Planners Workshop
- Computer Integrated Manufacturing Workshop

Michael can be reached at mpelphrey@roadrunner.com.

Acknowledgments

I acknowledge the support of my wife, Terrie, and family members, Mark, Karen, Brock, Isaac, Jordan, Meghan, Brennan and Emily Pelphrey, Bert, Donna and Karen Harrington, Julie, Marissa and Sarah Valdivia, Cliff, Nancy, Clifton and Carey Tarpy, Steve, Nancy, and Marshall Pelphrey, Steve, Rose, Tom, Suzanne, Christopher, Lindsay, Nicole, Michael, Roy, Jessica and Garrett Chimick, Christy, Andrew, Ava, Lilly and Ian McKenzie, and Dennis DeBorde.

I especially thank Bill Walker for an exceptional job of critiquing and providing insightful nuggets of advice to guide the development of this book.

I salute my professional and personal colleagues, who worked with me, encouraged me, and participated in my personal development on ERP, including Larry Barman, Ron Johnson, Preston Blevins, Adrian Messer, Herb Langthorp, Bob Lacy, Tom Davis, Dick Bourke, Bob Gilson, Bob Honaker, Harold Cook, Gunnar Sundstrom, Mel Stuckey, Tom Scheele, Warren Hinze, Mark Wurzel, Craig Wibby, Ron Marinella, Frank Wilhelm, Terry Fearn, John Petitclair, Tom Kaspar, Steve Campanelli, Randy Madison, Ellis Camp, Steve Senard, Tony Padilla, Stuart Baker, Len Scaff, John Dougherty, John Strang, Dave Waggoner, Dave Loomis, Matt Porter, Dave Porter, Edward Hawkins, Sarah Mitchell, Suzanne Musick, Marvin Bleiberg, Randy Kroeger, Dick DeShong, Guyscott Hickey, Rama Radhakrishnan, Dan Avila, Daniel Aleman, Mark Britton, Gani Shaikh, Izzy Kotton, Greg Heath, Moshe Segal, Al Loebel, Craig Birt, Craig Heilman, Brian Owen, Joe Coleman, Bob Davis, Bob Galten, Arya Farinpour, Bill Figini, Carl Coppel, Dan Baker, Denis Axtell, Dennis McCormick, Ed Chew, Dick Evans, Don Trutwin, Tom McClung, Gene Reahl, Hal and Terry Rose, Herb Anderson, Joe Tansy, John Helvie, Kelly Keller, Rudy Morales, Randy Graves, Rick Kahlbau, Sue Yant, Ron Kemper, Steve Rhorer, Kenny Roesler, Ray and Vicci Anderson, Richard Trifan, Howie Berger, Harlow Bumstad, Cameron Hand, Dave Nerran, Ray Duran, Dan Maloney, Fred Alegria, Bill Friese, Gary Forde, Jim Weigel, Bryan O'Neill, Bill Verne, Bill Greenya, Tom Clark, Dave Shea, George Monge, Larry Rosalez, Anna Jones, Dave Davis, Lester Guillory, Craig Johnson, Chad Murff, Leslie Whitney, Jim Dellinger, Justin Cunningham, Ted Smith, Larry Banks, Jack Robb, Bob Flemming, Joe Rosa, Keene Bridgeman, Ric Brown, Bob Jobe, Marty McLaughlin, Reggie Barrow, Christy Bracher, Mike Butler, Mike Rosiak, Rojae Charity, Paul Dies, Nick Derrico, Jay DeVries, Gene Dowell, Connie Egerdahl, Jay Engle, Mark Fitzgerald, John Garrison, Greg Garza, Pam Giesegh, Kirsten Gomez, Brian Greiner, Michelle Griggs, Howard Hetrick, Gwen Irby, Vincent Johnson, Shirley Juden, Pat Kessler, Denish Kumar, Candy Robinson, Marc Weston, Jack Lindsey, Melanie Zwernemann, Charles Margolin, Brian McCalip, Bill Wild, Sohil Mody, Dan Mullen, Terry Munroe, Mer Parhizkary, Fred Mengini, Dale Perry, Kenny Phillips, Bob Phillips, Ron Bernard, Megan Porter, Scott Pressey, Don Purcell, Chris Raffetto, Dennis Beeson, Sheila Tresvant, Lee Robinson, Jeff Sakata, Bahram Djahad,

Colleen Green, Lenny Schilb, Ed Shamdas, Mark Shocklee, Vince Simon, Kevin Stevens, John Chan, Linda Van Der Baan, Jeff Vibert, Vicki Baker, Christy Blackman, Steve Lemire, Lindsey Beard, Bob Jones, Scott Nickerson, Mark Bohlman, Kimberly Cook, Mark Davis, Sergio Cortez, Joe Compositor, Athena Burns, Tom Eng, Sherrill Fellows, Nan Guyse, Nicole Hanes, Beth Hilbing, Raymond Ho, Greg Howard, Lauro Jaime, Joe Lopez, Mike Maceyko, Ken Marquart, Kathy Maxwell, James McIntyre, Mark Morocco, Nate Nalven, Dan Perez, David Pessin, Adam Pierce, Tim Reese, Kenny Roberts, Dave Torchia, Stan Yip, Jim Waltz, Rodney Warner, Debora Wright Henley, Mike Testa, Paul Linder, Art Lofton, Guy Votta, Greg Wilcox, Owen Filbey, Stacy Virga, Nancy DeFranco, Greg Carmichael, Julie Irvine, Scott Elmassian, Mark Colunga, Paul Easterling, Vince Mauro, George Umlauf, Don Coates, Terry Tett, Chuck Jones, Annette Simons-Petersen, Hasib Husain, Arnold Swan, Frank Chavez, Chavelle Ward, Ron Aday, Joyce Lau, Chris Borden, Barry Paul, George Earle, Brandon Wright, Mike Reed, Sourajit Guha, John Irvin, Dillon Simmons, Fred Barnett, Linda Burdett, Danny Ortega, Larry Rivas, Roger Smith, JP Batache, Martin Petersen, Al Perez, Craig Ashley, Celeste Anton, Roger Bowen, Sam Badwan, Charles Budde, Lisa Bob, Scott Boehme, Dan Bucowski, Ron Baez, Richard Benevidez, Bernie Macam, Trish McHugh, Ken Knapp, Bill Knaffi, Michelle Varney, Dean Westcott, Chris Wegner, Bill Stone, Jack Tobin, Ryan Wollner, Greg Villaroel, Mike DeRosa, Jack Cooley, Mike Garrett, Al Nowak, Ray Happe, Iris Watanabe, Mike Seley, Mike Santy, Jennifer Holbrook, Domas Vailokaitis, Yoli Strickland, Jeff Reed, Tim Krantz, Dennis Hedding, Jerry Baird, Jerry Bowman, Manny Gochico, Dick Kust, Ken Niggemyer, Brad Dixon, Jeff Dingwall, Don Trojan, Dave Auxier, Gene Averill, Gene Bailey, John Ball, Brian Bastow, Mike McLane, Don Werter, Charles Bennett, Joe Cardoni, Karen Ladika, Gus Tepper, David McCament, Cyndi Flaugher, Dick Kiernan, Jim Shane, Tom Clippard, Margaret Poorman, Karen Ladika, Jeff Caldwell Admad Jafari, Wilson Crider, Mike Postle, Chris Lord, Doug Meyer, Mike Gibbons, Jack Knudsen, Scott Myhre, Phil Philbin, Carey Butler, Alan Rosenbaum, Michael Souder, Aphrodite Caserta, Paul Jonas, Shawn Keller, Mike Benway, Joey Bietdashto, Ed Birch, Ray Blinde, Steve Blyth, Jim Bonnel, Lois Brandt, Jim Breslauer, Bill Brezel, Burt Caldwell, Carlo Campo, John Carlson, Mike Buchner, Ray McMinn, Ray Chavira, Vic Childs, John Choate, Don Church, Roy Cirunay, Jimmy Zepeda, Greg Clark, Virginia Colwell, Joe McGrath, Terry LaRock, Jim Stromsness, Karen Jacoby, Ed Leckliter, Bob Krist and Fran Smith and Jim Varner, Tom Wardrup, Rich Nakamoto, Tom Owens, John Power, Bill Staley, Bruce Swenson, Deborah Taylor, Jesse Connor, John Barch, Rick Robinson, John Sage, Doug Cook, Ed Cusack, Lee Dapper, Brian DeHart, Kurt Fulle, Larry Dossett, Matt Feider, Greg Fordham, Bob Galten, Carlos Granda, Jim Greathouse, Vince Guess, Jerry Hancock, Warren Hanke, Mike Haupt, Glyn Homer, Doug Howardell, Chuck Drake, Tony Kuczynski, Barry Rowland, Vic Janulaitis, Kelly Jones, Jim Kensock, Court Koep, Lionel Laurin, Al Arnold, Steve McGrath, Buck McNichols, George Miller, Dan Murphy, Scott Myhre, Nelson Lee, Earl Newsome, Gene Oster, Dan Pavelich, John Pepper, John Proud, Frank Scavo, Eric Schubert, Ray Spears, Dan Steele, Tom Stevenson, Dane Sullivan, Bill Swan, Joe Trino, Terry Tuttle, Suzanne Wade, Lynn White, Bruce Wilson, Mavis Winter, and Steve Wood.

I also acknowledge Indumathi S., project management executive at Lumina Datamatics Ltd., whose team did an outstanding job in editing and improving the quality of the book.

Annotated Table of Contents

Section I—This section discusses the best practice environment and sets the stage for success. This includes engineering a plan, generating of requirements, and obtaining a results-oriented commitment.

- *Chapter 1*—Creating a project plan is realistically consistent with the company culture. The proper engineering of a planning roadmap depicts high-level deliverables that may be further decomposed into weekly deliverables, which then help the project team members achieve their expected statement of work (SoW).
- *Chapter 2*—The requirements generation process defines the functionality of desired results as well as the engineered process changes essential for an order of magnitude of operational performance improvement. The attributes of requirements definition include categories such as mission critical, essential, and nice to have, which then establish the baseline for a traceability matrix that flows through the project phases, including design, prototyping, customization, testing, piloting, and delivery.
- *Chapter 3*—It discusses the commitment of senior management and the company movers and shakers to expected results ... a two- or three-day off-site meeting where the "blood mobile" draws "pints of blood" in the form of committed expected system results. This is a process where specific systems results are cataloged, including such areas as inventory reduction, customer service improvements, product cost reductions, and yield improvement, and a variety of other expected results are committed to and reasonable time frames are defined. A rough scorecard is developed so that the rules of engagement (what is in-scope and what is out-of-scope) are clearly understood, and tracking may begin at the onset of the project.

Section II—This section addresses the practical deployment framework essential for success and includes a variety of tools that position the company for success.

- *Chapter 4*—This is "the minimum acceptable quality level for transactions, job functions, work processes, and ultimately the resulting information." Without a standard there is confusion regarding work expectations. The Information Workmanship Standard (IWS), such as financial, quality, and a variety of other standards, clearly defines the expectations associated with information. In addition, it fully develops the nested internal customer and supplier of a service framework to define process-based performance metrics that reflect in the trenches end-to-end business process activities. Engineering "process-based" metrics allows players from the entire organization to understand their specific contribution to profitability, which is lacking in traditional hierarchical metrics.

- *Chapter 5*—Test driving the blending of functionality reflected in the design (features, functions, and capabilities) as well as processes (how the design is configured) and the environmental structure (policies, procedures, and performance metrics, within the players' culture). This chapter not only pursues project team piloting but also demonstrates senior leadership piloting, customer/supplier piloting, and other business partnership piloting.
- *Chapter 6*—The backbone to system success involves the entire user community exhibiting the competence and mastery of the new system. Contrary to popular practice, this does not occur through osmosis or attending a couple of training classes. Like all competency processes, this must be achieved through proper design and fulfillment.

Section III—Ensuring the project hits on all cylinders requires proactive involvement by the project core team, executive sponsors and stakeholders as well as the working level systems champions (functional experts of the current system).

- *Chapter 7*—Ensuring the fulfillment of project tasks and commitments, reporting the status, and invoking Steering Committee guidance and day-to-day issue/decision management are the tried and true practices of good project management. However, merely doing these things is inadequate for a best practice deployment. This chapter discusses these essential tasks but also exploits the practices that transform a good project into a best practice project. Things such as behind the scenes salesmanship, removing risk barriers, executive ownership process practices, monitoring rules of engagement, and other differentiating elements are discussed.
- *Chapter 8*—This chapter peels back the layers of opportunity and explores realigning measurements on an end-to-end process basis, which allows the entire organization to understand the importance of every job function and gives the working-level tier of the organization the ability to measure their individual contribution to profits.
- *Chapter 9*—Experience has shown that many ERP projects just are not successful. This chapter addresses how to convert potential failure attributes into critical success factors. It explores such topics as follows:
  - GO/NO GO voting decision—looking at the technical review and recommendations, functional review and recommendations, open issues, cutover plan, transition to production strategy, and other criteria for successful cutover
  - How to tell when the project is going off the rails
  - How to decide and prioritize what aspects of the system need tuning
- *Chapter 10*—Keeping sanity yet achieving exponential results on time and on budget.

Executive Summary

Too many ERP implementations fail. This does not have to be the case. There is a plethora of publications that discuss approaches to successful ERP implementation efforts, yet facts prevail that the significant investment in ERP has yielded small, if any, ROI. And many projects outright fail.

There are a variety of reasons that companies miss the mark, not the least of which include the following:

- Unrealistic expectations
- Lack of clarity of vision associated with ERP goals and objectives
- Lack of addressing process changes essential for project success
- Inadequate key performance indicators
- Lack of up-front commitment by senior management to detail specific expected results, for example, don't just say reduce inventory, but rather commit to a 40% reduction in inventory
- Lack of strict adherence to a best practice implementation methodology
- Inadequate system requirements definition
- Inadequate project system engineering

The purpose of this book is to provide a proven roadmap that gives the ERP implementation a best practice process to improve the odds of system implementation success.

This book discusses essential planning ingredients that are frequently omitted from the ERP implementation start-up. Without a solid planning framework, and meaningful and rigorous expected results, the project monitoring and execution process tends to result in flaccid results. This includes the need to engineer comprehensive requirements, which may be chained through all phases of the ERP implementation steps. Early on, end-point system expected results need to be clearly defined and results may be monitored throughout the implementation.

Once an effective planning framework is engineered, the book will elaborate proven foundational methods and principles that position the company for a successful implementation. This is like tree roots building a structure, which not only supports tree growth but allows for factors such as winds, floods, and other environmental impacts to maintain structural integrity of the tree. An ERP implementation must include similar elements to help ensure structural integrity of the entities, attributes, and relationships essential for sound business practices. These principles and practices include the IWS, comprehensive prototyping through multilevel CRP flights, and education, training, and deployment of a best practice implementation framework.

The capstone, to the frameworks discussed to this point, focuses upon project monitoring and ultimately realizes the project and business process performance that yields significant ROI.

Every ERP implementation project experiences multidimensional drivers and dynamics, which influence success or failure. Few companies are adept at weaving an alignment of resources and engineered framework into their ERP project plan, their deliverables, their education and training deployment, and other essential process changes, while balancing their culture and persistence to achieve success.

Many of the best practices documented in this ERP implementation roadmap are applicable as best business practices transferable outside ERP projects. For example, the Section 1.2 on SoW may be used in the product engineering process to assist design engineers achieve their design deliverables on time and within budget.

Section I

This initial phase relies upon the client engineered processes that set the stage for success and includes the following:

- Creating a project plan that clearly describes the ERP implementation process that will be followed
- Delineating the expected results pursued
- A roadmap exhibiting goals, objectives, and short-interval schedules that lead to their attainment and including commitments from the operational resources for such deliverables as exceptional delivery performance, inventory optimization, and significant cost reduction
- A creative approach to generating the requirements essential for ROI attainment. This process spans the traditional bulleted features, functions, and capabilities pursued, as well as generating narrative statements for the Pareto class A functionality expected. Breaking down bullets into narratives tends to elicit the formulas and logic and complex methods that are essential in a robust design standard. Other by-products of a comprehensive requirements generation include the following:
  - A framework for a comprehensive traceability matrix that may be tracked through all the implementation phases—design through deployment
  - The ability to highlight end-point triggers such as activity-based costing drivers and other elements that are central in experiencing exceptional ROI

Chapter 1

- Planning roadmap of deliverables
- SoW—Managing expectations through project life cycle
- Managing change

Chapter 2

- The art of generating requirements
- Categorization of requirements
- Requirements generation life cycle
- Traceability matrix—integrity between project phases: design, prototyping, customization, testing, piloting, and delivery

Chapter 3

- Two- to three-day off-site session to review the significance of the ERP investment
- Rules of engagement agreement
- High-level review of requirements
- Aligning requirements traceability to committed expected results and assigning accountability and timetable for achieving results
- Accepting resignation of any resource refusing to agree to commitment
- Agreeing upon measurement scorecard

Section II

This central phase relies upon a good foundation that sets the stage for success and includes the following:

- Ensuring that an IWS is clearly defined within the organization and providing little doubt as to leadership's information quality expectations.
- Delineating performance expectations on an end-to-end process basis.
- Creating a successful roadmap for defining performance goals, objectives, and accountability.
- Providing a prototype environment for testing
  - Software
  - Data
  - Processes
  - User competence
  - Policies
  - Procedures
- Fostering a healthy environment for clearly understanding what ERP is about, its importance to the business, and the need for change, while creating a roadmap that will enable the user community to become functional experts using the new tools, processes, and business acumen.
- Engineering a best practice implementation framework and roadmap, which enables the project team and stakeholder community to define critical success factors, audit project performance, and realize nested improved business performance simultaneous with achieving project completion results.

Chapter 4

- Definition of an IWS
- Criteria for an IWS
- Performance measurements for transactions, documents, and files
- Data accuracy
- End-to-end process
- Performance goals and objectives
- Performance accountability
- Managing performance expectations
- Certification
- Systems champions

Chapter 5

- Definition of a conference room pilot
- Structuring the conference room pilot
- Deliverables resulting from an effective conference room pilot
- General points of awareness

Chapter 6

- Structuring an education and training program
- Implementation framework

Structuring an Education and Training Program

- Perspective
- Commitment
- Setting goals
- Achieving quality education
- Planning for new ERP system education
- Achieving overall efficiency
- Instructional methods
- Ongoing education

Implementation Framework

- Overview
- Project planning and control
- Education plan
- Implementation audits
- Implementation success factors

Implementation Plan

- Define and establish project implementation team
- Define and establish the project Steering Committee
- Prepare and execute project team education plan
- Develop project objectives
- Develop project milestones
- Develop and receive approval for the project charter
- Communication plan (Comm)
- Risk management plan (RiskMgmt)
- Project health (ProjHealth)
- Present executive seminar
- Define existing in-house systems
- Install software
- Develop detailed education and training plan (TrainingPln)
- Conduct module definition review for phase I modules

- Develop detailed implementation plan
- Execute detailed education and training plan (TrainingPln)
- Develop functional specification and test interfaces
- Develop functional specification and test conversion programs
- Plan and execute conference room pilot
- Develop test plans (testing)
- Environment and performance testing
- Develop final production system definition
- Develop user manual (UserDoc)
- Develop and execute production pilot
- Review results of the production pilot
- Conduct it post production pilot audit
- Develop and execute production conversion plan
- Conduct postimplementation audit

Section III

This final phase relies upon a good foundation that sets the stage for success and includes the following:

- Ensuring the fulfillment of project tasks and commitments, reporting the status, invoking Steering Committee guidance and day-to-day issue/decision management are the tried and true practices of good project management; delineating performance expectations on an end-to-end process basis.
- Delineating the practices that transform a good project into a best practice project.
- Fostering a healthy project environment by infusing behind the scenes salesmanship, removing risk barriers, executive ownership process practices, and monitoring rules of engagement.
- Creating a roadmap that peels back the layers of opportunity and explores realigning measurements on an end-to-end process basis, which allows the entire organization to understand the importance of every job function and gives the working-level tier of the organization the ability to measure their individual contribution to profits.
- Fostering accountability at the data level, the internal nested customer and service provider level, and at the end-to-end process level.
- Providing a best practice framework for optimizing organizational performance.
- Creating a roadmap that converts potential failure attributes into critical success factors.
- Discussing the deployment of ERP for the first time and the ways to avoid derailment.
- Providing real-life examples of ERP projects that experience snags, traps, and black holes with remedial prescription on how to avoid them.

Chapter 7

- Visionary
- Innovative
- Flexible
- Ingenuity
- Agile

- Exceptional throughput
- Nimble
- Project plan
- Documenting AS IS
- Project schedule

Chapter 8

- Definition of process performance management (PPM)
- Criteria for PPM
- Job functions require a PPM
- Data accountability
- Data accuracy
- Performance goals and objectives
- Performance measurements for optimal "in-the-trenches" results
- Performance accountability
- Managing performance expectations
- Process performance measurement
- Retooling information resource management
- Organizational perspective
  - Parochial performance objectives
  - A better perspective—Value streaming
  - Refining, streamlining, and reducing cycle time
- The "vision" of the business process
  - Dreaming a bit
  - Future orientation
  - Developing action plans that deliver effective and orderly change
- Prioritizing
- The emerging natural work teams
- Natural work teams
- Process-based performance measurements
- Performance-based compensation
- Committing to the journey

Chapter 9

- Software
- Hardware
- Business process
- The big package deal
- Lack of ownership
- Cross functional, matrix management, and stalls
- Managing third-party relations and SoW
- Portfolio management
- Project management
- ERP performance management
- Timely decision management

- Data accuracy
- Resource commitment breaches
- ERP for the first time
- Miscellaneous
  - Critical success factors while approaching GO LIVE
  - Stabilization
  - System tuning
  - ROI tracking

PLANNING AND PREPARING FOR ENTERPRISE RESOURCE PLANNING SUCCESS

I

Overview

This section discusses the best practice environment and sets the stage for success. This includes engineering a plan, generating requirements, and obtaining a results-oriented commitment.

- *Chapter 1*—Creating a project plan is realistically consistent with the company culture. The proper engineering of a planning roadmap depicts high-level deliverables that may be further decomposed into weekly deliverables, which then help the project team members achieve their expected statement of work.
- *Chapter 2*—The requirements generation process defines the functionality of desired results as well as the engineered process changes essential for an order of magnitude of operational performance improvement. The attributes of requirements definition include categories such as mission critical, essential, and nice to have, which then establish the baseline for a traceability matrix that flows through the project phases including design, prototyping, customization, testing, piloting, and delivery.
- *Chapter 3*—It discusses the commitment of senior management and the company movers and shakers to expected results … a two- or three-day off-site meeting where the "blood mobile" draws "pints of blood" in the form of committed expected system results. This is a process where specific systems results are cataloged including such areas as inventory reduction, customer service improvements, product cost reductions, and yield improvement, and a variety of other expected results are committed to and reasonable time frames are defined. A rough scorecard is developed so that the rules of engagement (what is in-scope and what is out-of-scope) are clearly understood and tracking may begin at the onset of the project.

Chapter 1

Creating a Project Plan

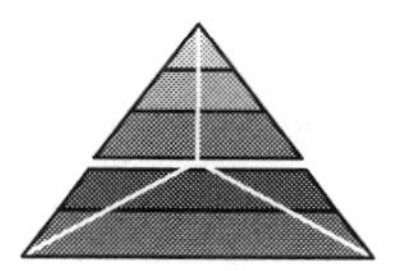

The importance of project planning cannot be overemphasized. Without adequate planning, projects tend to sway and wobble, and frequently get deflected and go down unwanted rabbit trails.

I recall one project; I was called on to turnaround, which had derailed, whereby the statement of work (SoW) was not clear to a variety of support resources.

Project background: This commercial off-the-shelf (COTS) deployment was categorized as one of the top three "mission critical" projects in the project portfolio. Because it was a COTS solution to reduce cost, management leadership chose to assign a junior project manager. This individual was competent in COTS deployments that required no system modifications, but was inexperienced in managing software "mods" and had limited skills in resource management. Management leadership misread the importance of the modification process to the success of this project.

Essentially, the core resources were a bit clued in as to their deliverables, but the support resources had no idea that they had deliverables due at a specific time. The deliverable results are as follows:

- Software engineering (SE)—The SE "core resource" thought that another group, systems engineering (SyE), had the lead and was responsible for the primary design document and that their role was merely a "consultive" role.
- SyE—The SyE resource had no idea that they were responsible for any deliverable on the project and committed no resource to the design deliverable.
- Leadership did not recognize that the software modification process was a critical success factor (CSF) on this project.

- Both the schedule and the cost became delinquent by over a three-month variance.
- There was minimal slack time and reserve built into the scheduled deliverables, primarily because it was a COTS deployment.

Project recovery: Due to subsequent related projects' impact, this three-month schedule delay forced a mitigation strategy to recover two of three months of delay. The schedule recovery involved enlisting outsourced resources (at over triple the cost of in-house resources). These outsourced resources functioned a "second shift." By employing an overlapping (concurrent processing) recovery method and adding the "second shift" outsourced resources, the project mitigation strategy only missed the original schedule by 2 weeks. However, the cost variance was over $300,000, which was nonrecoverable.

Lessons learned: The derailing of the project schedule postmortem concluded the following:

- Leadership needed to triage the scope of modifications and evaluate the critical skill set requirements before assigning the project manager, regardless of COTS deployment.
- The design toll gate review was inadequate on the modification aspect of COTS deployment SoW.
- The modification aspect of the SoW needed earlier and more focused attention.

The challenge is to create a project plan that is realistically consistent with the company culture. The proper engineering of a planning roadmap depicts high-level deliverables that may be further decomposed into weekly deliverables, which then help the project team members achieve their expected SoW:

- Planning roadmap of deliverables
- SoW—Managing expectations through project life cycle
- Managing change
- Risk management

1.1 Planning Roadmap of Deliverables

The planning roadmap may be represented in the following high-level visual aid (Figure 1.1):

- The technical environment
- The business operating environment
- The user community
- The project environment

Inasmuch as an enterprise resource planning (ERP) implementation tends to permeate virtually the entire company, the project plan must address the impacts accordingly. Let us look at the four elements just listed in some detail:

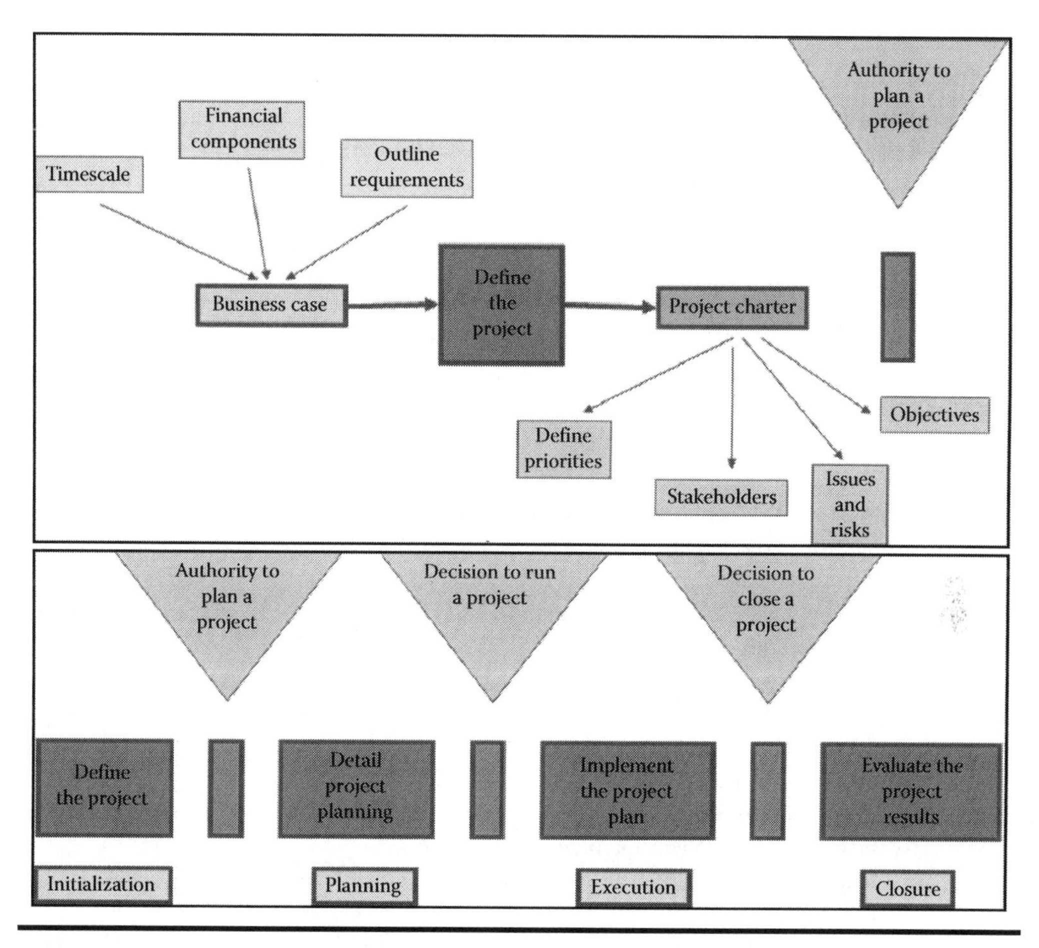

Figure 1.1 High-level planning life cycle.

1. The **technical environment** impacts will depend on the degree of change and budget defined by leadership. At a minimum, it will include ensuring that the following ingredients are assessed:
   a. Is there any hardware, network, telecommunications, web, or other infrastructure impacts? A proper technical plan must assure that a business continuity strategy is well documented including such issues as whether the environment will be fault tolerant or not. This might also include updating the architecture diagram.
   b. The user service-level agreement (SLA) includes system user response time, number of users supported, geographical span, whether it is a 24/7 or other support span, and break/fix commitments. The SLA must also define system maintenance windows (for hardware/software/network updates).
   c. Technical roles and responsibilities may need to be documented, depending on the size of the technical staff. This artifact defines the various technical job functions by who is accountable (A), responsible (R), consulted (C), or informed (I)—(also called a RACI chart). In smaller companies, this may not be necessary, or it is combined as an overall

company-wide document. A similar artifact, for larger companies, is a collaboration checklist, whereby tasks and functionality intersections are documented to diagram upstream and downstream interactions and accountability.

Responsible (R): The doer—the individual who actually completes the task

Accountable (A): The buck stops here—the individual who is answerable for the activity or action

Consult (C): In the loop—small and medium-sized enterprises, the individual who needs to be consulted prior to a final decision (a two-way communication)

Inform (I): Keep in the picture—the individual who needs to be informed after a decision or action is taken (a one-way communication)

d. Are hardware, software, or other assets going to be retired at the successful implementation of the ERP project? Documenting an asset retirement task at the onset of the project helps ensure cleanup of clutter later and nipping the "lingering old system" in the bud.
e. Agreement on the number of databases that will be required to support the production, test, and training activities. This number of instances may also be influenced by the author of the ERP package selected for implementation. Hand in hand with the number of database instances is the question of how frequently to back up each database instance. Again, the ERP package author may influence this as well as the budget for storage space and other factors.
f. Readiness toll gates may be needed, if the company wants to be especially risk averse and keep leadership and stakeholders engaged. This may be a stand-alone technical environment approach or consolidated with functional users' combined reviews. Toll gates may include, but not limited to, the following:
    i. *Project plan review* (validates/synchronizes the project activities as related to budget, business annual operating plan, schedule, etc.)
    ii. *Requirements review* (validates the magnitude of requirements versus project funding)
    iii. *Design review* (validates/synchronizes leadership and user requirements to technical strategy)
    iv. *Test readiness review* (validates integrity of ERP requirements, design, configuration, and test plan activities)
    v. *Production readiness review* (validates that the Technical Environment is fully ready to GO LIVE and may include a GO/NO GO Voting artifact)
    vi. *Sustainment readiness review* (validates that postproduction support model is fully operational)
    vii. *Problem log and issues review* (validates that adequate progress and tracking are being completed on technical concerns)
    viii. *User training review* (validates that users have a minimum acceptable competence level in their use of the ERP system tools, menus, reports, and typical day-in-the-life-of [DILO] tasks)
    ix. *User desktop procedure review* (validates that users have created a minimum acceptable level of support documentation whereby a "temporary employee" may step into any job, with minimal training or preparation, and use the ERP system tools, menus, and reports to fulfill a typical DILO task for every support job function)

2. The **business operating environment** will benefit greatly from the ERP implementation. However, managing expectations and engineering realistic deliverables is a key to obtaining

exceptional results. For example, if a stated goal of leadership is to "concurrently install a new ERP package, re-engineer every nested business process and attain 100% Software Certification and Competency levels all in a 3 month time period," the project is likely doomed to failure. Therefore, the project plan must include such elements as risk management (every project has risks, which need to be defined up-front and mitigated), rules of engagement (a charter of what the ERP implementation will deliver and what it will NOT deliver, e.g., what is in-scope and what is out-of-scope), and fully documented expected results. In addition, defining the project health parameters early will help the project core team navigate when conflicts arise. Each of these elements will be discussed in Section 7.8.1, but they are essential critical success elements and need to be defined early in the project life cycle. One other aspect is management commitment. Leadership must not only fund the project but must roll up their sleeves, remove barriers to success, contribute daily to project endeavors, and provide the necessary guidance in a timely (almost instantaneous) manner. A similar breakout of ingredients listed in the technical environment needs to be configured for the business operating environment (as well as the user community environment and the project environment), or at least shared with the technical environment.

3. The **user community** will also reap substantial benefits from a successful ERP implementation. As discussed above, the user community expectations need managing, ensure that business processes are properly engineered and realistic deliverables defined, and so on. Few companies have the resources to fully dedicate "a cast of thousands" team members to an ERP project implementation. Most companies must operate the business concurrently, while performing the project tasks for the ERP effort. Consequently, the user community must buy into their SoW, fully understand the rules of engagement, and rigorously participate in risk management/mitigation activities.
4. The **project environment** is typically a matrix structure, with project resources having line responsibility to another organizational entity and not under the control of the project organization. This structure then provides its own challenge in managing tasks, resources, schedules, and deliverables. Equally important are the rules of engagement, risk management, and attendant elements discussed above.

Now that we are beginning to appreciate the various stakeholder dynamics, the project plan will need to be created in sufficient detail to guide the project team to success. The overall project schedule may consist of the specific deliverable tasks on a date-specific horizon. These discrete tasks may be further decomposed into a series of short-interval schedules for more finite monitoring. Typically, a short-interval schedule spans about 5 days, which is usually finite enough to help monitor in sufficient detail to help ensure task completion on time. In its simplest definition, then, a project plan documents the resource requirements, the span of control for the project, roles and responsibilities, risk management, approval latitudes (grants of authority), and change/configuration control parameters. According to Wiki,* "It [Project Plan] is used to guide the project team in execution and project control activities and facilitates communication. It also defines the scope, cost and schedule." A core ingredient of the project plan is the project schedule, which defines task deliverables, milestones, and the resources needed by which the task is successfully completed.

The key ingredients of the project plan are defined in subsections 1.1.1 through 1.1.18:

---

* Wikipedia, http://en.wikipedia.org/wiki/Project_plan.

1.1.1 High-Level Acceptance Criteria (Accept1)

According to the Project Management Body of Knowledge (PMBOK),*

> Acceptance criteria represents specific and defined list of conditions that must be met before a project has been considered completed and the project deliverables can and will be accepted by the stakeholders. Customarily the acceptance criteria should be outlined in specific detail before work on the project has commenced and a very careful timeline should be set forth to make sure that all parties are onboard. Acceptance criteria may include certain essential requirements that must be met within the final deliverables themselves, or specific conditions that must be met during the process in which those deliverables are assembled and completed. In providing a series of acceptance criteria to the stakeholder, the project core team should, when possible, prioritize the acceptance criteria. In the event that a series of acceptance criteria is not met, or is met only partially, the final set of deliverables can either be refused for acceptance outright or, in some cases, it may be assigned the status of conditional acceptance, that being, an acceptance pending modification or correction to better meet the acceptance criteria.

1.1.2 End-Point System Expected Results (ToBeResult2)

As an early ingredient of the ERP implementation project effort (ideally, even before the ERP software is chosen), the executive leadership and movers and shakers within the company will collaborate (hopefully off-site to minimize distractions) and agree upon the expected results. These expected results will be detailed, not generalized (e.g., 45% inventory reduction, 60% customer satisfaction improvement, 30% unit cost reduction, etc.). These expected results of the end-point system are somewhat nested with the high-level acceptance criteria. For example, we want to ensure that the ERP software module design contains the necessary software capability to deliver the expected results to the degree of the stated commitment. At this point, it is essential to recognize the importance of this element … There would be no reason to make a significant investment in an ERP implementation project if there were only mediocre expected payback. This is a key executive leadership project participation deliverable and it needs to keep clarity of focus through the project life cycle and hold the business entity accountable for delivering these expected results. I believe that this should be an agenda item at every leadership staff meeting conducted, with progress reported to the executive group regularly. This topic will be covered more thoroughly in Chapter 3.

1.1.3 Rules of Engagement (RulesOfEngage3)

The rules of engagement is a list of what the ERP implementation will deliver (in-scope) and, maybe even more importantly, what it will NOT deliver (out-of-scope). Managing leadership expectations is crucial to success. This topic will be covered more thoroughly in Chapter 3.

1.1.4 Risk Management Plan (RiskMgmt4)

The risk management plan is a list of project risks (distractors) that may result in a negative impact on project results. It includes mitigation strategies used to manage the risks. To be discussed in more detail in Section 1.4.

---

* PMBOK_Guide, Project Management Institute, 2013.

1.1.5 Quality Assurance Plan (QA5)

Quality assurance (QA) confirms adherence to standards and provides visibility into the processes being used by the project and project deliverables. It provides management with appropriate reviews or audits of the project by independent personnel. It ensures the process steps and applicable standards that are being followed.

1.1.6 Requirements Management Plan (RqmtsMgmt6)

Requirements management is the process of documenting, analyzing, tracing, prioritizing, and agreeing on requirements and then controlling change and communicating to relevant stakeholders. It is a continuous process throughout a project. A requirement is a capability to which a project outcome (product or service) should conform (Wiki*). This will be discussed further in Chapter 2.

1.1.7 Configuration Management Plan (CM7)

Configuration management (CM) identifies and defines the process to establish and maintain the integrity of the project components and the configuration throughout the project life cycle.

1.1.8 Training Plan (TrainingPln8)

The planning team and functional stakeholders evaluate the technical, procedural, process, and managerial aspects of the project and determines the training needs of the team members and user community. The approach for fulfilling the training requirements is also documented, including the method, source, time frame, and any associated costs. It includes a skills assessment and training plan that is resourced by ERP functional experts.

1.1.9 Collaboration Coordination Plan (CollabCoord9)

The planning team creates a plan for communicating the commitments between various resource disciplines. This plan is used for coordinating and tracking the work associated with performing interdisciplinary coordination, and for identifying critical dependencies.

This plan is reviewed and agreed to by all the representatives of the participating disciplines. This plan should do the following:

- Communicate the commitments and deliverables between nested disciplines
- Identify the critical dependencies between these disciplines
- Document the methodology to be used to track critical dependencies and other interdisciplinary commitments, issues, and deliverables
- Coordinate and track the work associated with performing interdisciplinary coordination

1.1.10 Project Health Reporting Plan (ProjHealth10)

Identify project and performance metrics—Objective project metrics to be used to track progress and guide management decisions for interactive leadership.

---

* Wikipedia, http://en.wikipedia.org/wiki/Requirements_management.

The project should include the measures from at least these five categories:

1. Schedule
2. User testing
3. Risk
4. Training
5. Technical environment

Additionally, large projects may want to collect and monitor the level of change activity for requirements and configuration items. The project should be able to identify the level of change per unit time (volatility) for requirements changes. For configuration items, the project will log changes and may choose to add a measure of volatility where there is a potential for a diverse set of sources involved in change activity. Volatility measures can provide early warning indicators of a need for replanning and should be reviewed for root cause and potential project impact (cost and schedule).

The project plan will identify the specific metrics that will be tracked as well as their collection and reporting frequency. The format of the reports used to communicate these metrics will also be identified. They will be part of the metrics package developed to control the project and the critical components for ongoing project status reporting. An example of a project health report card is shown in Figure 1.2.

I recall one project I was assigned where one of the management resource leaders was constantly delinquent in making the needed decisions to keep the project flow on course. After numerous formal meetings as well as behind the scene prodding, I was unable to get the needed progress so I elevated the status to a Red (unsatisfactory) health report for schedule risk and technical environment. Because all levels of leadership focused attention on these report cards and no manager ever wanted to see a Red status, this caused a stir among the ranks. The culture at this particular company, especially within the project managers, was fearful of ever promoting the health report to Red (it always garnered a lot of senior management "assistance"). However, there are times when Red breaks logjams and facilitates project integrity. The director (sponsor) was furious with me for elevating the status to Red. I reminded him of the number of meetings and behind the scene prods I'd given him with lack of responsiveness on his part. He suggested that I resign from the project and I suggested we escalate the issue before a management review. An escalated management review would be detrimental to the director, inasmuch as it would show his lack of commitment to schedule integrity. He could not fire me from the project and did not want the management review; therefore, he was forced into making the decision the project team needed. I further reiterated the need for agility in the decision process, not just by his resources but by all project stakeholders. We agreed to change the rules of engagement to include decision agility as a key factor. Although furious with me and what he viewed as project manager insubordination, we ultimately became rather close working partners over time. To assist in this transition of "attitude," I recognized that the director had an engineering background and I decided to detail and frame each decision requirement in a way that would make him comfortable in "pulling the trigger" in a timely way. Even though that framework was a role (SoW) for one of the directors' staff assistance, I found it to be the proper elixir to remove conflict and gain the "support" essential for project schedule integrity.

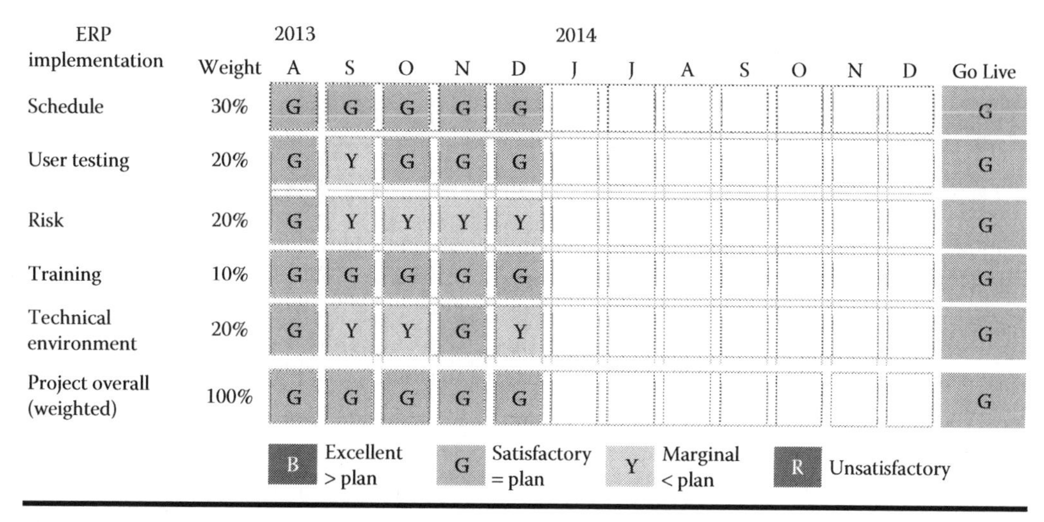

ERP implementation	Weight	2013 A	S	O	N	D	2014 J	J	A	S	O	N	D	Go Live
Schedule	30%	G	G	G	G	G								G
User testing	20%	G	Y	G	G	G								G
Risk	20%	G	Y	Y	Y	Y								G
Training	10%	G	G	G	G	G								G
Technical environment	20%	G	Y	Y	G	Y								G
Project overall (weighted)	100%	G	G	G	G	G								G

B Excellent > plan G Satisfactory = plan Y Marginal < plan R Unsatisfactory

Figure 1.2 Project health report card.

1.1.11 User/System Documentation Plan (UserDoc11)

The project core team identifies what is included and the standards for any user or system documentation that needs to be developed (this may default from corporate standards if available). System documentation includes collection of documents that describe the requirements, capabilities, limitations, design, operation, and maintenance of the system. It is archived for historical purposes and ongoing support of the system, as well as future upgrades. User documentation comes from operating practices, work flows, and desk instructions essential for day-to-day job activities.

1.1.12 Knowledge Transfer Plan (KT12)

The project core team assesses the need and creates a plan and schedule for knowledge transfer activities including the supplier/customer as well as in-house personnel. Knowledge transfer activities occur throughout the project life cycle and may include formal training and working with suppliers/customers on shared and self-service activities.

1.1.13 Communication Plan (Comm13)

Communication among the project team and with leadership and stakeholders is essential to project success. Frequency, format, and content of communication are defined in the communication plan (see Appendix A.1).

1.1.14 Plan for Reviews (Toll Gates) and Walkthroughs (TollGate14)

The detailed project plan includes review points to help ensure that information about the project is communicated to leadership and customer management and that the deliverables conform to all applicable policies, procedures, and standards. Typical reviews include management reviews, stakeholder reviews, requirements reviews, design reviews, interdisciplinary reviews, technical reviews, QA reviews, peer reviews, and walkthroughs. The number and type of reviews is based on the project size, the relative risk, and the operating style of leadership.

1.1.15 Contractor Agreement Management Plan (CAM15)

Contractor agreement management is essential when an outsourcing decision is made to contract any portion of the project scope. These contractor activities are identified and included in the project schedule. This also includes location where agreement will be performed and methodology that will be deployed. It may also include milestone payment points based upon the percentage complete. Because a contractor's work effort is typically beyond the company's day-to-day control, this effort needs to be monitored aggressively. This effort may have particular cost/schedule risk impacts.

1.1.16 Test Strategy (Testing16)

The test strategy, within the project plan, describes the goals and objectives for project deliverable testing. It also defines the exit criteria and user acceptance methodology.

The test plan (developed at a later phase) provides the following supporting details:

- Testing approach (types, phases, e.g., unit testing, configure integration testing, system integration testing, regression testing, process testing, user acceptance testing)
- Testing tools
- Testing resources
- Testing environment
- Test data and frequency of refresh

1.1.17 Business Information Assurance Plans (BIA17)

1.1.17.1 Information Systems Continuity Plan

The project core team will determine the criticality and value of the system and the work effort and specific tasks will be included on the detailed project schedule. A CSF is to determine the financial impact of a disaster and to identify vital applications, datasets, and system/application interdependencies. Based on the financial impact, if it is determined that the system is mission critical, an information systems continuity plan is prepared to develop an off-site backup and/or alternate processing strategy.

Note: This may already be in place as part of a corporate policy or mandate.

1.1.17.2 Fault-Tolerant Plan

Based on mission critical status and stakeholder influences, a fault tolerant, or high-availability capability may be necessary to support the user community (based on SLA). However, many companies include this as part of the information systems continuity plan discussed in subsection 1.1.17.1. The role of the project stakeholders is to evaluate the magnitude of change from the former operating environment to the new ERP operating environment and to recommend changes, if needed.

1.1.18 Software Implementation Strategy (SWImple18)

This section defines the implementation strategy for the project. Normally, the implementation strategy includes the software implementation process, be it phased or a single rollout. During the software implementation phase, the project will complete the work activities and deliverables, adhere to any toll gate reviews or other leadership audits, and conform to any defined transition to production checklists, such as Go/No Go voting process. The implementation strategy will also include agreed-upon stabilization activities after implementation.

1.2 SoW—Managing Expectations through Project Life Cycle

The SoW is at the heart of project integrity pursuit (cost/schedule/return on investment [ROI]) and consists of agreed-upon criteria associated with the project deliverables. Like requirements generation (discussed in Chapter 2), the SoW relies upon leadership guidance to elicit the stakeholder community and internal/external service providers to develop the list of deliverable tasks, quality tolerances, and expectations achieved from the project effort. The SoW covers the scope of the work, technical goals and objectives, user goals and objectives, identification of stakeholders and end users, imposed standards, assigned responsibilities, cost and schedule constraints and goals, dependencies between the project and other organizations, resource constraints and goals, and other constraints. It is also used to describe the criteria defining what is meant by "ERP implementation success." Rather than relying upon the whims of all attendant players, the SoW is an engineered instrument to define project latitudes, CSFs, quality of deliverables, scope, and other attributes that influence ERP success in the minds of leadership. When designed robustly, the SoW becomes a defining best practice that facilitates the achievement of stellar ERP ROI results.

There are a couple of points of emphasis regarding the SoW, which are as listed below.

- The SoW provides a crisp understanding as to **who** does **what** and **when** according to an attendant specification. Although typically used to ensure that the rules of engagement for external entities are clearly spelled out (software suppliers, implementation service providers, hardware suppliers, etc.), it is easily adapted to internal project activities as well. The content may include the following:
  - Definition of project team roles, responsibilities, and deliverables (this function may be included in a project charter, however)
  - Definition of expectations and deliverables from noncore project resources
  - Definition of associated business units where resources will be periodically used
  - Definition of expectations from bargaining unit users
  - Especially difficult activities or high-risk tasks that need special attention and very diligent monitoring
- The SoW provides a boundary of expectations. Similar to the rules of engagement, it includes what will be delivered on the project effort as well as what will NOT be delivered. This approach to management by specification adds clarity to expectations. As will be discussed

repeatedly, managing expectations is a core principle to successful project management. The SoW is an ideal medium to facilitate the management of expectations because it is a definition framework ... The entire project community (internal and external) clearly understands what is expected in the form of deliverables.

Wiki* defines the areas a typical SoW addresses as follows (paraphrased):

A statement of work typically addresses these subjects:

- *Purpose:* Why are we doing this project or effort activity within the project? A purpose statement attempts to answer this.
- *Scope of Work:* This describes the work to be done and specifies the hardware, software and process boundaries involved.
- *Location of Work:* This describes where the work is to be performed, including the location of hardware and software and where people will meet to do the work.
- *Period of Performance:* This specifies the allowable time for project/activities, such as start and finish time, number of hours that can be consumed per week or month, where work is to be performed and anything else that relates to scheduling.
- *Deliverables Schedule:* A schedule specifies and describes what is due and when.
- *Applicable Standards:* This describes any industry/company specific standards that need to be adhered to in fulfilling the effort.
- *Acceptance Criteria:* This specifies how recipients (internal customers) will determine if the product or service is acceptable, usually with objective criteria.
- *Special Requirements:* This specifies any special hardware or software, specialized workforce requirements, such as degrees or certifications for personnel, travel requirements, and anything else not covered in the contract specifics.
- *Type of Contract/Payment Schedule (External):* The project acceptance will depend on if the budget available will be enough to cover the work required. Therefore a breakdown of payments by whether they are up-front or phased will usually be negotiated in an early stage.
- *Miscellaneous:* Many items that are not part of the main negotiations may be listed because they are important to the project and overlooking or forgetting them could pose problems for the project.
- *Performance Clause:* In the event that the SoW owner falls behind on their commitment to deliverables, there needs to be a statement of consequences ... for example, this might be the guideline for a 3rd Party Consultant SoW Schedule Performance clause ... In the event that Consultant falls behind in their commitment to the agreed upon Project Plan deliverable dates, Consultant billing invoices shall be placed ON HOLD and not paid until schedule integrity is reinstated by fulfillment of the past due deliverable(s). In addition, the Project Manager will be provided a detailed Corrective Action Plan which details the remediation roadmap, as well as, daily status updates on remediation progress. If it is an internal resource area delinquent, the follow may be suitable ... In the

* Wikipedia, http://en.wikipedia.org/wiki/Statement_of_work.

> event that an internal resource center falls behind in their commitment to the agreed-upon Project Plan deliverable dates, Project Health Status shall be Red until schedule integrity is reinstated by fulfillment of the past due deliverable(s). In addition, the Project Manager will be provided a detailed Corrective Action Plan which details the remediation roadmap, as well as, daily status updates on remediation progress.

As we discuss deliverables, we must recognize that some resource areas tend to commit to deliverables casually, without giving the critical thought necessary to ensure a quality and timely deliverable within a well-managed resource plan. The SoW certainly helps in clarifying the expected deliverable result. To enhance the probability that the deliverable, when delivered, meets the requirements specification to a tee, I have found that working with the resource areas to engineer a roadmap by which delivered results are acceptable (quality, timely, and comprehensive). This process is equivalent to creating a detailed mini-project schedule for each deliverable. There are a variety of benefits associated with this drill down technique:

- It helps ensure that there is consistent clarity of vision between what is needed (quality level) by the project and what the resource will, in fact, deliver.
- It helps the fulfillment resource better understand the level of effort essential for completing the task. Most resources work on a variety of deliverables concurrently; therefore, giving critical thought to a success path for the project deliverable helps ensure timely results.
- It helps the fulfillment resource think through potential barriers to successful completion. The earlier barriers (obstacles) are identified, the better the chances that their removal will not adversely affect schedule.

These detailed roadmaps should be an ingredient in the project documentation library for potential use on future project (time saver) and used as an artifact for the traceability matrix (see Chapter 2).

A best practice associated with deliverable fulfillment is as follows. *Every deliverable* requires that the fulfillment resource send an artifact to the project core team. This artifact

- May be stored in the project documentation library for audit purposes and/or part of the traceability matrix,
- May allow future projects to benefit (be more efficient) from adequately detailing the process roadmap,
- Will be used by the project core team to affirm that the deliverable meets the quality needs and spirit of the defined task.

1.3 Managing Change

One of the key contributors to ERP implementation success is the proper use of change management. ERP implementation is a journey, not an event. Consequently, as a company progresses during the journey, there are unexpected circumstances that may occur, which tend to distract project team focus or impede progress ... One thing for certain is that change will occur some time during the journey. To best handle aberrations is to, up-front, at the onset of the project, the prudent project team will plan for change to occur.

In companies that design and build products, there is already in place the practice of CM and control to address product design or build anomalies. Some of the CM practices require elaborate and/or sophisticated CM processes, including one or more control boards that evaluate change impact. Depending on the impact (cost/schedule risk), these practices could take hours, days, months, or years to fulfill the life cycle. However, in all cases, the CM activities are supported by disciplined, repeatable, and visible processes that were specifically engineered for CM.

Recognize that the goal of CM is to

- Support business, stakeholder, and project core objectives, and maintain integrity of the project requirements, schedule, and budget guidelines.
- Establish a best practice change environment that facilitates the generation of pertinent information in support of quality and timely decisions.
- Optimize resources (both internal and external resources), assets, and efficiency of the support organization.
- Minimize the severity of adverse impacts to the ERP implementation.

Let us take a closer look at how we might engineer an effective best practice change management process to help ensure that the ERP implementation project keeps on course and maintains cost/schedule integrity.

1. *A lean project change management framework.* We need to engineer a nimble, agile, and quick response change framework that flows well within the ERP implementation processes. Depending on the magnitude of the change, we need to develop approval workstreams that align within the grants of authority established within the company. This will require enlisting executive level leadership and operational level and project level appointees to be on board and postured to approve changes on an almost instantaneous (agile/nimble/quick response) basis. Therefore, accommodations need to be in place for both periodic and ad hoc (24/7) convening of approval members and subsequent communication broadcast. Due to the needs for agility, Delegation of Authority (DoA) or member backups need to be appointed and a clear escalation path defined so as to not bog down the approval process. In addition, there needs to be clarity of vision for the assigned roles and responsibilities, processes, tools, and procedures associated with the project change management framework.
2. *The change log (may be part of the traceability matrix).* Depending on the formality desired by leadership, some form of change proposal, log, evaluation, and approval mechanism needs to be designed so that when a change needs to be made, the end-to-end process may quickly facilitate fulfillment. The simpler the better, this may include, but not be limited by, approval board/entity, forms, log, workflow, document repository (library), and so on.
3. *Lean change execution.* Once the change is approved, there needs to be a lean change implementation process in place. This may be as simple as changing the project schedule. However, there needs to be created attendant change approval artifacts that are consistent with company practices for audit control as well as supporting project integrity. In addition, if there is a large volume of change activity, there needs to be some agreed-upon metrics in place to alert management of root causes for scale and types of change. “Affecting change” costs money, may dilute resource capacity and is a **best practice** defining ingredient to ERP implementation success.

Fortunately, an ERP project change framework does not typically need the rigors of a product change process. However, as stated above, change will happen at some point and the ERP project change framework needs to be in place to avoid scope creep, escalating cost/budget/risk impacts, or impacts associated with diluting expected ROI.

1.4 Risk Management

One final ingredient of the project plan needs elaboration, namely, risk. Regardless of the simplicity of the ERP implementation effort, there will be risk associated with implementation. Therefore, it is essential to develop a risk management plan, which documents the risk associated with ERP implementation.

It is important to document all known risks early in the ERP implementation process; in fact, it might be a deliverable from the **Senior Leadership Collaboration Workshop** (discussed in Chapter 3). Risk inputs should come from all areas where the ERP implementation will have impact. A sample risk management log (including mitigation plan and contingency plan) as well as a risk action plan is discussed in subsection 1.4.2.

A sample risk management strategy is discussed in subsection 1.4.1.

1.4.1 Risk Management Strategy

Managing risk is a foundational precept for a successful ERP implementation process. Risk management is the process used to identify, quantify, and rectify issues that can adversely affect the success of a project. To that end, the following framework will be used to identify and mitigate risks:

- Risk management plan
  - Document known risks, then prioritize.
  - Identify the root cause of the risk.
    - Cost, schedule, technology, requirements, capacity, and process
  - Assess probability and severity that an undesirable activity will occur.
  - Document consequences of risk.
  - Develop a mitigation strategy.
    - Avoidance, control, transfer, monitor, and acceptance
  - Review risks/mitigations on a regular basis.
  - Manage leadership perceptions regarding risks.
- Reporting risk status
  - Make risks, plans, actions, concerns, exchanges, forecasts, and progress known, particularly at the project management reviews.
  - Help ensure the visibility of risk information to the senior staff.
  - Enable all project members to be aware and participate in defining and managing risks.
  - Help ensure an understanding of risk and mitigation plans.
  - Establish an effective, ongoing dialog between the management leadership and the project team.
  - Help ensure that an appropriate level of attention is focused on issues and concerns.
- What risk management can do
  - Minimize or eliminate many risks.
  - Highlight areas of uncertainty and false confidence.

Figure 1.3 Dilbert risk analysis cartoon. (Data from DILBERT © 1997 Scott Adams. Used By permission of UNIVERSAL UCLICK. All rights reserved.)

  - Help to decide best course of action.
  - Provide early warning of many problems.
  - Provide a basis for plan changes.
  - Increase management and customer confidence.
- What risk management cannot do
  - Identify all risks.
  - Minimize or eliminate all known risks.
  - Guarantee the best course of action is always chosen.
  - Provide early warning of all problems.
  - Compensate for poor planning.

1.4.2 Sample Risk Management Log, Mitigation Plan, Contingency Plan, and Risk Action Plan

Depending on the company, the type of risks, and the impact of the risks, risk management may include sophisticated quadrant cubes that analyze each risk and alert leadership when risk factors become a reality. See Appendix A.2 for a sample risk management log, mitigation plan, and contingency plan, and Appendix A.3 for a sample risk action plan.

As denoted in the Dilbert cartoon (Figure 1.3), risk may be perceived differently by stakeholders. How a company approaches risk management is a matter of company culture and style; however, it is essential to the integrity of the project effort.

Developing a risk management strategy and mitigation plan is a leadership **best practice**.

Chapter 2

Requirements Generation

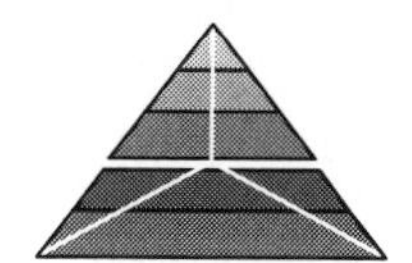

The requirements generation process defines, in detail, the system functionality as well as the engineered process changes essential for an order of magnitude improvement in operational performance. The attributes of requirements definition include categories such as "mission critical," "essential," and "nice to have," which then establishes the baseline for a traceability matrix that flows through the project phases including design, prototyping, customization, testing, piloting, and delivery.

No matter what the size or type of a project, the enterprise resource planning (ERP) project implementation journey cannot begin without defining, at least, initial requirements. Performing a comprehensive job of distilling good ERP project requirements (needs and expectations) is one of the hardest tasks within a project and functions as a best practice in the ERP implementation process. Defining clear and incisive requirements is an art and those ERP project teams that are very effective at it, early in the project, have few costly changes during the project life cycle while facilitating risk mitigation. According to Wiki* (paraphrased), "A requirement is a defining capability central to a project outcome (product or service)." Therefore, the entire listing of all the requirements is a key factor in the composite of project outcomes. Requirements come from a variety of sources and it is essential to maintain the ability to trace each requirement back to the source, which is called requirements traceability.

A **best practice**, in generating good requirements, is the ability to tie the business drivers (unique aspects differentiating our company from our competitors) in such a way that the resource base (users), benefitting from ERP-delivered functionality, may excel in their individual job performance. When viewed as a means to optimize functionality, as a competitive differentiator, requirements generation is an art.

* Wikipedia, http://en.wikipedia.org/wiki/Requirements_management.

In recalling a requirements related project failure ... I was called in to conduct a formal requirements generation process as the company's mitigation strategy. After the company had poorly documented their requirements, they purchased software that did NOT meet their business need. During the implementation process, the key stakeholders recognized the deficiency and were faced with a decision as to whether to modify the ERP software to include the functionality or to seek better fitting software. The missing functionality was deemed mission critical. After evaluating the cost to perform the necessary modification (and the ongoing customization cost at version upgrade events), it was determined that the best mitigation strategy was to conduct a proper requirements generation process and purchase a better software fit. Fortunately, the software suppliers recognized that continuing to enforce the purchase contract would reflect badly on their product and they offered a 50% refund to void the contract. In addition to the recognized mission critical missing functionality, upon conducting a rigorous requirements generation, the project team and leadership uncovered a handful of other missing capabilities that would have resulted in a total ERP implementation disaster if they had decided to continue with this software. The mistake was not the software company's issue ... The software selected was excellent and the company had a great reputation in the marketplace, but was inadequate for the company's business. It was just a bad requirements generation process. The net impact of this false start failure was over a quarter million dollars of out-of-pocket expense and close to a year project implementation delay. Once good requirements were generated, the company purchased software that better met their business needs and ultimately were successful at implementation.

Lessons learned:

Lessons learned is the process of documenting tasks, processes or ideas that we learned during implementation of the project which, if used on future projects, may improve results of those projects.

- Conduct a thorough requirements generation process and fully document the "unique" functionality of their business needs.
- Ensure that a broad audience (whole company) participates in soliciting requirements.
- In the software selection scorecard used to select the best fitting solution, ensure that the "unique" needs have the proper weighting factor compared to routine feature/function capability weighting.
- If a software modification will be necessary to augment any software deficiency business need, ensure that the total ERP life cycle cost is evaluated, not just the cost of the modification effort.

2.1 Requirements Source

Requirements come from a variety of sources, including internal leadership, stakeholders, users, and business operating factors. They also come from external sources such as customers, competitive proficiency, contract clauses or specifications, innovation, and creative team members, as well as other sources such as regulations or standards. However, a typical ERP project primarily receives its generic feature/function requirements from the ERP-selected software supplier product. Note that a good requirements definition needs to be done first and is used to solicit the best fitting software candidates in the software selection process.

A significant amount of customer/stakeholder angst, emanating from selecting a less desirable ERP solution, originates from the requirements definition phase whereby one or more incomplete facts, incorrect facts, inconsistencies, misplaced requirement, omission, ambiguity, or other factors are generated. Therefore, a contributing factor for ERP implementation failure may be linked to requirements generation.

There are various types of requirements, some of which are as follows:

- *Functional*—An essential capability that the product/service must perform by defining the task, action, or activity that must be accomplished.
- *Technical*—Characteristics, size, dimension, form, fit, function, color, reliability properties, performance, and process that is expected from the product/service/system. It may include such things as architecture, structure, stress, behavior, or other like attributes or constraints.
- *Customer*—Statements of facts and assumptions that define the expectations of the system in terms of business objectives, environment, constraints, and measures of effectiveness and suitability (Wiki).
- *Nonfunctional*—Requirements that specify criteria that can be used to judge the operation of a product/service/system, rather than specific behaviors.
- *Derived*—Requirements that are implied or transformed from another requirement.
- *Interface*—Requirements that specify what external operating products will be nested within the ERP software product. These requirements define how the various products will function together, the data flows (one-way or bidirectional), precedence, upgrade frequencies, and a host of other compatibility and characterization specifications.
- *Process*—Requirements that specify how the need function will operate independently and together as a nested whole. In addition, process requirements should specify not only how it impacts a certain function, but also the end-to-end interactivity of the business dynamics.

The characteristics associated with *good* requirements include the following:

- Comprehensive
- Compulsory
- Consistent
- Crisp and concise
- Describes what, not how
- Explicit
- Lack of escape clauses
- Nonredundant
- Proper
- Sensible
- Simple, not complex
- Traceable
- Understandable by all stakeholders and users
- Verifiable and/or testable
- Viable

When implementing a commercial off-the-shelf (COTS) ERP product solution, the challenge becomes "what portion of this vast functionality am I going to implement?" Hopefully, a thoroughly detailed requirements generation process was used as a basis for selecting the ERP list of candidates. If that is correct, then the requirements baseline to be used for the ERP project becomes a blend of the original solicited requirements *and* the added functionality you choose to implement that comes standard with the ERP software product. In addition, there needs to be a definition for customizations, modifications, interfaces, and other elements that make the requirements definition reflective in a comprehensive way. There are other considerations as well, which are as follows:

- Most ERP software solutions allow the buying company to customize various functionalities. These are in the form of parameters, switches, default values, and so on. This customization then must be part of the requirements specification and, when the application installation is completed, will have the artifact as attendant to the documentation library as a deliverable. They should also be included in the traceability matrix spanning the technical work effort (requirements, design, prototyping, customization/configuration, testing, piloting, and delivery).
- System setups are to be documented as part of the requirements specification as well and requires attendant artifacts to be posted in the documentation library. Whether setup values will flow as part of the traceability matrix or not is optional. The downside to NOT including in the traceability matrix is that, during testing, if a failure (defect) arises, then troubleshooting research may take longer to determine that the fault was a setup value. The trade-off is the span of time needed for problem diagnosis and resolution.
- System integrations, tying your ERP software to third-party software, may be as simple as engaging an application programming interface (API) object (or hook), which is fully compatible with the ERP software. Or, at the other end of the continuum, it may require somewhat rigorous and, at times, complex programming. Yet a third option is called middleware that sits in between an API and full-blown programming. As discussed in system setups above, whether or not to include it in the traceability matrix, as a stand-alone entity, is optional, with the span of time needed to troubleshoot defects lying in the balance.
- If it is determined that the software does NOT have critical/unique functionality necessary to competitively run the business, an ERP software modification will likely be necessary. Whether that modification is purchased, fulfilled by resources within your company, the ERP software firm, or a third-party company, rigorous requirements generation and traceability matrix is essential for ERP implementation cost/schedule integrity.

As illustrated in the cartoon of Figure 2.1, a given requirement may be viewed differently, depending on the specification offered by the role the source plays. Therefore, it is essential that critical requirements be elaborated properly, analyzed thoroughly, and validated across the entire stakeholder community (internal and external) to help ensure that it meets the various roles within an organization.

2.2 Requirements Generation Life Cycle

The requirements generation life cycle (Figure 2.2) begins with collecting requirements from various input sources (stakeholder, customer, supplier, user, etc.), validating the requirement via a requirements analysis process ("determining whether the stated requirements are clear, complete, consistent and unambiguous, and resolving any apparent conflicts"—Wiki), requirements flow-down (decomposing requirements in a more finite or lower level), generating a functional specification, and then verifying and validating the process (Does it meet the original source inputs needs and does it conform to standards, is it correct? Does it conform to minimum acceptable quality level? Does it have adequate level test/use cases [UCs]? Does it correct ambiguities and vagueness?).

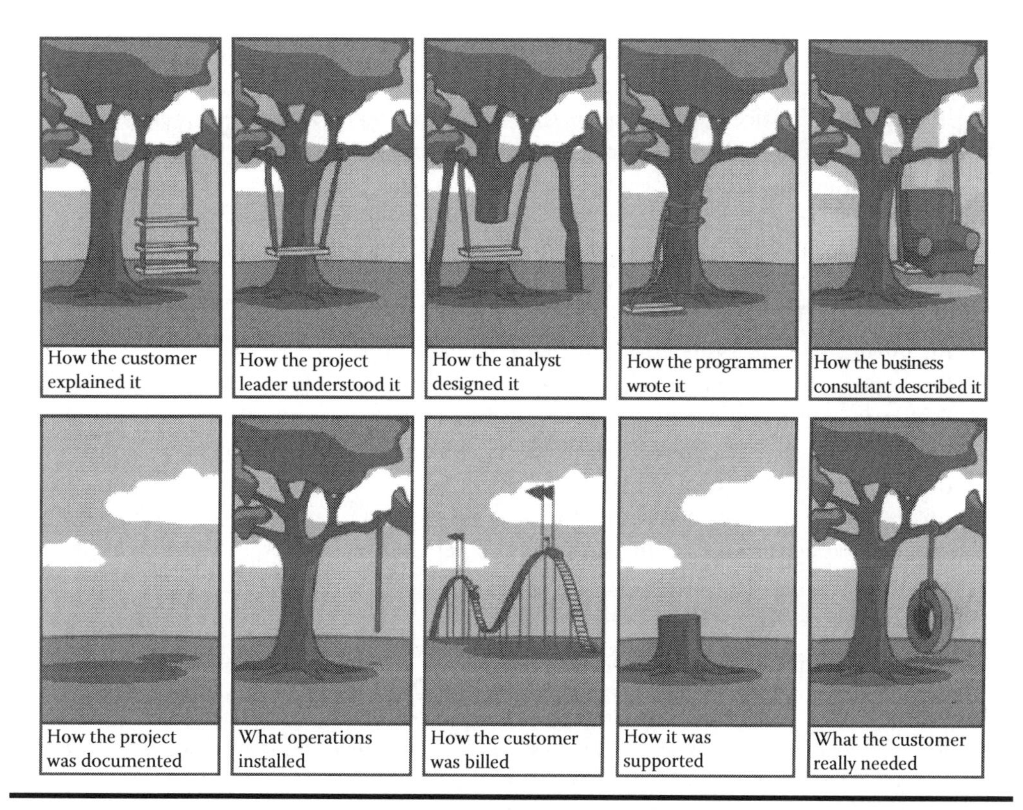

Figure 2.1 Requirements viewed differently. (Data from Tree Swing Cartoon compliments of projectcartoon, http://www.projectcartoon.com/about/. Original author unknown, permission granted by projectcartoon.)

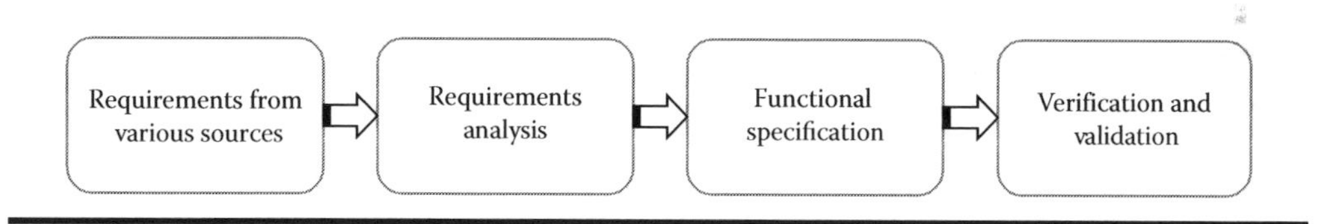

Figure 2.2 Requirements generation life cycle.

There are some common mistakes that distract from the integrity of a GOOD requirements document:

- Defining business rules without nested perspective to their processes, resource requirements, data points, end-to-end process management, human factors, roles and responsibilities, key performance indicators (KPIs), company culture, and strategic business goals and objectives
- Omitting key functionality such as interfaces, metadata, and failing to test veracity of requirements
- Lack of proper requirements gathering guidance, core competence, and technical review
- Not providing both functional (what it does) and performance (how well it does it) criteria
- Documenting bad assumptions for the requirement and/or not documenting sufficient breadth of expected assumptions

- Lack of the correct mix of resources in cross-functional collaboration (balancing review span)
- Not collecting goals and objectives associated with the complex or very unique needs
- Lack of tying service-level agreements to requirements performance, as well as executing sufficient audit process to help ensure improving performance over time
- Documenting realization elements instead of requirements
- Requirements specify a need, realization specifies an end result or a how to
- Failure to ensure that requirements processes fit well with project methodology
- Describing operations instead of documenting requirements
- Operation—"The operator shall be able to turn the machine on and off"
- Requirement—"The system shall provide a manual on/off switch"
- Not layering business, functional, and technical requirements ... Emphasis needs to be balanced
- Failure to keep requirements simple (yet comprehensive), adhering to a simple template to help ensure that requirements are consistent in form and use of visualization techniques (pictures are worth a thousand words, many times)
- Failure to have end-to-end traceability to source inputs, KPIs, key attributes and functionality, proper documentation, training (minimum acceptable quality level), realized deliverables, and realized expected results
- Not using accurate or precise terms
- Examples—support, but not limited to, (the qualifier not limited to is vague, it needs to be specific)
- Requirements omission or incomplete list (lack integrity)
- Ensuring that requirements are actionable and not too precise nor too ambiguous
- Requirements not focused upon future state ... Many current system requirements tend to be lethargic, rather than nimble, agile, and lean oriented
- Failure to align strategic goals with annual operating plan with project scope with requirements portfolio
- Over engineering requirement
- Ask yourself "What is the worst thing that could happen if this requirement were missing?"
- Adhering to a clear requirements generation process methodology and toolset
- Absence of specifying tolerances
- Ask for absolute values rather than a tolerable range

There are a variety of requirements management products (tools) available in the marketplace that may be used to help gather, store, and categorize requirements, if such a tool is desired. For a COTS ERP solution, these tools are likely an overkill and they tend to be pricey. However, they are available if desired.

Now that there is better clarity on what good requirements consists, there are a couple other aspects that need further elaboration, some of which are as follows:

- Attributes of requirements
- Process engineering
- Traceability matrix

2.3 Attributes of Requirements

There are elaboration categories of requirements ... The *first* implies comprehensiveness. It is important that everyone recognizes that a requirement is an engineered specification. It denotes what is essential for each and every data element in all ERP modules pursued to function according to the needs of the stakeholders and user community. Because it is an engineered specification, it will need to include sufficient attributes to properly define essential ingredients. For example, if the requirement size is important, that attribute must be included. If temperature tolerances, humidity factors, or any other defining element is essential, these attributes need to be included. Therefore, a requirement may be a one-line descriptor, a multiple element descriptor, or multiple page detailed description of the need. Each ERP implementation project will need to develop its requirements to the level of detail sufficient to ensure integrity of specification.

The *second* category implies criticality. Criticality refers to importance of need. Is the requirement mission critical, essential, or merely nice to have? It is important to specify criticality for activities such as evaluating merits of various software offerings in the software selection process. In this activity, criticality may be given different weighting factors, which will help choose the best fit software candidate. It may also be used, if the requirement will affect a system modification. Any system modifications must be developed using a stringent development process to help ensure timely and cost-effective solutions, which align to project goals and timing.

The *third* category implies whether the requirement is business unique or common and used in most businesses. This category covers matters such as patents, intellectual property, or other like capabilities that differentiate this company needs from its competitors. This particular requirements category typically needs disclosure protection such as nondisclosure agreements or other instrument to help ensure secrecy. Typically, these type requirements are classified as mission critical and include exhaustive levels of detail.

The *fourth* category spans the ancillary aspects of the project implementation effort, namely, support activities. A typical ancillary attribute may be training. Training is likely a statement of work (SoW) or task on the project schedule to bring internal/external users into a proficiency level associated with usage of the ERP tool. However, it may require a broader specification. Training may imply being an integral component of product deliverables. An example might be a supplier fulfillment aspect of the software whereby the supplier is the resource transacting activities on your behalf within your ERP system (a special firewall is created to accommodate external self-service activities). Another example might be part failure/repair transaction processing or customer collaborative design activities.

As discussed earlier, implied in attributes is performance. Therefore, as requirements are distilled into the requirements artifact, performance expectations need to be specified.

Reflecting upon a software selection project in my past, I recall one company who did such an outstanding job in detailing their business needs ... They not only focused upon their "unique" needs, but gave emphasis to detailing their perceived "competitive advantages." While evaluating the final software selection candidates, there were two ERP suppliers in a neck-to-neck tight competition. As it turned out, the defining factor and primary differentiator was that one ERP supplier's product development life cycle strategy was aligned to my clients' product life cycle. Although the finalists' ERP product functionality were scored about the same, the software selection team saw the value of aligning their final candidate's life cycle to their own, and this alignment netted the contract award. My client had established a critical success foundational element, by extracting superior requirements from the company, and leveraged these to select an ERP software partner. They continued this exceptional performance throughout the implementation process to become a premier ERP software user.

2.4 Process Engineering

At the onset of the project, there needs to be a specification as to which business processes will be reengineered and to what extent. (Note that this might be a great deliverable from the senior leadership collaboration, which will be discussed in Chapter 3.) If the ERP project is merely a software product version upgrade, there may be few, if any, process changes. However, if this project is the first time that sophisticated ERP capabilities have been deployed, there needs to be significant analysis performed as to the degree of process change to be undertaken. This is one of those project activities that may derail a project if not engineered and managed properly. Just like data element mapping which is important to features and interfaces, process mapping is essential to understand the magnitude of change that is expected in day-to-day business processes.

Whether simple process changes are expected or a broader swath, processes need to be engineered, expected results agreed upon, and the proper team of resources deployed to make it happen. Note that the process engineering effort may be a large SoW and may require bringing in technical experts from outside the company. In many companies, processes have not been properly engineered from the onset. Therefore, this activity may require a long period of time to properly deploy. To that end, if the process engineering is significant, it should be separated from the software portion of the ERP implementation. This will depend on the senior leadership's timetable for expected results.

We will spend a good deal of focus upon the importance of process engineering in Chapters 4 and 8. Needless to say, process engineering is a very important aspect of ERP implementation success.

2.5 Traceability Matrix

The ability to trace requirements flow from their source (originator), through the various project phases (design, prototyping, customization, testing, piloting, and delivery) is a requirements generation's **best practice**. It helps ensure that the integrity of the requirements is maintained, that change impacts are handled properly, and that the results are attained as expected.

According to Wiki,*

> A traceability matrix is a document, usually in the form of a table, that correlates any two baselined documents that require a many-to-many relationship to determine the completeness of the relationship. It is often used with high-level requirements (these often consist of marketing requirements) and detailed requirements of the product to the matching parts of high-level design, detailed design, test plan, and test cases.
>
> A requirements traceability matrix may be used to check to see if the current project requirements are being met, and to help in the creation of a request for proposal: (1) software requirements specification, (2) various deliverable documents, and (3) project plan tasks.
>
> Common usage is to take the identifier for each of the items of one document and place them in the left column. The identifiers for the other document are placed

* Wikipedia, http://en.wikipedia.org/wiki/Traceability_matrix.

across the top row. When an item in the left column is related to an item across the top, a mark is placed in the intersecting cell. The number of relationships are added up for each row and each column. This value indicates the mapping of the two items. Zero values indicate that no relationship exists. It must be determined if a relationship must be made. Large values imply that the relationship is too complex and should be simplified.

To ease the creation of traceability matrices, it is advisable to add the relationships to the source documents for both backward traceability and forward traceability. That way, when an item is changed in one baselined document, it's easy to see what needs to be changed in the other.

The attributes of an effective traceability matrix include the following:

- It is easy to understand.
- Requirements ID is a unique key.
- It has the ability to cross-reference and search.
- It has ties to comprehensive testing (unit level test, integrated system test, user acceptance test), training, and deliverables.

From the Wiki, a sample traceability matrix is discussed subsequently ... It shows down the vertical plane the various test cases and reference numbers (i.e., 1.1.1, 1.1.2, etc.) and across the horizontal plane the requirement (REQ1), whether it is a UC or a technical requirement (Tech) ... The "x" indicates the matrix intersections. This matrix tool will chain the interactions across the requirements and project life cycle (design, prototyping, customization, testing, piloting, and delivery).

In practice, the traceability matrix might list on the horizontal plane the life cycle:

- Unique or business requirement (e.g., BR001)
- Functional requirement (FR001)
- Design (TR001)
- Process (PR001)
- Modification (MR001)
- Configuration or setup (CSR001)
- Verification (SIT001—system integration test)
- Validation (UAT001—User Acceptance Test)

Each of the above inclusions may have a series of attribute columns that describe the feature and designate who it is assigned to, expected completion date, or other reference data. An example is shown in Figure 2.3.

Business/Unique Requirements					Functional Requirement			Technical Design Requirement					■ ■ ■
Requirement ID	Requirement Desc	Requirement Type (Functional, Technical, Process)	Requirement Status (New, In Process, Approved, Deleted)	Notes	Function ID	Functional Desc	Assigned To	Tech ID	Desc	Assigned To	Design Approval Date	Notes	■ ■ ■
BR001	Item Number	Functional	Approved		FR001	Create Item	Jim Hooper	TR001	Entity	Jon Bridger	1/3/2013		

Figure 2.3 Traceability matrix tool.

Sample traceability matrix															
Requirement identifiers	Reqs tested	REQ1 UC 1.1	REQ1 UC 1.2	REQ1 UC 1.3	REQ1 UC 2.1	REQ1 UC 2.2	REQ1 UC 2.3.1	REQ1 UC 2.3.2	REQ1 UC 2.3.3	REQ1 UC 2.4	REQ1 UC 3.1	REQ1 UC 3.2	REQ1 TECH 1.1	REQ1 TECH 1.2	REQ1 TECH 1.3
Test cases	321	3	2	3	1	1	1	1	1	1	2	3	1	1	1
Tested implicitly	77														
1.1.1	1	x													
1.1.2	2		x	x											
1.1.3	2	x											x		
1.1.4	1			x											
1.1.5	2	x												x	
1.1.6	1		x												
1.1.7	1			x											
1.2.1	2				x		x								
1.2.2	2					x		x							
1.2.3	2								x	x					
1.3.1	1										x				
1.3.2	1										x				
1.3.3	1											x			
1.3.4	1											x			
1.3.5	1											x			
etc.															
5.6.2	1														x

Figure 2.4 Sample traceability matrix. (Data from *The Definitive Guide to Requirements Traceability*, p. 8, Accompa [e-book].)

The vertical plane would list every requirement by ID. A pattern may be used to differentiate the type of requirement:

- Business requirement—BR prefix
- Functional requirement—FR prefix
- Technical requirement—TR prefix
- Customer requirement—CR prefix
- Nonfunctional requirement—NFR prefix
- Derived requirement—DR prefix
- Interface requirement—IR prefix
- Process requirement—PR prefix
- Configuration/setup requirement—CSR prefix

Figure 2.4 shows an example of a traceability matrix.

2.5.1 Requirements Documentation

Once the compendia of requirements have been collected, analyzed, validated, and rationalized, they are ready to be codified and published. The publication approach must conform to company culture, standards, and management style. For example, in a small business, the requirements document may simply be a spreadsheet of one-liner requirements specification, mostly derived from the library of their selected ERP software publisher, and may consist of 5–10 pages. For larger organizations, and especially those who embrace capability maturity model integration or similar standard, the requirements are published in-depth and very formal and may consist of one or more artifacts with labels such as concept of operations, system requirements document, and functional specification. Regardless of the level of detail, documentation style and expense of data included, formal requirement documentation is a **best practice** for companies wanting to excel in their ERP implementation. A good requirements document removes the guessing as to what will be achieved from a needs assessment. A well-engineered requirements document will be used as the baseline for ERP solution deliverables. It renders stakeholders and users the ability to know precisely what will be delivered. As mentioned earlier, there will likely be a need to change requirements as the project progresses. That's OK, in fact, depending on your company's technical operating practices; if they adhere to scrum (agile) practices, requirements definition occurs continuously with quick response and small deliverables (about a month's worth). Regardless of the style or operating practice, change must be managed properly and the cost of change clearly evaluated and incorporated in the project budget.

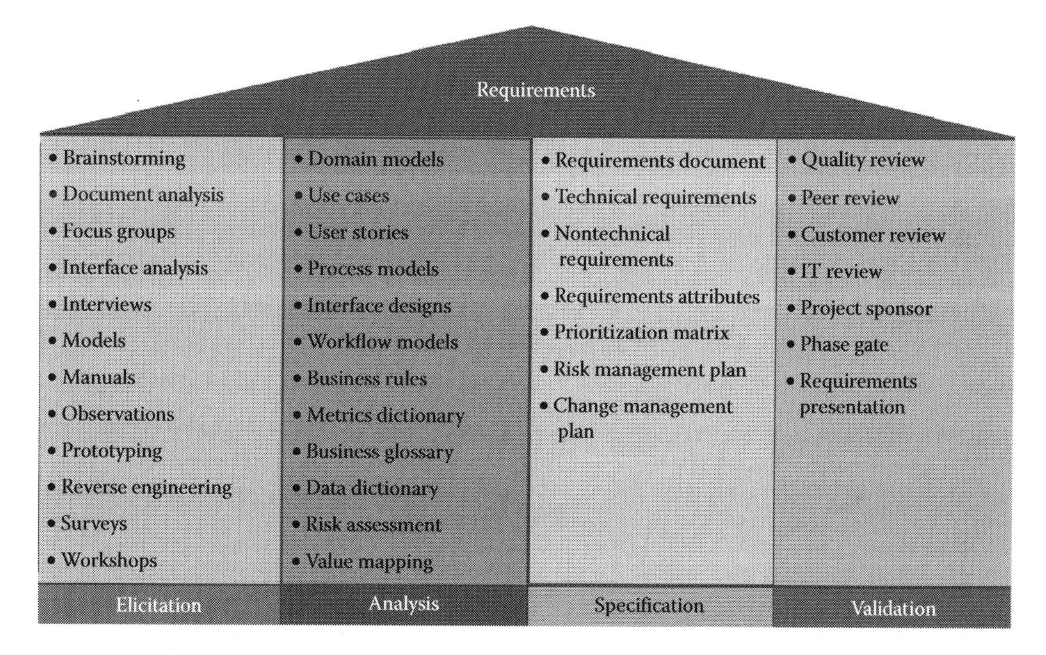

Source: Gathering Business Requirements presentation by Nikita Atkins (nikita.atkins@au1.ibm.com) at the IBM Cognos Forum.

2.6 A Final Comment about Requirements Generation

It should be clear that requirements generation sets the stage for a successful ERP implementation. Adhering to an engineered process defines the expected feature, function, and capability to be delivered as a result of the ERP implementation effort. Chapter 3 will discuss the senior leadership collaboration workshop where committed expected results will be documented. Suffice to say, the requirements generation will reflect the structured framework and process roadmap whereby an order of magnitude value of results may be achieved. As the organization evaluates their committed expected business results, there will be a variety of tools/practices, such as activity-based costing, projected future variances, and cost of change, explored which will evoke and stretch the leadership team to migrate from the menial achievements to the extraordinary business results arena. These extraordinary results do not occur by accident. Rather, senior leadership must set the stage and lead by example to make the hard decisions and blaze the process changes needed to realize an order of magnitude of business benefit (return on investment). Achieving stellar results requires a commitment to the needed change, embracing the **best practice** framework and relentless pursuit to excellence as promised in most annual reports to stockholders. The venture begins by defining exemplary requirements. If best practice requirements cannot be defined and managed, then the subsequent ERP implementation process will be destined for, at best, menial, or perhaps derailed results and subsequent failure.

Chapter 3

Senior Leadership Collaboration Workshop

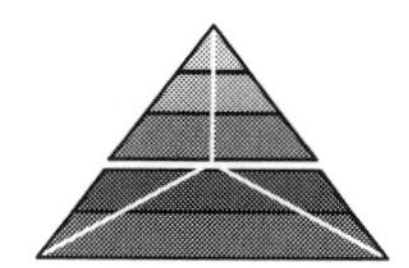

This is the forum where senior management, along with the company movers and shakers, commit to enterprise resource planning (ERP) implementation expected results ... a two- or three-day off-site meeting where the "blood mobile" draws "pints of blood" in the form of committed expected system results. This is also the forum where specific systems results are cataloged, including such areas as inventory reduction, customer service improvements, product cost reductions, and yield improvement, and a variety of other expected results are committed to and reasonable time-frames are defined. A rough scorecard is developed so that the rules of engagement (what is in-scope and what is out-of-scope) are clearly understood, and tracking may begin at the onset of the project.

An agenda that is appropriate for the workshop is as follows:

- Two- to three-day off-site session to review the significance of the ERP investment
- Rules of engagement agreement
- High-level review of requirements
- Visionary functionality
- Align requirements traceability to committed expected results and assign accountability and timetable for achieving results
- Accept resignation of any resource refusing to agree to commitment
- Agree upon measurement scorecard

I recall an ERP turnaround project I was called in to recover a derailed implementation effort. As the discovery questions, of what went wrong, were pursued, it was obvious that the stakeholders really didn't understand what the objective of the project really entailed. As it turned out, senior leadership sent a couple of working-level emissaries to a workshop to gather information about approaching the ERP implementation. Essentially, they reported back that the IT department should lead the effort and bring functional users into the effort on an "as-needed" basis. It was also clear, from mid-level leadership, that this project should not interrupt day to day getting the job done. I quickly decided to have a face to face with the CEO and understand his perspective. He had a completely different objective than what formed as the project framework evolved into. I recommended an off-site three-day workshop to drill into details and create the framework of what would be necessary to get the project on track and deliver real business results. We essentially followed the agenda listed earlier. The topics we delved into will be discussed in separate sections … The actual results of the workshop will be presented Section 3.4.

It was clear from discovery that there was inadequate commitment from all levels of the business management. Therefore, to begin with, we had to define the rules of engagement from which the project would operate.

3.1 Rules of Engagement

Rules of engagement are directives that define the latitudes within which the project will operate. They are different from a charter inasmuch as they specify not only deliverables but also what will NOT be delivered. An ERP implementation project tends to be rather comprehensive because it encompasses a very broad segment of the business. To help manage expectations, the rules of engagement clearly define, at a high level, what the project will deliver (inclusions and exclusions). These are specified so as not to have to deal with each sponsor and stakeholder individual expectations. Another way of saying it … The intent of the rules of engagement is that it is a contract of project deliverables. It is better to specify rather than assume … A rule of engagement specifies the behavior of the project regarding deliverables, expectations, and core attributes of the project.

A sample rules of engagement is described in subsection 3.1.1.

3.1.1 Project Rules of Engagement

The expected deliverables from the ERP implementation project are as follows:

- Project team to deliver the production environment (hardware, software, process, training, resource capacity, policy, and procedures) without business interruption.
  - If post GO LIVE issues arise, they will be remediated as quickly as possible according to mission critical triage priorities.
  - An issues log will be maintained with all issues resolved within reasonable time frame.
- IT to maintain concurrent production, development, demo, training, and test environments.

- Documentation deliverables.
  - Project team to publish documentation standards.
  - IT to fully document customizations, setups, and modifications.
  - Users to update/create desk instructions and procedures for every business process.
- The business fit/gap analysis option pursued is to adhere to a plain vanilla ERP software deployment and defer enhancements to phase II of the project (after GO LIVE).
- Many operational resources are frequently single-threaded and time-constrained; however, commitment has been made to give best effort to support the project Go-Live Date.
- A communication plan has been formalized and deployed. It is expected that all facets of the business (executive, operational leadership, key users, and casual participants) will be thoroughly briefed on project status accordingly.
- Project change control will be approved at the weekly/biweekly project briefing.
- A risk management plan will be published and tracked through the tenure of the project.
  - Mitigation strategies and contingency plans will be developed and tracked.
- Project health will be evaluated and posted monthly.
- An ERP requirements document will be generated up-front and used as a key input to the ERP software selection process.
  - Business unique requirements will have a 60% weighting factor for best-fit package selection.
  - Functional weighting (40%) will be weighted as follows:
    - Supply chain management—25%
    - Production operations—25%
    - Finance—10%
    - Engineering—10%
    - Logistics—10%
    - All other functionalities—20%
- Project schedule and budget will be managed tightly with less than ±5% variability to target. If variability exceeds 2% in a given month, the variance and corrective action strategies shall be an agenda item at leadership staff meetings and quarterly Board of Directors meeting. Greater than 5% deviation will have an impact on leadership performance compensation.
- Delegation of Authority (DoA) will be assigned to all project decision approval assignees. In the absence of DoA, the vote of the attendees prevails, no arbitration after the fact.
- The return-on-investment (ROI) commitments made at the senior leadership collaboration workshop will be tracked from the onset of the project. Greater than 5% deviation from the agreed-upon ROI (amount and schedule) will have an impact on leadership performance compensation. Any extraordinary events that might impact the ROI must be approved by the Board of Directors.

Creating and managing a project rules of engagement is a leadership **best practice**.

3.2 High-Level Review of Requirements

Once the rules of engagement have been agreed upon, senior leadership, sponsors, stakeholders, and other "movers and shakers" that form the intellectual critical mass of the company's business performance need to be aware of, and agree upon, the critical business requirements that the ERP implementation will address. As discussed in Chapter 2, excelling in the definition of requirements is a critical success factor (CSF) in the ERP implementation. To that end, the collaboration workshop attendees must be on the same page regarding the importance of the implementation effort achieving the deliverables from these requirements in a stellar manner. The workshop becomes the attendee's commitment to achieve these specified requirements as a minimum acceptable project quality deliverable.

Attendant to the requirements is an "expected result" associated with improved business performance regarding these requirements. Therefore, there needs to be an owner responsible for every critical requirement and its expected result. Ownership at this level should be assigned to an individual at the vice president level. Ownership also implies that their performance/compensation package should be tied to achieving the committed results. Commitment, then, must be focused as an attitude of all out emersion (effort, budget, resources, policy, etc.) to achievement of results.

It is important to recognize that company scale (small, medium, or large organizations) will have influence on how the company facilitates ERP implementation. Small-scale organizations do not have the depth of resources that medium- and large-scale organizations have at their disposal. Each deliverable associated with the ERP implementation may be evaluated based on scale. Figure 3.1 below contrasts deliverables by scale. Small-scale organizations may not have the budget to tackle a large portfolio of deliverables as do medium- and large-scale organizations. As one views the scale of organization, the list of impact shown in Figure 3.1 would be representative.

Deliverable	Small Scale	Medium Scale	Large Scale
Requirements ownership	Each key resource has a multitude of owned requirements	Small group of requirements owners	Large group of requirements owners
Expected results commitment	Each key resource has a broad portfolio of expected results	Broader group of key resources has a limited portfolio of expected results	A very broad group of key resources has very limited portfolio of expected results
Responsibility/accountability	Each key resource has a broad portfolio of responsibilities and accountabilities	Broader group of key resources has a broad portfolio of responsibilities and accountabilities	A very broad group of key resources has a broad portfolio of responsibilities and accountabilities
ABC costing	Not likely	Finance owns tool with limited accountability for generating results	Decentralized ownership and accountability for generating vast cost savings
ROI targets	Each key resource has a broad portfolio of ROI targets	Broader group of key resources has a broad portfolio of ROI targets	A very broad group of key resources has a broad portfolio of ROI targets
Cost segmentation	Not likely	Sales/marketing owns tool with limited accountability for generating results	Decentralized ownership and accountability for generating significant profit improvements

Figure 3.1 Deliverable scale chart.

3.3 Visionary Functionality

If it wasn't completed during the requirement definition, this would be an ideal time to "stretch" the organization, a bit, by exploring some expanded or visionary capability of the ERP functionality.

3.3.1 Projected Future Variance

Instead of merely reporting and managing variances, you may want to enhance the reporting to PROJECT FUTURE VARIANCES. Projecting future variances allows the operating team to take early corrective action to avoid unfavorable variances. This preemptive strategy may be used on a variety of variances as described below:

1. *Purchase price variance (PPV)*—A look ahead at future variances might allow the ability to combine purchase events in the future to obviate unfavorable PPV. If looking out into the future progressively, if given enough time, a project could be formed to reengineer the part in a manner that would reduce costs.
2. *Material usage variance*—If you previously experienced unfavorable material usage, resulting from a completed work order, if there is a replicated work order planned in the future, an early warning trigger could be generated to notify the accountable department ... The accountable party may then take the necessary steps to obviate a recurrence of an unfavorable material usage variance before beginning work on the planned work order.
3. *Labor efficiency variance*—In a like manner to material usage described above, if you previously experienced unfavorable labor efficiency, resulting from a completed work order, if there is a replicated work order planned in the future, an early warning trigger could be generated to notify the accountable department ... The accountable party may then take the necessary steps to obviate a recurrence of an unfavorable material usage variance before beginning work on the planned work order.
4. *Material substitution variance* (same as variances 2 and 3)
5. *Labor usage variance* (same as variances 2 through 4)
6. *Material standards variance* (same as variances 2 through 5)
7. *Routing variance* (same as variances 2 through 6)
8. *Labor standards variance* (same as variances 2 through 7)
9. *Machine efficiency variance* (same as variances 2 through 8)
10. *Machine utilization variance* (same as variances 2 through 9)
11. *Labor rate variance* (same as variances 2 through 10)
12. *Machine rate variance* (same as variances 2 through 11)
13. *Yield variance* (same as variances 2 through 12)
14. *Scrap variance*—The difference between the amount of the authorized scrap and the actual amount of scrap generated on a work order
15. *Outside processing variance*—The difference between the authorized cost and the actual cost of outside processing
16. *Setup rate variance*—The difference between the authorized setup rate and the actual setup rate
17. *Setup cost variance*—The difference between the authorized setup cost (allotted time) and the actual setup cost
18. *Overhead rate variance*—The difference between the authorized overhead rate and the actual overhead rate

19. *Overhead cost variance*—The difference between the authorized overhead cost (allotted time) and the actual overhead cost
20. *Configure to order (CTO) variance*—The difference between the authorized CTO cost (allotted time) and the actual CTO cost

3.3.2 Cost-of-Change Analysis

There are a plethora of opportunities to enhance the integrity of decision making associated with cost of change. This analysis technique is seldom included in the ERP solutions. Essentially, a cost-of-change analysis permits the modeling of the cost impact and gives a pro-forma financial effect before executing the change. Similar to the projected future variance capability discussed in subsection 3.3.1, the simulation of the cost impact allows decision makers to make intelligent decisions, regarding cost. Many times, there are attributes, in addition to cost, that may be modeled to improve the quality of the decision-making process. This preemptive strategy may be used on a variety of cost-of-change impacts as described below:

- *Product (engineering) changes*—The evaluation of the impact for adding, deleting, or changing product design
- *Process (engineering) changes*—The evaluation of the impact for adding, deleting, or changing process design or routing
- *Material substitution changes*—The evaluation of the impact for adding, deleting, or changing material composition (work order parts list)
- *Labor utilization changes*—The evaluation of the impact for using alternate labor work centers or routings
- *Machine utilization changes*—The evaluation of the impact for using alternate machines or equipment on the work order
- *Make/buy changes*—The evaluation of the impact for alternating from a *make* or *buy* posture to the alternate posture

Note: One or more of the projected future variance listing of variances (see subsection 3.4.1) may be calculated for cost of change if the magnitude of activity warrants.

3.3.3 Triggers, Drill-Downs, and Simulations/Projections

As stakeholders and users are afforded to "dream" of ideal systems features, there is an unlimited array of productivity improvement opportunity at their disposal. To assist in early warning or preemptive alerts, triggers may be used to keep tasks, schedules, costs, and so on "on course." Triggers may be deployed, virtually anywhere, when goal achievement, production rates, or other measureable activities are essential. Associated with triggers are escalations, broadcasts, and other communication tools that are bent on achieving successful operations. Not only is an individual alerted, but a nested team of participants may be included in a workflow … If deployed before unfavorable results occur, this tool becomes an essential capability to attain the goals and objectives associated with such things as annual operating plans, individual performance achievements, group achievements, or company-wide achievements.

In addition to triggers, there are other productivity tools that may be deployed, one of which is drill-downs. Drill-downs broaden the scope of data and frequently open a user's perspective from merely data analysis to a more "information"-related awareness. There are a variety of drill-down capabilities that may be explored, including the following:

- *Activity-based costing*—The value associated with drilling down to the essential "cost drivers" frequently leads to the development of strategies to significantly reduce cost.
- *Cost segmentation*—A valuable capability, for many companies, is the ability to drill-down and analyze individual product and/or customer contribution to profitability. This analysis is frequently overlooked as a meaningful tool for achieving improved profitability. Frequently, I discovered that this has never been done, or is performed so infrequently, that significant reduction opportunity eludes the operating units. This activity should become a regular review task assigned to an accountable individual and regularly measured.
- *Productivity dashboard*—This is the ability to drill-down and readily monitor selected functions on a minute-by-minute basis. The value of measuring production rates, quality assurance, or a variety of other important measures in minute details help prevent runaway costs and minimize potential productivity disasters.

One final tool capability that might be pursued is the use of simulation technology and/or artificial intelligence (neural network) enhanced projection logic to help monitor or predict task results. If modeled properly, these tools may function as an early warning trigger to help offset or obviate potential unfavorable impacts. Similar to the production dashboard discussed above, these tools may be deployed in the background rather than real time, yet with near real-time effectiveness.

Regardless of the visionary tools deployed, the purpose is to stretch the envelope in a manner that will allow the organization to gain exceptional performance enhancement capability that will move the user from merely lumbering along with minimal results to significant results. Use of this capability is a reporting **best practice**.

3.4 Align Requirements Traceability to Committed Expected Results and Assign Accountability and Timetable for Achieving Results

One of the most important deliverables from the collaboration workshop is agreement, by participants, on the results that will be forthcoming from the successful implementation of the ERP product. To most companies, the ERP project will likely be one of the most costly projects ever pursued. Because ERP typically encompasses such a broad impact to the business, the investment is significant. Some large organizations spend significant amounts on the hardware, software, consulting, software modifications, internal resources, and ongoing cost of their ERP implementation. Because of the investment significance, it is right to help ensure that there will be a reasonable ROI by the business unit (executive leadership, sponsors, stakeholders, managers, users, etc.). Therefore, at the forefront of the project, the workshop attendees need to quantify the expected results and commit to their attainment within a reasonable time frame. Each ERP project will be targeting their focus on "business improvements" in different areas. However, there is a common thread across all organizations, which needs to be mandated ... The improvements cannot be merely lip service, rather they need to be a commitment tied directly to participant's performance management process. In other words, if the company makes the ERP investment, that investment will be foreshadowed by realized (actual) results.

Popular ROI targets include the following:

- Inventory reduction
- Customer delivery performance improvements
- Product cost reduction (material and labor)
- Overhead cost reduction
- Improved equipment utilization
- Improved labor efficiency (direct and indirect)
- Improved supply chain performance (across tiers of supply chain)
- Improved total cost of ownership
- Improved utilization of fixed assets
- Improved throughput
- Improved demand management
- Reduced personnel turnover rate

There are a variety of ROI modeling tools readily available on the Internet, and many are free. The important result from this ROI commitment is to be specific in the expected results (percentage or dollar savings, specific rate improvements, etc.). There are a couple of keen precepts regarding these "specifics": One is agreed upon the starting baseline (basis from which we measure results) and the other is a credible method for validation. For example, if inventory is to be reduced by 60%, what is the starting inventory value at the onset of evaluation period? The baseline needs to include rules of engagement; for example, if the actionable deliverable is merely to claim that 60% of the inventory is obsolete and, therefore, throwing away obsolete inventory achieves the expected results, this is not netting any improved business results and foolish. A much better approach might be to segment inventory into two categories: existing inventory and future purchase inventory. The goal might be to redeploy "obsolete" inventory (through engineering obsolete parts into future products, exploit after-market products, etc.) while reducing future inventory purchases by reducing the number of parts (reengineering the products) and working with the tiers of the supply chain to help tier suppliers reduce their costs (better processes, better utility of tooling, better labor utility, etc.). Each ROI commitment target area needs to follow a similar process.

The last precept is the results validation process. Many organizations agreed upon a scorecard that transcends ROI commitment targets whereby each area is rationalized into agreed-upon measures that are "fair" across the organization. As discussed above, the measurement, and results validation, should begin at the onset of the project and incrementally tallied as the project timeline "% complete" is evaluated. The reason for incremental evaluation is to allow for implementation process changes to be invoked keeping the results realization on target while the project progresses. A CSF of the validation process is to gain agreement on the scorecard by all participants in the beginning. This will help ensure that each team member not only facilitates their committed expected results but also helps the other team members meet their objective (synthesized collaboration).

The **best practice** in this area is to engineer an agreed-upon process that yields an order of magnitude of business improvement results to transform the business into a lean, mean, high-productivity business operation. If the process is well engineered, then it will gain unanimous acceptance by all participants. Unanimous means that executive leadership must "accept the resignation" of any resource refusing to agree to their accountabilities regarding this commitment.

3.5 Agree upon Measurement Scorecard

We briefly discussed the measurement scorecard for the expected results (ROI) from the ERP investment. Depending on the project, there are likely other scorecards needed to properly manage the ERP project. In Chapter 1, there was the project health (**ProjHealth10**), which is a measurement scorecard for the overall health of the project. Other scorecards that might be used may include requirements generation progress, integration testing progress, or any pertinent deliverable that might benefit from measurement and status reporting. The determination of which measurement scorecards to use should be an agenda item at the project kick-off meeting, which would consist of executive leadership, sponsors, stakeholders, and other key team members of the overall project effort. Figure 3.2 shows an example of an ERP expected results scorecard. Using the scorecard, to rigorously track results, is a reporting **best practice**.

> I recall a project that initially had failed to incorporate a measurement scorecard as part of its deliverable portfolio. At one Steering Committee Briefing, an executive asked the segment team member a question regarding progress and the criteria used for reporting results. After a variety of stutters, hems and haws, and a trip down a rabbit path, it was decided that there was a need for a variety of measurement scorecards, with clarity regarding criteria, accountability, reporting frequency, and a whole host of other pertinent attributes. The project core team members determined the number of scorecards needed to properly monitor the project status, defined the criteria and other attributes essential for each, and obviated the need for future measurement-related diversions during the Steering Committee Briefing.

To conclude the discussion from the beginning of this chapter where there was a need to turnaround a derailed project by conducting an off-site collaboration workshop. There is a distinct difference between what the CEO expectations were and how the project began. During the

ERP implementation	Commitment amount	Year 1 Q1	Q2	Q3	Q4	Year 2 Q1	Q2	Q3	Q4	Year 3 Q1	Q2	Q3	Q4	End of project
Inventory reduction	40%↓	GR	GR	GR	GR	GR	GR							B
Day sales outstanding	80%↑	GR	YL	GR	GR	GR	GR							B
Customer delivery improvement	65%↑	GR	YL	YL	YL	YL	GR							GR
Product cost	38%↓	GR	GR	GR	GR	GR	GR							GR
Labor efficiency	30%↑	GR	YL	YL	GR	YL	GR							GR
Improved throughput	50%↑	GR	GR	GR	GR	GR	GR							GR
Etc. ↓														

B Excellent > plan; GR Satisfactory = plan; YL Marginal < plan; R Unsatisfactory

Figure 3.2 ERP expected results scorecard.

workshop, it became clear that a significant gap existed between the CEO and the operating management team. The CEO intended to only spend an hour or so at the session and return to work (on more important things). As we progressed and started to scribe the expected results opportunity within the business, the CEO recognized that this project would likely result in a completely different business operating model and this interested him greatly. As we developed the criteria for the scorecard and started quantifying the commitment, both the operating layer and the executive leadership layer got excited about this project for the first time. This cohesion quickly recognized that without their active engagement and commitment, there would be limited business results. The CEO urged that aggressive goals be pursued and he tagged the HR director to determine how these results could be weaved into the performance bonus process and the total team compensation. The cost accounting group was tagged with monitoring the expected results and to align the project monitoring with expected results monitoring including a monthly briefing going to the Board of Directors. Needless to say, this company was really committed to improving their bottom line as a result of their ERP effort.

WRAP-UP I

This concludes Section I where we examined the importance of creating a planning environment by engineering a realistic project plan, diligently gathering system requirements, and conducting a senior leadership collaboration workshop to distill the "expected results" (return on investment [ROI]).

We developed a roadmap of project deliverables that impacted four environments (technical, business operating, user, and project), discussed readiness toll gates, and developed the best practice ingredients essential for a reliable project plan. We also talked about project health and importance of the statement of work (SoW) to the integrity of the plan, and discovered the value of change management and risk management to a successful project.

We discussed the importance of generating pristine requirements (functional and process), accounting for attributes of requirements as well as process engineering. In addition, we looked at the value of a traceability matrix that tracks the requirements through the project life cycle.

Finally, we discussed the merits of conducting a senior leadership collaboration workshop where all the movers and shakers within the organization come together on agreements such as rules of engagement, scorecards and their criteria, and most importantly, a commitment to the bottom-line performance improvements which will result from implementing enterprise resource planning (ERP).

Obtaining improved business results will align with doing a good job implementing the ERP solution. However, obtaining stellar results (even order of magnitude improvements) will only result from managing critical success factors and adhering to best practices, which include the following:

- Developing a risk management strategy and mitigation plan is a leadership best practice
- Generating and adhering to a meaningful SoW
- Populating a broad library of deliverable artifacts
- Fostering lean change management practices
- Documenting a robust requirements list (including business drivers and a traceability matrix) aligned to expected results tracking
- Creating a robust library of documentation that supports the requirement and process deployment
- Using out-of-the-box visionary tools

- Obtaining a commitment to expected results and validating their accomplishment
- Using a scorecard to track and help ensure attainment
- Creating and managing a project rules of engagement is a leadership **best practice**

Now we'll begin discussing the topics of Section II.

FOUNDATIONAL PRINCIPLES, TOOLS, AND STANDARDS

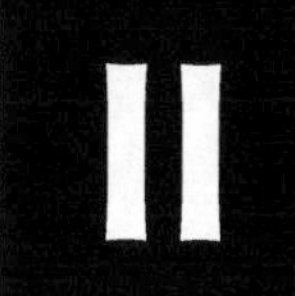

Overview

This section addresses the practical deployment framework essential for success and includes a variety of tools that position the company for success.

- *Chapter 4*—This is the minimum acceptable quality level for transactions, job functions, work processes, and ultimately the resulting information. Without a standard there is confusion regarding work expectations. The Information Workmanship Standard, such as financial, quality, and a variety of other standards, clearly defines the expectations associated with information. In addition, it fully develops the nested internal customer and supplier of a service framework to define process-based performance metrics that reflect in-the-trenches end-to-end business process activities. Engineering "process-based" metrics allows players from the entire organization to understand their specific contribution to profitability, which is lacking in traditional hierarchical metrics.
- *Chapter 5*—Test driving the blending of functionality reflected in the design (features, functions, and capabilities) as well as processes (how the design is configured) and the environmental structure (policies, procedures, and performance metrics within the players' culture). This chapter not only pursues project team piloting but also demonstrates senior leadership piloting, customer/supplier piloting, and other business partnership piloting.
- *Chapter 6*—The backbone to system success involves the entire user community exhibiting the competence and mastery of the new system. Contrary to popular practice, this does not occur through osmosis or attending a couple of training classes. Like all competency processes, this must be achieved through proper design and fulfillment.

Chapter 4

The Information Workmanship Standard

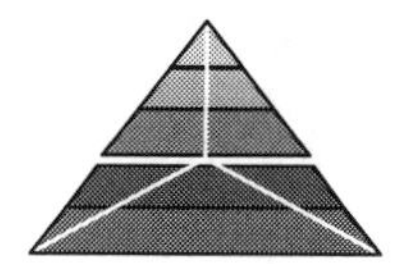

The Information Workmanship Standard (IWS) is defined as "the minimum acceptable quality level for transactions, job functions, work processes and ultimately the resulting information." Without a standard there is confusion regarding work expectations. The IWS, such as financial, quality, and a variety of other standards, clearly defines the expectations associated with information. In addition, it fully develops the nested internal customer and supplier of a service framework to define process-based performance metrics that reflect in the trenches end-to-end business process activities. Engineering "process-based" metrics allows players from the entire organization to understand their specific contribution to profitability, which is lacking in traditional hierarchical metrics. The IWS may be deemed the "Pinnacle in productivity optimization."

This topic is one of the most elusive precepts in the enterprise resource planning (ERP) implementation journey. As I reflect back on ERP implementation efforts and their results, the IWS becomes one of the most ignored areas, yet a foundational critical success factor (CSF) for a rewarding ERP implementation conclusion. On the surface, ERP project effort leadership teams tend to believe that the information quality within their organizations is adequate. Consequently, the ERP implementation project team members seldom challenge the need for an information quality standard.

I recall a project, where the program I worked on, used the external "customer's" data repository (fully collaborative subsystems). Because the customer controlled the data environment, I asked the question "what is the Information Workmanship Standard?" and received no reply. From experience, I know that if there is NO STANDARD, the record quality is at the whim of the beholder and becomes questionable. The customer was completing a new subsystem repository for critical logistics records and determined to GO LIVE with the existing quality of records, without a review for accuracy. Inasmuch as these records were a "contractual deliverable" and record quality would impact mission critical readiness, the implied record accuracy was in the 99%+ realm. As a skeptic and with NO STANDARD to fall back upon, I conducted an audit of the newly launched subsystem data repository. The result of the first audit was deplorable, at less than 50% accuracy. In conducting a root-cause analysis, there were a variety of factors that contributed to the poor record quality, including missed records, inaccurate records, and a variety of other contributing factors. Corrective action required over six months of collaborative effort, with dozens of combined (both internal and external customer) resources, to establish reasonable integrity of these mission critical records.

Another precept, associated with the IWS, is the engineering of the internal customer and service provider relationship standard (service-level agreement [SLA]). At the core of the nested internal customer/service provider relationship are the following attributes:

- The need for high-quality information flow
- Agreed-upon service delivery—SLA
- Agility, nimbleness, and fast action responsiveness between nested players

Let's look at the IWS in more detail.

The outline of topics is shown in Figure 4.1.

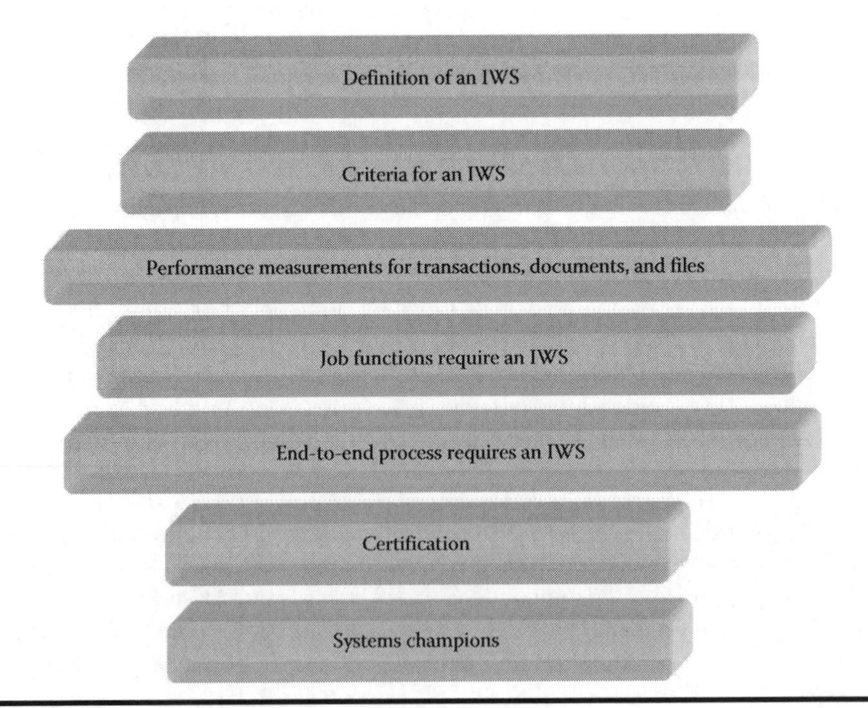

Figure 4.1 Outline of IWS topics.

4.1 Definition of an IWS

An *IWS* is essential to businesses that are serious about maintaining a high degree of accuracy in their record-keeping activities and want exceptional, optimized operational performance results. An IWS defines the *acceptable quality performance level* necessary to achieve record-keeping excellence and functions as the *internal SLA* between the internal customers and their service providers. The IWS allows the business process functionaries to know what is expected of them, to manage information on a timely basis and deliver information excellence. The IWS is the information workflow **best practice**. In short, the IWS is a method to synthesize information flow across business processes and account for performance accuracy between the internal customers and their service providers on an end-to-end process basis. Today's fast-paced and competitive environment requires timely information as an integral asset in the day-to-day decision process. Companies that maintain high-quality and responsive business information systems use information as a competitive weapon. Timely, accurate information is the tool of astute executives in their pricing decisions, cost variance management decisions, scheduling activities, and customer performance excellence assessments.

Like so many companies' policies and procedures, an IWS without management care and feeding over time, becomes stale. An IWS should be a living and constantly updated tool if it is to reach its potential as an integral weapon within a company's competitive arsenal. Traditionally, there are various types of workmanship standards within an organization. For example, there may be a workmanship standard used for documentation within a design engineering function, or a workmanship standard for various manufacturing processes, or a workmanship standard for product assurance and testing. One of the most commonly followed workmanship standards emanates from the financial side of the house in the form of accounting standards. These standards are present primarily due to generally accepted accounting principles (GAAP) fueled by the annual review by a company's external auditors.

However, few companies have any semblance of an IWS. Information that feeds the accounting system and affects the quality of book numbers is typically lacking a standard. Because the sources of the accounting information lack a standard, how accurate can the accounting data be?

Note: Accounting uses information from the rest of the company as its source. Establishing standards downstream is important to ensure that the closing values follow consistent procedures, practices, and cutoffs. However, if the source of the information upstream is not equally managed by a standard, the quality of downstream information can be no greater than the upstream source. For example, let's assume that the inventory value booked by accounting is based on the computer's on-hand inventory valuation from the manufacturing system. Let's further assume that the accuracy level of on-hand inventory is 80%.

In lieu of an IWS, companies must compromise. Accounting can base its book valuation upon the following possible sources:

1. ERP system (80% accurate)—way too low
2. Take a monthly physical inventory—costly validation technique
3. Creative management accounting—a technique of interpolating information and adjusting based on certain assumptions and expertise

None of the above alternatives yields the desired (100% quality level) results. A company employs its best judgment in order to close its books on a monthly basis, yet each month introduces the same dilemma. The dilemma cannot be overcome until the "source" causing the discrepancies is addressed. The source involves documents, initiation and processing standards, individual employees' standards, data auditing, a closed loop feedback, measurement, and corrective action techniques. These issues will be developed further below as "criteria" are discussed.

It is an interesting dichotomy that companies either already have or are in the process of implementing computer-based systems. However, the standard that should be in place prior to implementation is prominent by its omission at point of conversion. This omission is further salient by the fact that it is not unusual for a company to spend 4%–5%, or more, of a sales dollar for an information (computer)-oriented support budget. Yet something as basic as the IWS is not in place. Subsection 4.2 will describe the criteria for an effective IWS.

4.2 Criteria for an IWS

The IWS comprises the following ingredients:

- Each transaction, document, and file within the system will have a performance measurement associated with it (e.g., on-hand balance accuracy, purchase order [PO] accuracy, etc.).
- Each job function within an organization shall have a uniquely defined IWS associated with it.
- Each end-to-end process IWS should incorporate a composite of all the job-level IWSs within its business process, and the performance measurement of the business process as an entity. In addition, the end-to-end process audit guidelines are necessary ingredients for the total system-wide IWS.

One of the missing elements that many companies lack, in rolling out their corporate performance objectives, is the **vision** on how the performance is to be achieved. In addition, recognition that a company consists of various business processes synchronized together to achieve common objectives. However, performance excellence does not occur by osmosis but requires **focused and diligent effort** to realize the expected results. Agreeing upon **acceptance criteria** is the origin of the journey toward performance excellence. An acceptance criterion removes any doubt as to what is expected between the SLA team members. Once the acceptance criterion is established, **meaningful metrics** may be defined and then the **roadmap** on how to achieve these results may be finalized. Any of these CSFs omitted will result in suboptimized performance.

An essential "quality" component of *criteria* is information validation and auditing. The auditing/validation ingredient should **not** consist of bureaucratic approvals or other methods to delay information processing. On the contrary, the process of auditing and validation should be engineered to facilitate more timely updates. Information validation may be accomplished by use of one or more of the following techniques:

1. *Data element editing*: Predescription of the authorized range of values acceptable for each data element. For example, a company's unit of measure table may only consist of "ea.," "ft.," and "gm." All transactions dealing with an on-hand value will be edited to accept only the units of measure specified. Transactions with any other unit of measure will not be permitted. As another example, a company's part number may be edited to allow only numeric characters, with the size of the part number restricted to nine digits. Likewise every data

element should be reviewed for the possibility of defining an electronic acceptable range edit. This editing should take place real time, when feasible.

2. *Data element continuity editing*: The preestablishment of an acceptable combination (configuration) of two or more data elements. Once the acceptable range for every data element is established, some combinations of data elements may be illogical. The electronic validation of the combination of two or more data element values may be accomplished real time (at data entry) or scanned via a deferred processing audit report. **The timelier the validation, the better.** An example of a multiple data element continuity edit is as follows: establishing a reorder policy with a minimum value of 100 and a maximum value of 50. It is illogical. Another example is as follows: establishing the part type as "pseudo" (blow-through) yet a bill of material component type (used in) for the part as "standard" (not blow-through).
3. *Editing for reasonableness*: The preestablishment of a reasonableness range of values for critical data elements. For example, it is probably unreasonable for a part that has component values of cents and less dollars to have a unit cost of millions of dollars; therefore, inputting a value of $1 million would be electronically challenged. Every data element should be likewise reviewed to determine if a reasonableness test is appropriate.

Note A	This is usually an alternative method of editing when an acceptance range cannot be established (as in 1 above).
Note B	This method usually affects only numeric and date fields.
Note C	The reasonableness challenge is usually set up to allow the input operator to "override" the challenge (warning) rather than preventing input as described in (1) above.

4. *Controlled document edit*: Certain control documents (such as PO invoice numbers, payable check numbers, etc.), which are not key field values, may be edited to ensure that they are within a proper range. This editing is especially helpful for document control purposes. For example, if the current valid PO number range being used were 70001 through 79999, the input of 35001 would not be acceptable (or at least would be electronically challenged).
5. *Missing data edits*: Input transactions should be electronically reviewed (prior to input processing) to ensure that all required data are present. For example, an issue for a stock transaction should include an edit for the proper unit of measure from the part file.

   The above editing occurs with the assistance of computer review at the time of input. Individuals preparing input transactions should manually perform a quality review on each document to ensure that it includes all the necessary information and that it is legible. In addition, the transacting individual should also ensure that it is performed on a timely basis, especially sensitive to cutoff dates and times. (This should be part of the individual's IWS.)
6. *Internal customer acceptance*: By far, the best audit is not necessarily a computer review at all, but the acceptance and approval of the *internal customer*, who uses or benefits from the quality of the data. *Internal customers and their service provider team members are at the heart of the IWS philosophy.* Internal customers negotiate a quality and service performance level from each of their process service providers. This negotiation concept will be discussed in Section 4.8.

As we critically think about accuracy, there is a difference between "control" data (major impact on decision process) and textual data (minor impact on decision process). Emphasis needs to focus upon the control data, yet it needs to be recognized that all data reflect upon the company image and reputation. Scale of accuracy is also an important criterion ... If your company embraces the Six Sigma concept, then the expectation for defects is 3.4 defective units/million. Six Sigma techniques were founded on the quality of manufacturing defects on manufactured products. However, this concept may be applied to any aspect of the business, from inventory accuracy, sales order accuracy, PO accuracy, to a minute aspect of a process. It is important for the company to specify the meaning of accuracy in all metrics. This is a core ingredient to the SLA and defines the performance measurement acceptance criteria.

Now that the scope has been defined, a review of the above ingredients in light of the scope is appropriate.

4.3 Performance Measurements for Transactions, Documents, and Files

The computer editing of data described above (1–5) should constrain a vast amount of faulty data resulting from data entry errors. However, it will not totally inhibit erroneous data.

Tools such as bar coding and digitized scanning improve data entry substantially. However, there are still large amounts of data being entered via keyboard and mass transaction processing. Therefore, it is necessary to establish an acceptable quality performance level for each transaction, document, and file. For example, an acceptable quality level for the inventory balance file may be 99.9%. Validating the achievement of that level requires the following:

1. A reasonable means to validate the accuracy level (e.g., cycle inventory programs).
2. A valid method to identify the root cause of an error (e.g., reconciliation procedures).
3. Preventive measures to head off sources for data contamination (e.g., bar coding, scanning, and an effective ongoing education and training program).
4. Management intervention through corrective action (e.g., making it an agenda item during staff meetings, including data accuracy goals in individuals' performance review criteria, etc.).
5. An aggressive awareness **process** is needed, which clearly specifies the *acceptable quality levels*. This awareness **process** may include one or more of the following:
   a. Conspicuously posting achievement levels outside work areas, cafeterias, and so on
   b. Quality-level performance in monthly progress reports to top management
   c. Quality-level performance in the company newsletter or other house organizational literature
   d. Quality-level performance in the employment *job description* so new job applicants understand the standard when they apply for a job
   e. Having contests and awards for consistently achieving a preestablished level. This may include such things as a monthly catered luncheon if goals were attained, certificates, and company meetings that announce attainment

Information data files become accurate and sustain accuracy by establishing goals and measuring performance to these goals. Within these data files, inaugurating similar attention and review to the component elements that influence the accuracy level, namely, transactions and documents, influences record accuracy. Focus upon the proper entry of all relevant data is necessary for input forms, if achieving accuracy goals at the output level (files) is to be attained. (Accuracy of each data element insures accuracy of the whole document, and consequently accuracy of the file.)

4.4 Job Functions Require an IWS

An IWS for a job function should be an integral part of a job description; however, the premise of an IWS is different than the premise of a job description. *A job description typically broadly defines the scope of responsibility and tends to relate to global issues.* In comparison, an IWS is very finitely focused on specific goals and clearly defined objectives. *The IWS defines a specific performance level, within a process, on a document-by-document basis and individual-by-individual basis, as an increment of an end-to-end business process within the value chain.*

Note: A document may be a form, a computer process, a drawing, a triggered exception, or any other media that convey workflow tasks needed to achieve business performance excellence.

The context of a receiving associates IWS includes the combination of subsections 4.4.1, 4.4.2, and 4.4.2.1.

4.4.1 Documents

1. Receiving memo
    a. *All receipts* will be processed into the system within 15 minutes of physical receipt of goods, measured by the date/time stamp of the packing slip, and monitored by a weekly review from the department supervisor.
    b. The data entered will achieve 99% accuracy level on the following data elements:
        i. PO number
        ii. Item number
        iii. Quantity
        iv. Unit of measure

    All other data elements will be within 97% accuracy. The level of accuracy will be based upon the following data:
        i. Packing slip data
        ii. PO data
        iii. Item master data
        iv. Accounts payable data

The accuracy process will be audited by the combination of the following activities:

- Online editing
- Receiving inspection review (internal customer)
- Receiving inventory exceptions (internal customer)
- Accounts payable three-way match (internal customer)
- Internal spot audit of receiving activity

2. Receiving memo process certification
    To achieve receiving memo process certification, it is mandatory to receive personnel to process transactions at the *minimum acceptable quality level of 95%* prior to being allowed transaction update privileges. Certification should be an essential ingredient for new employees to pass the 30–90-day probation period. The inability to become certified within a 30-day period should be grounds for immediate dismissal.

4.4.2 Return-to-Vendor Credit Document

The documents for the receiving associates IWS include subsections 4.4.1 and 4.4.2.

1. All return-to-vendor (RTV) credit documents will be processed within 30 minutes of receipt of document from purchasing, measured by the date/time stamp entered on the document.
2. The data entered will achieve 99% accuracy level on the quantity returned data element.
3. The level of accuracy will be based on the following data:
   i. PO data
   ii. Item master data
   iii. Debit memo data

The accuracy process will be audited by the combination of the following activities:

- Online editing
- Receiving inspection review (internal customer)
- Debit memo reconciliation (internal customer)
- Internal spot audit

4.4.2.1 RTV Credit Document Certification

To achieve RTV credit process certification, it is mandatory for responsible personnel to process transactions at the *minimum acceptable quality level of 95%* prior to being allowed transaction update privileges. Certification should be an essential ingredient for new employees to pass the 30–90-day probation period. The inability to become certified within a 30-day period should be grounds for immediate dismissal.

4.5 Data Accuracy

The receiving associate will maintain receipts in process on a receipt-by-receipt (item level) basis of 99% accuracy level. This will be audited by a receiving inspection physical count, on a sample basis, and a stockroom 100% sample review.

Within certain job functions, it may be necessary to tailor an individualized IWS for each person depending on the knowledge level, grade, and/or password security level of the individual. For example, an employee who has functioned within the job for over 12 months should achieve a higher performance level than an employee with one-week experience. However, a caution is necessary here. If your objective is 99%–100% process accuracy level, even recently hired employees must function at peak performance if the goal is to be attained.

How can this be accomplished? Simple! As described in the example above, an individual should not be allowed to perform transactions on documents against the "production database," without having achieved the specified *minimum acceptable quality level* on the conference room pilot (CRP) (training) (see Section 5.2) database. This process is frequently termed *certification* or *software certification.* It is analogous to the process of obtaining a vehicle driver's license. For example, in most states you must pass both a driving test, hands-on, and a written test before you can obtain a drivers license. A certification program is very similar. For certification, an individual must pass a hands-on (online) and written exam in the CRP database before being allowed security access to the "production database." If the individual *cannot* pass both the hands-on and the written test, they should be reassigned into a responsibility area that does *not* involve data input (or creating any source document) or they should be considered for dismissal. Anything short of this process rigor is, by default, endorsement by management that data contamination is acceptable behavior by its employees.

The certification process needs to be tailored by job (and/or employee) and requires recertification whenever an employee is hired, has a job transfer, or is promoted. Recertification may also be required when new software modules are added. An annual review for recertification provides an increased insurance policy for individual proficiency.

At this point, many readers are saying:

"Our company will never buy this concept ... it is too expensive, it is too time consuming, it is too" ________ (you fill in the blank). However, I will challenge you:

What is the annual cost of obsolescence?
What is the annual cost of lost customers?
What is the annual cost of lost sales?
What is the annual cost of inaccurate product costs?
What is the annual cost of employee turnover?

The frustration level, decisions, and actions from day-to-day operations are all influenced by information accuracy. Some, if not all, of the above "What is the annual cost?" items are influenced by whether a company has information standards or not. Once a company assesses the *cost of quality* (cost of rework, poor decisions, etc.) resulting from inaccurate information, it will be clear that it is massive compared to the cost of providing a *certification program.*

The companies surviving in the next decade will be those companies who have and use information as a competitive weapon. The difference of a 1%–2% accuracy level may be that edge which elevates one competitor over another. Consequently, defining these performance measures on a job function and/or individual level will be essential in the future. (Appendix D.1 illustrates the representative IWSs by job function.)

4.6 End-to-End Process

The end-to-end process IWS comprises the composite of every job function IWS within the business process value chain, and any additional goals and objectives for the business process itself. Synergy (the whole is greater than the sum of the parts) promulgates the direction that standards should pursue.

For example, let's continue to amplify the receiving associate's IWS and integrate it within the "quote-to-cash business process IWS" based on the receiving associate's job function IWS. Within the quote-to-cash process IWS, a receiving memo will require a 99.5% process accuracy level and all receiving documents will be transacted within 15 minutes of the time of receipt. Any performance failure to the 15-minute transaction processing standard is recorded based on 15-minute increments, the reason for processing delay (e.g., system down, unavailable purchase documentation, etc.), and the corrective action taken.

4.7 Performance Goals and Objectives

By nature, most individuals are more comfortable being a part of a group than being alone. Therefore, it is natural to obtain consensus on the group's commitment and process performance level before acquiring any individual commitments. Research has shown that *higher quality results* are obtained when goal-setting sessions are performed as a group before individual (or job function) goal setting occurs. Also, workmanship standards goals should be constantly changing (at least annually), consistent with company goal changes. The specific attributes of goals are as follows:

1. Goals should focus upon constant tightening of tolerances as time goes on.
2. Goals should be achievable and a source for "pride" within the department.
3. Goals need to be championed by individuals if success is to be attained.

To relate this "goal-setting" methodology to the end-to-end process IWS, each business process work element should establish its IWS objective (see Appendix D.2 for an example of end-to-end process IWS objectives):

1. The performance attainment should be based on the objective within the business process and, when appropriate, stepped up (with tighter tolerances) since the last review.
2. The end-to-end business process goals should be a subset of the company-wide IWS goals and related performance measures. Company-wide measures are not normally attainable if all business process goals do not focus upon and add up to company-wide objectives. Many companies now extend the end-to-end business process concept through the entire expanded supply chain (customers and suppliers including across multiple tiers).
3. The logical hierarchy is that there are company-wide IWS objectives that are disaggregated into end-to-end business process IWS objectives. These end-to-end process objectives represent the internal customer and service provider team's nested element objectives (SLAs), which are then disaggregated further into job function IWS objectives. Ultimately, the IWS objectives are personalized to the individual. The IWS hierarchy is shown in Figure 4.2.

There is typically a one-on-one relationship between a physical organizational structure and an IWS structure (although an IWS structure is usually more modularized along functional lines).

The end-to-end business process (internal customers and their nested service providers) IWS becomes the catalyst for encouraging improved performance and, ultimately, tightening tolerances. The business process is the lowest operational level of accountability that maintains a cohesive association among a variety of job functions. Each business process team leader should be chartered with setting the tone for the *minimum acceptable information quality level.* It is normally at the *business process level* where operating procedures dovetail (are synchronized) with an IWS. Because business processes typically cross organization departmental lines, there may be a bit of conflict between the departmental and the business process objectives. End-to-end business process performance management is an information management **best practice.**

This conflict is most easily resolved by establishing an *independent,* senior management-level sponsor for the business process itself. Independent implies that the senior management appointee does not have any direct influence upon any individual's performance evaluation within the end-to-end business process itself. Therefore, the independent senior manager is able to objectively adjudicate any conflicts between departmental and business process *objectives,* being removed from

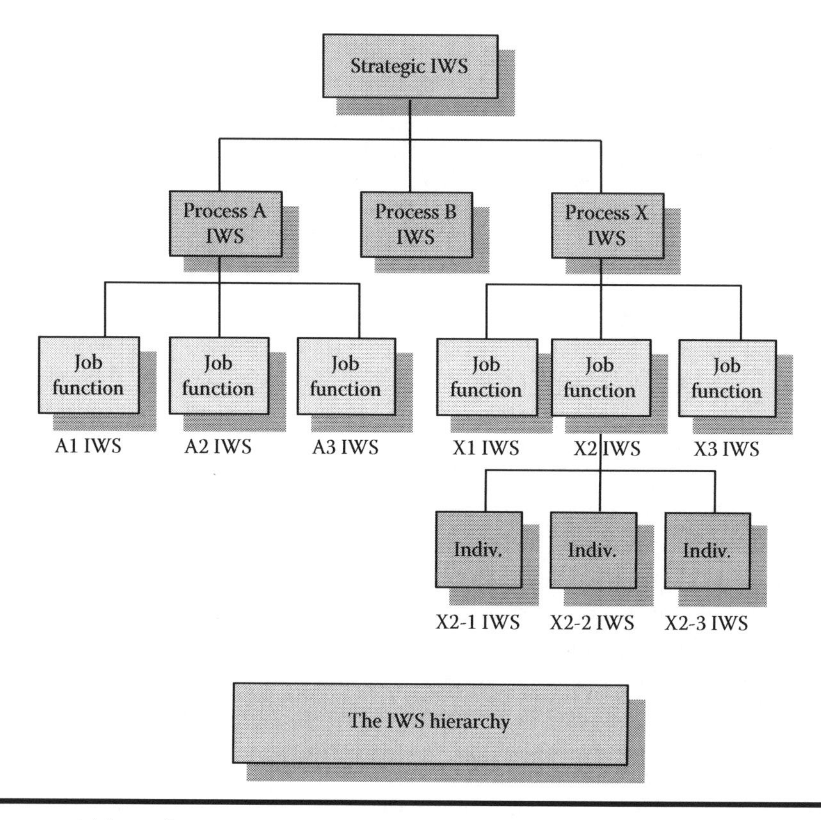

Figure 4.2 IWS hierarchy.

line authority. This independent executive sponsor precept is another information management **best practice.**

4.8 Performance Accountability

To digress a bit, those committed companies that aggressively pursue an IWS typically take the position that **every employee should understand their "contribution to profit" within the organization.** Given that the concept of a workmanship standard functions as the means to assess individual performance toward attainment of profitability, the IWS serves as its integral component. *With profitability as a baseline, and using* ***information*** *as a competitive weapon in the marketplace, the IWS becomes a mechanism to measure the individual's performance to profitability results.* There can be no organizational performance without business process performance and no business process performance without individual performance. Companies consist of people working together toward the common goals of the organization. The business process organizational structure facilitates the establishment of objectives and the monitoring of performance. **The workmanship standard serves as the means to express the composite company-wide objectives in a small enough increment to help ensure accountability, and contribution to profitability may be established at the lowest possible organizational level.**

The IWS provides management the confidence level (quality of information) from which they may base day-to-day decisions. The business process level, then, becomes the hub for information accountability.

At this point, it is essential that we discuss the process performance relationship more fully. As a company migrates from a parochial entity (one with tall, monolithic, silos) to an integrated process-oriented entity, there are various dynamics activated as well. First, the isolated individual begins to function as a member of a team, rather than merely autonomously. Second, the teams and individual performance measures need to be aligned to provide balance and harmony. Third, accountability (taking ownership through fulfillment), as well as responsibility, becomes essential ingredients. As discussed a bit earlier, the ultimate goal is to drive accountability into the heart of the organization at the process working level. Not only accountability but also decision authority need to be given as well. When we are serious about process-based accountability, we must empower the internal customer and their nested service providers to define the performance expectations for quality and delivery.

These expectations then become the IWS guidelines from which to assess process fulfillment. To amplify further, if the individual team (internal customer and nested service provider[s]) is the most finite accountability entity, then their process performance measures become the spark plug for end-to-end business process performance. Further, departmental objectives, in harmony with end-to-end business process performance objectives, radiate the synergy that ultimately determines organizational performance. It is simple to conclude then that if we are maximizing the internal customer/service provider performance elements along profitability lines, we will then be capable of optimizing profitability across the organization. By defining the internal customer/service provider performance criteria at a level understood by those required to perform the process, these individuals can begin to clearly perceive how they are contributing to the profitability of the organization. These process-based performance metrics will be further detailed in Chapter 8. However, suffice it to say, engineering performance metrics, which will lead to an individual's contribution to profitability, is a performance management **best practice**.

Process-based measures are far superior to mere financial measures.

- Everyone in the organization can understand them and, individually, relate to how to achieve their success.
- They are much more real-time than financial measures.
- People can readily act on the information from a process metric.

A financial measure, such as *return on net assets* (*RONA*), may be understood by an elite sophisticated segment of the organization. However, the vast majority of the organization (in the trenches) can neither understand it, nor can they relate to how **they personally** are able to contribute to RONA achievement. When people in the trenches agree upon internal customer/service provider measures which are synchronized to maximize profitability objectives, the broader financial measures, such as RONA, will also be maximized. The IWS then serves as the performance measurement bridge between working-level components of a process and the combined end-to-end process itself.

The recognition that information quality has been and/or is being compromised serves as the genesis of a truly effective IWS. When a company does not admit that there has been information quality compromise, it is like a person who is sick and either refuses to see the doctor or refuses to heed the advice of the doctor. Without acknowledging the *data integrity illness*, corrective action and future prevention are unlikely. Once we are able to get to this baseline, we may begin the process that will enable companies to finally realize their potential return on investment from their business systems.

4.9 Managing Performance Expectations

Managing expectations is a cornerstone of the senior management tasks, as they commit to implementing this IWS process. Companies seriously committed to the IWS process ensure that the internal customers have a significant say within the service provider's performance evaluation process. Inasmuch as the internal customers view performance achievement continuously (at least daily), they must be considered competent to evaluate their nested service provider's performance. However, any given individual may likely be both an internal customer of some people and a service provider to others. Being consistent, across organizational boundaries, is an essential senior management theme, if the IWS process is to achieve its impact potential across the organization.

The agreement between the internal customer and their nested service provider team must clearly specify the quality and delivery performance expectations. Once the SLA is clearly defined, the internal customer must work closely with their nested service provider(s) to help ensure that the measurements are doable and regularly measured, that both parties agree upon an approach to measuring progress, and that the actual measurement is consistent. Senior management must actively participate in the removal of any barriers that would inhibit the successful discharge of these performance execution tasks. Senior management must also remove the *politics* from interfering with the trench-level execution. The final senior management involvement becomes an ongoing task, namely, ensuring that continuous improvement is encouraged, that the IWS process becomes the second nature to every individual (acculturated), and that the process teams are continually motivated and inspired with strong leadership and commitment to the change process.

One last comment is appropriate here. To properly establish natural internal customer and service provider work team measurements, the internal customers and their nested service providers must spend quality time each day assessing performance fulfillment, identifying ways to improve the process, and creatively thinking of better, or more impactful, approaches. On at least a biweekly basis, they should meet with their process team leaders and senior manager sponsor to discuss the merits of the proposed changes, assign a relative priority, and determine the impact it will have across the entire business process. Then they must plan to incorporate these improvements by detailing a change strategy and developing a work plan (roadmap) to help ensure that these proposed changes become a reality, with a sense of urgency.

> It must be remembered that there is nothing more difficult to plan, more doubtful of success, nor more dangerous to manage, than the creation of a new system. For the initiator has enmity of all who would profit by preservation of the old institution and merely lukewarm defenders in those who would gain by the new ones.
>
> **Niccolo Machiavelli, The Prince**

One might ask how the IWS would be deployed by, say, the size of business organization. Once the IWS is embraced by an organization, it typically becomes a guiding principle and ingrained in the operating culture promoted by senior leadership. Because the IWS is "process based," the scalability is aligned to the number of business processes functioning within the organization.

As will be discussed in Chapter 5, the IWS may certainly be deployed and is encouraged to be deployed among supply chain trading partners. The larger the shared data, processes, and resources between trading partners, the more effective and efficient is the end-to-end nested efforts and likely lowered costs. This would logically also pertain to tiers of supply chain trading partners, exploiting maximum collaboration and throughput potential.

4.10 Certification

Webster's Dictionary defines certification as "to attest as being true or meeting a standard."* It further defines a certificate as "a document certifying that one has fulfilled the requirements of and may practice in the field."† Establishing a standard in the form of certification accomplishes the following:

1. It reflects management's seriousness about information quality.
2. It instills a pride of accomplishment in the individual having achieved the standards of certification.
3. The pride of accomplishment develops into a personal source of ownership. Establishing a "personal commitment" is so powerful, which actually has unlimited potential toward goal attainment. Those companies attaining goals are ones making money and achieving excellence.

Employees want and expect their leadership and the organization to specify the level of performance expected of them. The absence of "expected" performance tends to

1. Put the emphasis of establishing expectations upon the individual (roll your own).
2. Introduce a looseness that is perceived by the individual as "okay," but is counter to the natural need for vigorous organization established by a person's mind.
3. Become the essence and point of propagation for frustration.
4. Lead to other behavioral bad habits such as procrastination, laziness, and insensitivity. It sets the tone for the organization's culture and priority for achievement.

Establishing a standard, helping employees achieve that standard, measuring the performance to the standard, and taking the necessary corrective action are the lifeblood of a healthy, vibrant organization. **Those companies, found on the leading edge of performance to profitability, are the ones who expect and get excellence from their employees.** These are the ones who have established doable goals and objectives. They are the ones who have clearly defined workmanship standards. The act of proactively expressing the desire to achieve "a minimum acceptable quality level," in the form of a certification program, is the capstone of excellence.

The ever-increasing concern, associated with employee loyalty, may be reckoned with quickly if certification standards are instilled. *The lack of aggressive standards setting has allowed atrophy to infest individual performance expectations.* Consequently, the result has manifested itself in employee turnover and a trend toward decreasing productivity. *Frustration and problem identification need to be replaced with a dose of pride, a standard, and the perception by employees that the organization cares about them.*

---

* *Webster's New Collegiate Dictionary*, G.C. Merriman, Springfield, MA, 1987.

† Ibid.

There is no better expression of "organizational caring" than the establishment of workmanship standards and acknowledgment of individual (and group) attainment of these standards. Continuing to treat the symptoms (such as high turnover, low productivity, and low morale) may bring short-term results but lacks the endurance necessary for long-term results. Reestablishing accountability and specifying performance expectations are the formulas for long-term success. This process begins by implementing a certification program. Certification reestablishes an individual's self-worth (basis for motivation). As individuals, we would demand that the organization invest that much in us. As organizations, we owe that to our employees.

Certification may be implemented by sanctioning one or more of the following:

1. *Software certification.* Exhibiting an expertise on a module-by-module basis related to
   a. Transaction inputs (transaction matrix)
   b. Processing logic
   c. Output and interpolating information to enhance decision-making reliability
2. *Procedural certification.* Exhibiting an expertise on the application and process of procedures affecting a given department. This includes the proper filling out of documents, understanding of the upstream and downstream effects of documents, and the data control aspects of documents/processes
3. *In-process certification.* Exhibiting an expertise for diagnosing interrelational activities that cross-organizational and procedural lines frequently involve multiple transactions. These include such activities as engineering change management, material review board rejection and disposition, obsolescence diagnosis, and supply/demand management activities

4.11 Systems Champions

The leading expert, who surfaces as the "most knowledgeable," may be labeled the "systems champion." These special individuals are the backbone of support for module implementation success.

- They are the aggressors challenging the "doubters" and the complacent (wishy-washy).
- They are the ones who go out of their way to educate the newcomer and slow learner.
- They are the ones who point out the saboteurs and individuals with poor attitudes.
- They are the ones who offer "solutions" rather than merely pointing out problems. These "informal" leaders serve a vital role in system success.
- They are the individuals whose only alternative is winning!
- They are frequently cautious, requiring a rigorous and successful CRP.
- They are the standard setters, who take it personal, when things go wrong.
- They are the frequently overlooked heralds of organization strength.
- They adapt the Marines' slogan, "The Few. The Proud," which provides the systems engineering where the rubber meets the road.
- They attempt to balance out the streamlining activities with the issue of sufficient control. Terms such as "bureaucracy," "company politics," and "building staff needlessly" are not part of their vocabulary.

The above dialog attempts to describe many of the attributes of systems champions. It is by no means an exhaustive list. It does point out that these individuals are the "movers and shakers" of the system implementation effort. They are the real people with the vision of how the successful

Without measurement

- ✓ Performance is *not* being managed.
- ✓ Priorities *can't* be identified, described, or set.
- ✓ People *don't know* if their productivity is off track.
- ✓ There *cannot* be an objective basis for rewards.
- ✓ There are *no triggers* for improvement.
- ✓ People *don't know* what's expected of them.

Figure 4.3 Without measurement.

implementation of systems can benefit their company as well as themselves personally—the glue that keeps the organization focused upon the important issues. Figure 4.3 is a rule of thumb precept of what transpires in the absence of measurement:

In reflecting back, I recall a project that had lost its FDA certification primarily due to record-keeping practices. There were a variety of contributing factors for losing their certification, but at the heart of it was the lack of an IWS. There were a variety of "system"-oriented improvements to be deployed and there were training issues needing resolve, processes needed refinement, and procedures needed to be polished. When the concept of an IWS was embraced by the leadership team, the turnaround project gained significant traction and their FDA certification restored.

4.12 Conclusion

The IWS is a key vehicle that drives accountability throughout the organization.

- It is a catalyst that consumes the necessary energy to achieve excellence in performance results.
- It addresses the transactional and data elements, the job function, and the departmental activities.
- It expresses the minimum acceptable quality level and exhibits the pride of achievement.
- It encompasses the accuracy level of all records, the accompanying procedures, and the related processes.
- The record accounting migration journey should be aggressive, yet achievable.
- The sensitivity to information accuracy becomes the second nature to the organization.
  The ultimate minimum acceptable quality-level goal for all information is 100%. The computer does not care whether information is accurate or not; however, management cannot live with inaccuracy. Therefore, it is recognized that information accuracy is not a system problem, but rather a management problem. Figure 4.4 shows the 100-cubed triangle (100% of the records, 100% accuracy, 100% of the time).

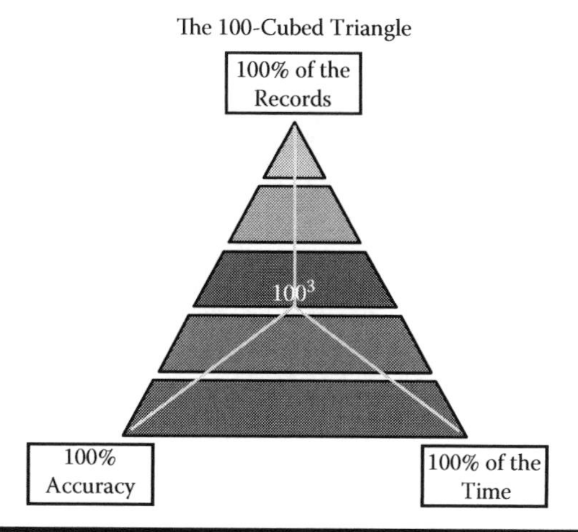

Figure 4.4 The 100-cubed triangle.

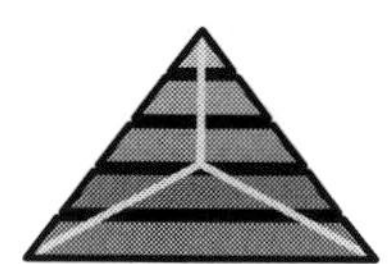

Note: The triangle is the Best Process Logo and represents both **best practice** and **best process** precepts.

- The surviving management teams in the twenty-first century will be the teams that cherish information and build information management positions into their strategic plans. They will be the organizations that understand that accurate, timely information is the new competitive weapon in the marketplace.

Associated with the IWS is the engineering of the internal customer and nested service provider relationship standard (SLA) leading to exceptional process-based performance management results.

Next we will discuss the CRP proving grounds essential to validate functionality, train users, and provide an end-to-end process integration test (hardware, software, process, procedures, and culture) acceptance foundation using the sandbox.

The goal of every company that is serious about vital information management must be to improve the quality and timeliness of information flow. The IWS provides the means to improve the quality of data across the business process and provides the means to help ensure timely information flow between the business process functionaries. The IWS establishes an agreement between the end-to-end process team members to help ensure that the information lifeblood of the company facilitates immediate improvement of business results.

Chapter 5

The Conference Room Pilot

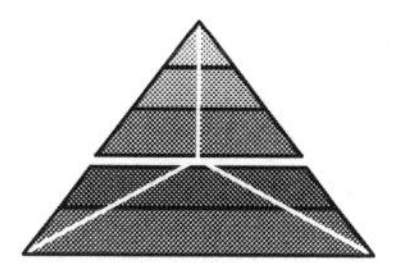

Test driving the blending of functionality reflected in the design (features, functions, and capabilities) as well as processes (how the design is configured) and the environmental structure (policies, procedures, and performance metrics within the players' culture). This chapter not only pursues project team piloting but also demonstrates senior leadership piloting, customer/supplier piloting, and other business partnership piloting.

The conference room pilot (CRP) is the proving ground (prototype framework) essential to validate software functionality, train users, and provide an end-to-end process integration test (hardware, software, process, procedures, and culture) acceptance foundation using the sandbox.

Previously, we looked at the importance of proper engineering a planning roadmap and the deliverables associated with the project plan. We also elaborated on the value of generating valid requirements and their associated traceability matrix. During the requirements generation discussion, we pointed out that the project life cycle included design, customization, and testing steps for a comprehensive depiction of an enterprise resource planning (ERP) project implementation. Now is the time to close the loop a bit and validate that our planning model will really work with the software selected, hardware infrastructure, and defined processes, policies, procedures, performance metrics, and other environmental dynamics. This is where the rubber meets the road, so to speak.

I recall a project that was on the edge of derailment due to a module interface not being rigorously tested. This particular ERP project implementation used a blend of software suppliers (it was called "best of breed" at the time). One of the unique requirements was to take an existing legacy financial system (general ledger, accounts receivable, accounts payable, fixed assets, human resource) and bolt on the newly purchased operational (engineering, supply chain, demand management, advanced planning, etc.) functionality. Because the financial system had been operational for decades, the leadership team wanted to retain this legacy functionality and consider integration at a later date. The objective, then, became interfacing the financial legacy system with the new ERP functionality. In reviewing the interface specification, there were a few missing pieces that were needed for high data integrity CRP activity. These omissions were corrected and the project team launched a rigorous CRP. In reviewing the comprehensive CRP results, the project core team recognized that there is still a missing element needed to sign off the legacy interface with new ERP functionality. The missing element was a transaction matrix that clearly mapped the operational-oriented transactions to the correct general ledger account at a very detailed basis. Once the transaction matrix was defined properly, the expected results and actual results were validated and the interface was properly completed. As it turned out, the CRP, when done properly, avoided an impending ERP implementation derailment.

Let's look at the CRP in more detail.

5.1 Definition of a CRP

One of the most fruitful activities and potentially the greatest software product knowledge learning tool is the CRP. There seems to be many different interpretations of a CRP. The CRP is frequently labeled a "training pilot" as well. The CRP can be used to

1. Test data elements and their relationships.
2. Educate and train users.
3. Validate policies.
4. Test user operating procedures.
5. Test issue resolutions.
6. Try something new out for the first time.

Let's look at each of these uses more closely.

5.1.1 Test Data Elements and Their Relationships

The project team must become familiar with all facets of each software module being implemented. This approach permits the project team to make an intelligent recommendation to management and the users as to the module features to be initially used. The purpose of testing each data element is to understand the functionality of the element as a stand-alone unit. This includes all input transaction effects as well as output. The element should then be tested on the effect it has as it interrelates with other data elements. Exercising each data elements singularly and interrelationally results in a keen understanding of the "logic" impact attained from the data element.

The CRP is an effective tool to test the functionality of software features. An example might be a scrap factor that typically inflates the demand for a component. A thorough understanding of how scrap interrelates with shrinkage or yield, which typically inflates the supply, is an important supply–demand balance principle. The compound effect of the two factors interacting together may have a significant impact upon intensifying potential shortage conditions or adversely affecting inventory levels.

A graphical example is shown in Figure 5.1.

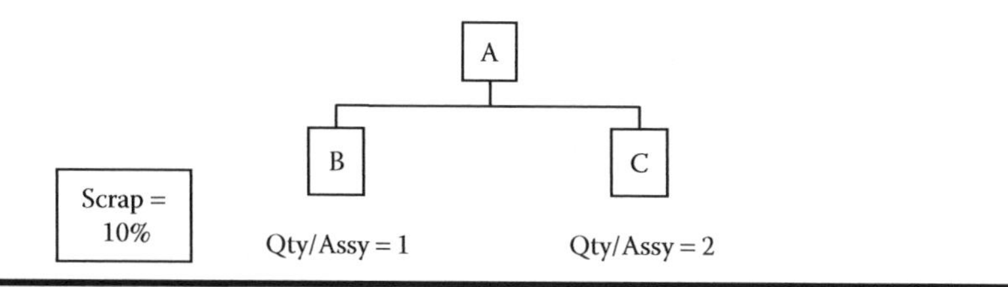

Figure 5.1 Bill of Material Baseline.

A manufacturing order for 100 As generates the following component demand:

	Without Scrap	*With Scrap*
B	100	110
C	200	200

The same relationship with a shrinkage rate for parent A of 10% generates the following component demand:

	With Shrinkage
B	110
C	220

The compound effect of scrap and shrinkage generates the following component demand:

	With Scrap and Shrinkage
B	121
C	220

The above example clearly points out the importance of understanding the systemic effect that each element has in and of itself. It is equally important to understand the compound effect of the interrelationship of the two factors working together.

Another example might be taking a close look at the effect of "default" values. In this example, let's assume that the issue control code is defined at four levels:

1. The system level
2. The item master level
3. The bill of material level
4. The order level (as depicted by the demand record)

The default logic is expressed as follows: When an item record is created, if not overridden, it will default to the preestablished value defined for the system level. Let's assume for this scenario that the issue control options are "manual" issue (the normal individual kitting withdrawal transaction) and "backflush" or automatic issue upon receipt of the parent (system does inventory control transaction). Let's establish the system-level default as manual. When a new item is created, it defaults to manual issue. Let's assume that when the bill of material relationship is created, it automatically defaults to the item level stored value, and when an order is created, it defaults to the bill of material level stored value. Now let's assume that all four levels of a relationship have respectively defaulted to manual issue. Let's assume further that the particular item discussed was created a year ago. It is a very active "common" part used in many different bills of material. These bills have been the source for generating many demand records within many orders. Let's also assume that management has recently specified that the common item should now be backflush issued rather than manual issued. How would the database administration go about updating the database to accommodate this decision? Clearly, the answer depends upon how the software handles the effect of changing the value. A software capability which has a "mass change" feature may locate "all occurrences," no matter what file or hierarchical level of definition, and make the change massively. A mass change strategy may surface a concern such as "how about already released orders will this mass change cause double issue?"

An alternative software capability may only mass change at a single level of definition. Then an approach may be to mass change each level one at a time.

Yet another software solution may not have mass change capability. How would the database administration go about accurately updating a large number of records? The documented result of trying each alternative in the CRP is a tool for determining an approach that minimizes operation risk. One final scenario highlights the misunderstanding of the software capability. For example, if the user assumes that by changing one level of definition, all existing records will automatically be changed, which may prove embarrassing. In practice, it may also appear as if the software were giving wrong answers. In our scenario, in fact, the software was providing the correct information, but it was being interpreted wrongly, (e.g., some item master files have over 100 data elements). This is only one data element. There are many data elements that may be candidates with equal complexity. Much of today's software has a significant level of sophistication. If the project team and/or user does not have a high degree of understanding and confidence as to the effort of procedural events, then "data integrity" will be suspect at the onset of a conversion. The proper place to establish a confidence is not on the "live" production database, after conversion, but rather on the "playground" known as the CRP.

5.1.2 Educate and Train Users

Education may be viewed as understanding of concepts and practices. Training, however, is the application of education into day-to-day events. Education should come early in a project. Training should occur just before cutover. These concepts will be further defined in Chapter 6. However, for purposes of our discussion, it is important to draw a distinction between the two activities. In the scenarios described in (subsection 5.1.1) above, the purpose of each event was to educate the project team and users so that the proper mix of features may be included at the conversion onset. However, it is equally important to keep the initial cutover simple (KISS) so as to increase the probability for early and decisive success. Therefore, it is just as important to proactively exclude certain features during the initial cutover phase. Then, go back later and add features once the user community is able to more aptly absorb the capability (sophistication). As a result of education sessions, the project team determines what features and functions are appropriate for inclusion at each phase. The training aspect comes much later as the working level employees are procedurally instructed as to what their jobs will entail using the new information capability. This juncture is where the rubber meets the road. This is also a crucial point for the "real people" to establish the confidence necessary for system success. Confidence displayed by the real people (who daily process massive numbers of transactions that affect the data upon which management makes decisions) is an integral ingredient in early success. If users are properly trained on the use of new forms, procedures, and input screens, their "resistance to change" may be significantly reduced. However, if users are not sufficiently trained, their level of frustration may be drastically increased heralding an impending disaster.

It is important to reemphasize that training typically occurs at a point just before cutover. At this point of the project, it is also an event's high activity level. It is a point where many projects are forced to take "shortcuts" if project deadlines are to be met. Would one of the shortcuts be manifested in inadequate training? The CRP should function as a focal point for establishing both software education and user training. As will be discussed from Sections 5.1.3 through 5.1.6, all facets of policy and procedures to be newly implemented or modified should be rigorously exercised in the CRP prior to adoption at cutover.

5.1.3 Validate Policies

One of the valuable ERP implementation by-products is the review and challenge of company's operating policy and practices. If "information" is to truly become a competitive weapon in your strategic arsenal, then the means by which data are obtained, used, and processed needs to be equally competitive. This competitive information strategy rightly belongs in the strategic business planning cycle. This strategic planning activity should constantly challenge current operating practices to ensure that policy does not become stagnant and become a latent source for losing competitiveness. The CRP may be an ongoing tool that allows management to simulate the effect of policy changes on the system before enacting the change. For example, what would the effect of removing a master scheduling time fence policy be upon the change of a master schedule? A small change at the top level of a product tends to cascade into a high level of activity at lower structure locations. The ability to simulate the effect of policy change in a "playground" tends to minimize the risk factor of changing a policy. This is not to imply that a long wait would be in store for policy changes. On the contrary, the CRP should provide the means to quickly simulate the effect so that management may exercise maximum flexibility to respond to changes in the marketplace.

5.1.4 Test User Operating Procedures

As policy changes, so should the correlating operating practices. Whenever the operating practice gets out of step with policy, there tends to be an embarrassing clash between top management and the rest of the business. The resulting frustrations are typically pooh-poohed as lack of communication. However, within some environments, policy tends to change upon the "whim" of its executive staff. Consequently, chaos and crisis tend to reign supreme. When confronted, management usually defensively responds by justifying their informal actions in words such as "flexibility" or "prerogative." However, the results usually lead to an intensifying of the problem and more profit erosion. Whenever policy affects procedures, these procedural changes should be simulated as to their effect, similar to the method described in subsection 5.1.3.

5.1.5 Test Issue Resolutions

A module definition review is the detail review of a given module's data elements, inputs, outputs, logic, edits, and job streams. The by-product of a typical definition review is a series of differences (issues) between the current (*as-is*) system and the new module (*proposed*). These issues should be written down, prioritized, and a work plan defined for their resolution. An ideal location to test out the effects of the issue is the CRP. Within this simulated environment, the project team can get a thorough appreciation for the full impact that the issue and solution will have. The normal issue resolution leads to one of these alternative actions: (1) change the current operating practices, (2) change the proposed system, and (3) defer capability to a later implementation phase. There is an inherent danger signal associated with alternatives (2) and (3). Let's look at the dangers associated with alternatives (2) and (3) in subsections 5.1.5.2 and 5.1.5.3.

5.1.5.1 Alternative 1 Danger: Change the Current Operating System

It is expected that the current operating practices will evaporate with the new or proposed system. However, in many companies, there is severe pressure to modify the new system to make it look like the old system ... Even without software changes, there is pressure to operate like we used to using new tools. This practice is ill-advised: where is the return on investment (ROI) in doing business like we used to? If you are going to make the investment in a new software solution, take as much new functionality as possible, as long as it adheres to the project time frame.

5.1.5.2 Alternative 2 Danger: Change the Proposed System

Quality packaged software currently available leaves few features unturned. In fact, much of the software is so feature rich that the mere sophistication stymies many project efforts. The introduction of system modification tends to add months and frequently years to the initial project effort. This makes the payback elusive at best. In addition, when thoroughly analyzed, many "must-haves" are merely taking the "bad habits" of the current system and contaminating the new system. With the exception of interfaces (which may be viable candidates and necessary for initial conversion), report writer requests (which are usually necessary), and the modification of certain default values, the general rule of thumb is to use the system "as is" as much as possible during the initial cutover. After gaining experience (usually one year), reconsider the real necessity of the modification.

5.1.5.3 Alternative 3 Danger: Defer Capability to a Later Implementation Phase

The real caution for this alternative deals with whether the feature is a "mainline" feature. For example, if your company's part number is currently 25 characters long and the software selected only supports 12 characters, there must be an up-front decision. In this example, the part number field is a mainline feature. Deferring the decision will cause significant risks to the project success factor. However, deferring a decision on a field such as "unit of weight" is not normally a mainline issue. It is important that the project team separate "mainline" issues and features from passive issues/features. The final decision on issue resolution should then be tested as to the impact that the decision will have on the database and use of procedures. This is especially important when dealing with forms, input transactions, and so on.

5.1.6 Try Something New Out for the First Time

Many software vendors provide a product enhancement program with the intent to keep the product "state of the art." These enhancements, blindly incorporated in a production database, could potentially spell disaster to the credibility established within the user community. Software companies take great effort to flush out bugs. However, they cannot test out all combinations and permutations that a given user could possibly configure. Consequently, it is the responsibility of the project team or database administrator to "try out" the enhancement, fix, or product release in a test mode. The ideal candidate is the CRP. Within the CRP, the promised capability may be verified especially as it relates to how your company is using the product. If you choose to modify the software, the CRP can be used to test the effect of the enhancement on these modifications. "Try something new" may also refer to incorporating the use of a feature postponed for later application. It is just as important to give a fresh look at the capability at a later date prior to incorporation as it was to try it out and decide to exclude it earlier in the project.

The CRP may be used for many different applications. Whether it is testing the effect of procedural issues or data elements, it is providing an important front-end simulation. It will not affect operations until the cause and effect is clearly understood within the "conference room."

When the CRP was described above, it presumed that the activity was "user" testing not "technical" or program testing. Poor cause and effects resulting as a default, due to "absentee management," is a user problem. The user must be active and directing the actions associated with system implementation. Being inactive and waiting for someone else to give it a shot (due to hectic schedules, low on resources, and other equally poor reasons) tend to add frustration and profit erosion to system implementation efforts.

5.2 Structuring the CRP

The reader is starting to get an appreciation for the power of the CRP. We will now turn our attention toward structuring the CRP so that it can provide maximum payback for time invested. It is important to recognize that most employees will be touched in some fashion or another by ERP. Therefore, it is fair to suggest that most employees will have an opportunity to "practice" in the CRP. Some operational functions, such as material planners and cost accountants, will have an opportunity to exercise the preponderance of system functionality, whereas job functions such as stockroom clerks may be more limited in their exposure. No matter what the extent of exposure, it is extremely important that the time spent in the CRP is quality time. Properly structuring the CRP is also essential. Some of the general guiding parameters that must be considered during the "engineering" phase of the CRP include the following:

What is the talent mix of the group?
How many personnel will be taught at a given session?
Who will get education sessions as well as training sessions? What is each employee's skill level going in?

Once the above questions have been determined,

1. A series of matrices may be formalized. Examples of education and training matrixes are given in Appendix C. The format used can apply to the CRP.
2. The next step is to define a specific "objective" and associated deliverables to be assigned to each session. Actively designing the purpose of each session tends to focus the student mind-set and energy upon bite-sized chunks. Overwhelming the participants tends to be

counterproductive. Soon the general population tends to lose "system enthusiasm" and many "mentally retire." It is important that the session be engineered concisely yet allowing the student sufficient time to interact thoroughly.

3. Each session participant should do his or her homework before class attendance by documenting, in writing, their "expected results" of the session. Actual session results are documented at the conclusion of the session. Then, *actual results* are compared to *expected results*. **Identifying the reason for the difference between "expected" and "actual" is where the real learning occurs.** At times, these differences lead to active and sometimes fiery dialog. It is important that the session facilitator be very familiar with the module so that confusion does not win out. Dissenting team members, having special difficulty, should be given independent study tutorial using documentation, training videos, and the training pilot to work through their problems. This process tends to develop "systems champions" who have a highly charged positive attitude toward the system.

Note: A log should be kept of actual results for future references. If anomalies are uncovered, they should be logged for corrective action.

4. The magnitude of the education/training program and the degree of aggressiveness of the project plan will determine the amount of parallel sessions running concurrently.

   Structured teams to consider for concurrent sessions include the following:

   a. Multidiscipline teams (all or a large mix of participants concurrently) understanding the interrelationship (cause and effect) functionality
   b. Specific homogeneous disciplines (limited mix) to rigorously exercise (becoming "expert") of their area of responsibility

   Each structure type has its merits and drawbacks.

   The interaction and problem-solving nature of the multidiscipline teams can be effective. However, the more "progressive system thinkers" group (a) tend to contaminate the session dialog, which frequently forces the lesser confident users into a shell (or further into a shell). It may be advantageous to team a "weaker" member with a stronger member to compensate.

   The intensity of the single discipline (b) session allows each discipline to fully shakedown certain features as it directly affects their jobs. The drawback of this type of session is that the participants are drawn to areas with which they already are familiar. It also permits them to shy away from interrelational areas they lack confidence in. This short changes their education and competence level. It also inhibits their understanding of the downstream effects that their action has upon other departments.

5. The quality time invested in engineering lesson plans, choosing the session talent mix (how about varying the mix frequently), analyzing results orally (have each team present their results and rationale for getting their answer), and debriefing can be the difference between the effective training pilot and just another "bullet" on the project plan. The sessions should be mentally enriching and challenging to the students.
6. Structuring the CRP encompasses scheduling online sessions (this may involve significant scheduling with computer operations), properly preparing the edits, defaults, control files, and so on. It also means to properly involve the job stream and minutely reviewing outputs.

5.3 Deliverables Resulting from an Effective CRP

The deliverables from an effective CRP include, but not limited to, the following:

1. Statements of issue resolution
2. Functional specification input document
3. Validation of policies and procedures
4. Software application certification
5. Enhancement shakedown
6. Training prior to cutover

5.3.1 *Statements of Issue Resolution*

The issue resolution task was discussed in depth earlier (subsection 5.1.5). The primary deliverable is a documented resolve on how the proposed system will function. Any procedural or organizational corrective action shall be incorporated early in the project process.

5.3.2 *Functional Specification Input Document*

A functional specification is a module-by-module "living document" describing how the users want the system to operate. A thumbnail sketch of a functional specification includes the following:

- Default values
- Interfaces (documented)
- Feature/function narrative
- Inputs
- Listing of source documents
- Suggested control file settings
- Expected response times
- Features "not to be used" at this time
- Edits
- Security matrix
- Logic and calculations
- Screens
- Report writer specs
- Job streams
- System down contingency plan

Output from the CRP provides practical documentation, which becomes a foundation for the functional specification.

5.3.3 *Validation of Policies and Procedures*

The policy and procedure handling was discussed in depth earlier (subsections 5.1.3 and 5.1.4). Essential ingredients in a successful system are the motivated cooperation and integration of the "people," operating practices, culture, and information tools. If an ingredient is not finely balanced with the others, it may result in frustration, poor results, and frequently crisis. It is important that the balance be multidimensional in nature. It should provide harmony between management levels and synchronized across organizational lines. This balance over time produces good business results. Policies and procedures are the guidelines that provide the means of attaining this balance. They must be formalized, adhered to by all employees, and updated when appropriate. They should be thoroughly understood by all employees and exceptions should also be discouraged.

5.3.4 Software Application Certification

One of the **most important** by-products of the CRP is the accumulation of the criteria to engineer a "minimum acceptable quality level" for information handling. Analogous to an individual not being able to operate a motor vehicle until he or she has passed a written and driving test, an individual should not be allowed to operate any portion of the information system until he or she has passed a minimum acceptable quality level. This qualification process is termed *software application certification*. It should include any individual who (1) inputs, (2) uses outputs, or (3) significantly affects data within the system. It should be tailored toward "job-specific" applications. The essence of the process is focused upon various levels of expertise. This was discussed in Chapter 3. Briefly, the process considers the following:

1. *New hire*: Without proper cultural and systems orientation, new hires must rely upon their understanding of the methods from their former employer. This may quickly introduce *information corruption*. Projects such as ERP require a significant amount of resource investment to achieve a positive level of performance. Why throw all that global organization effort aside by allowing someone to innocently destroy data integrity because of inadequate skills or training? *It is this author's opinion that a new hire should not be allowed to affect any portion of the system until he or she has passed a "minimum certification level."* It is very important to determine whether the new employee has the ability to be certified (known within two to three weeks of hire). If not, save the company pain, sorrow, and frustration by letting them go *early*. Don't allow the unqualified to (1) touch your product (your company's hard earned image), (2) speak with a customer or vendor (your friends and associates), or (3) interact with company data (your decision lifeblood) until they have culturally, procedurally, and software application wise passed a *minimum acceptable quality-level certification*.
2. *Transferees/promotees*: Equally important to potential data corruption is the employee who was recently transferred or promoted. He or she should not need a "cultural" education but should pass the software application certification credentials if he or she is to deal with data.
3. *Operational management*: If the daily decision makers are not familiar with the decision tools available, there tends to be differences in opinion (and information sources). It should have a fundamental business objective that everyone works from the same information source. To accomplish this requires at least a conversational awareness of the *input* and *output* tools used by the organization. The long-range information should keep the operating managers abreast of strategic decisions. A software certification credential should keep them day-to-day aware of the information quality level.
4. *Senior management*: The more successful leaders within an organization must establish a leadership direction by "example." If senior management is serious about obtaining the ERP-related benefits, then they must give more than lip service and funding. They must actively promote the system use, as well as use the system themselves. This requires an executive-level software application certification.

5.3.5 Enhancement Shakedown

The enhancement shakedown was discussed in detail earlier (subsection 5.1.6). It is important to note that the first place to try out a new software release, feature/fix, is the CRP, not the live production database.

5.3.6 Training prior to Cutover

The availability of scheduled personnel for training prior to cutover will be a major determinant as to the smoothness of the conversion effort. If people are not properly trained on the new system before cutover, they must be trained afterward. Resources may seem strained as you're heading toward the ***GOLD*** (*go-l*ive *d*ate). You may not have experienced a cutover ever before. In the "best" circumstances, your project team resource will be strained to the "limit" at cutover. This is not the time to ease up on training. Rather, it should be a time to intensify the training in order to fully understand the module and be properly prepared up-front.

The CRP model, so far, has been a framework for the project core team to pilot the hardware, software, processes, and environmental frameworks on an end-to-end integrated basis, primarily for end-user types. The framework model may apply to a multitude of other piloting opportunities as well. Let's take a look at how it might be adapted to other ERP implementation project stakeholders:

- *Senior leadership piloting*—As discussed in Chapter 3, the ERP implementation effort is likely one of the largest system endeavors ever taken on by the company. Chapter 3 also discussed the need to commit to significant expected results (ROI). For this ROI commitment to be achieved will require senior management to actively participate, even to the degree of a day-to-day involvement. As one astute leader said, "management devotes their time on the top priorities for the business." If you want to achieve "exceptional" results, the ERP implementation project must be a top priority for senior management. With that said, senior management should commit to optimize their ROI by embracing the new ERP implementation including tools, processes, and a host of ERP-related deliverables. To that end, senior leadership, as a group, must learn to adapt their operating practices in line with the ERP project deployment effort. To understand the impact of, say, deploying projected future variance capability, they will need to be educated on the new capability, may need to adapt to a new decision-making process, and become a cheerleader on how to "bank" the rewards. This adaptation comes from rigorous roll-up-the-sleeves participation (lead by example) and vision (guidance) on how the organization will operate accordingly. One of the best environments to understand the new framework and affect the changes needed for system success is the CRP sandbox. This may be accomplished by adding senior leadership flights or cycles to the end-user CRP. There is certainly value to integrating the senior leadership flights to the existing CRP cycles inasmuch as users and leaders alike learn the tools and processes together and learn from each other. However, this requires a significant time commitment by senior leadership and it may not be practical. An alternative is to appoint the number 2 senior leader as a full-time participant for the length of the project to represent the senior leadership team. This is also a significant time commitment with added value of buffering the leadership team from the vulnerability of exposing bruised egos. Yet a third option is to engineer a special CRP, with deliverables similar to user types, but expressly geared to the new "day-in-the-life" of an executive. The third option could either operate autonomously or be blended into user-oriented sessions when scenarios were overlapping. The senior leadership intense participation in the ERP Implementation effort is an ERP implementation **best practice**.

- *Customer/supplier piloting*—Every ERP implementation project is configured differently ... Each has its own business drivers, expectations, goals and objectives, and commitment. The broader the impact expanse tentacles, typically the longer the time. Small- to medium-sized organizations, limited by resource and budget, tend to narrow the expense due to their business constraints. Yet, these very same organizations tend to multitask their limited resource base facilitating more agility and quicker response to the statement of work (SoW) and are ideal candidates for covering a broader swath of impact than larger organizations. Regardless, an ideal opportunity, for any size organization, is to improve the productivity between their organization and their key supply chain partners. Only in the last decade or so has the technology been adequate enough to exploit a true partnering relationship. However, the benefit from those "true partnering" relationships is immense. Such activities as sharing product development (time, resource, and budget), tooling, processes, and technology tool deployments optimize these partnership productivity dynamics. Similar to senior leadership CRP, one of the best environments to understand the new partnering framework and exploit the leveraging of talent pools, sharing cost and pools is the CRP. Clarity of activities such as rules of engagement, SoWs, beginning baseline, and expected results need to be documented. However, if engineered properly, exploiting these opportunities may be the difference between menial ROI and an "order-of-magnitude" ROI results. Integrating customers and/or suppliers into your ERP implementation has numerous challenges yet opens a new dimension in potential rewards. Creating a "shared" CRP with "outsiders" may be as simple as limiting the impact to, say, capacity planning, whereby key supplier procurement schedules are integrated into the supplier delivery schedule, while managing capacity versus load factors for both internal and external work centers. This not only expands the utility of the ERP software effectiveness and span of control, but also seeds more effective and improved communications between partners. As discussed above, this "shared" CRP may expand into "joint product development" initiatives and/or expand the efforts through "tiers" of supply chain stakeholders to optimize the value chain across the partnership. These customer/supplier CRP activities may be accomplished by adding partnership flights or cycles to the internal end-user CRP or creating separate (autonomous) CRP events. Regardless of the approach taken, adding external partners to the company's ERP implementation effort may be a "game changer" in expected results. Extending the CRP to the customer/supplier framework is an ERP implementation *best practice*. The future "captains of industry" may be those companies taking great strides in this practice versus those taking merely baby steps.
- *Other business partnership piloting*—The concept of customer/supplier partnering may be exploited yet further as a business analyzes the potential of ERP implementation "beyond the borders" of the company. A list of potential opportunities may be created and their weighted merits of inclusion should be vetted to determine the likelihood of execution. A similar CRP framework, discussed in the **customer/supplier piloting** flights/cycles, may be engineered in a similar manner to any internal/external opportunity area.

5.4 General Points of Awareness

Understanding the multifaceted functionality of the CRP surfaces some provocative thoughts:

1. How many databases will be configured and supported?
   a. Production database (one or more)
   b. CRP

c. Technical database (accommodate maintenance migration test interfaces, test modifications, test new technical operating procedures, etc.)
d. Hands-on training database (one or more), which may be merged with the CRP database
e. Plain vanilla "vault" version
2. Will the CRP database be available whenever users need it? ***It's better, if it's to be an effective tool.***
3. Who participates in the CRP activities?
Hopefully, all users, to some degree. The participant rigors may be the following:
a. Project team/key players (continuously)
b. Operating management (actively)
c. Senior management (cursory to actively)
d. Clerical/transaction process ("Training only" to actively)

5.5 Conclusion

The time invested in an effective CRP will yield tremendous results at conversion time. It can be the difference between an orderly transition and mass confusion, activity as normal and crisis, and doing it right the first time versus having the opportunity to do it over (similar to rework, we don't have enough time to do it right the first time so we do it again!). The CRP is an invaluable asset to system success. Used properly, it can be the edge that allows your information systems to serve as a competitive weapon in the arsenal of business resources.

Next we will discuss in Chapter 6 the education, training, and implementation framework essential for ERP users to become prepared for the new environment.

Chapter 6

Education, Training, and Implementation Framework

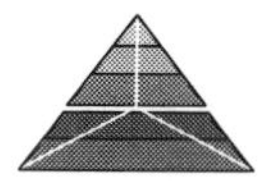

The backbone to system success involves the entire user community exhibiting the competence and mastery of the new system. Contrary to popular practice, this does not occur through osmosis or attending a couple of training classes. Like all competency processes, this must be achieved through proper design and fulfillment.

There is a host of information available addressing this topic area. It is the purpose of this book to provide essential ingredients that will allow the project core team to foster a successful enterprise resource planning (ERP) implementation project and, hopefully, in a way that yields exceptional results.

I recall a derailed project that I was called in to "resurrect from ashes" ... A significant investment had been made to send large numbers of key users to both generic education classes and the selected software training classes. On the surface, a person would conclude that this should have been an adequate approach to usher in a successful ERP implementation. However, when the cutover day arrived, the company found itself in a spiraling out of control disaster. There were a variety of misguided assumptions and inadequate framework that started the crisis, which including the following:

- There was no need for managers, supervisors, or senior leadership to attend any formal classes ... The brain trust had an adequate education and training foundation.
- Many of the end users were recent "new hires" and their former companies had prepared them sufficiently to the new system, so only rudimentary education and training were performed on a train-the-trainer basis.

- Trainers (train-the-trainer model) were selected based upon their time availability rather than training or organization skills.
- The vast majority of training was concluded months before the actual cut-over and there was no effort to conduct refresher courses.

It was obvious that the education and training was not engineered well, not end-user specific, insensitive to the job functions, and so on. The corrective action plan had to be engineered to respond quickly and the ERP project to get back on track; otherwise, the company would have to revert back to the old system. A senior leader and two key managers were assigned full time to the turnaround effort. The company went on a six-day working schedule and priority was given to this turnaround.

Let's take a look at a better approach that would have obviated this project derailment.

6.1 Structuring an Education and Training Program

6.1.1 Perspective

Recent history has generally seen a panorama of change in the level of technological development and application. Such development has benefited manufacturing and other industries by making their systems more productive. However, an associated cost trade-off also becomes realized in the limited availability of system-savvy employees. Dynamic unemployment (and underemployment) has not only become a very real sociological issue, but a very pressing competitive issue for business on a global scale. One strategy to maintain pace with technology is that firms are beginning to rely more and more on in-house education and training programs. The primary purpose behind these programs, therefore, has been to allow speedy and successful organizational adjustment to the persistent tide of technological evolution.

In addition to providing the means to meet the competitive challenges of structural adjustment, education and training programs tend, in general, to also accrue cost-related savings to the organization. For example, one class of savings is derived from increased employee loyalty and commitment, thereby resulting in reduced costs related to employee turnover (e.g., recruitment, selection, orientation) and redundancy training. Another example may be obtained through an employee's perception of a gained sense of control over his or her immediate work environment. Such employee satisfaction helps to instill a sense of goal congruence with the organization, which, in turn, has the effect of motivating each employee to produce more. This satisfaction operates by an employee's enhanced perception of his or her relative value to the organization as well as his or her personal sense of self-worth.

Effective education and training represents a worthwhile investment in the human resources (HR) of an organization (human capital investment), with both immediate and long-range returns. Education is the foundation of new system literacy. It is the process by which people begin to understand the concepts and philosophy behind the changes introduced with a new system. After the education of the concepts is complete, training may be performed as to the specific functions of the system and the "nuts and bolts" of the job task to be performed. Education should come early in a project, and training, just before conversion.

Employee training can be thought of in two ways. In a limited sense, training is concerned with teaching very specific and immediately usable skills. In a broad sense, training provides more general information designed for instilling knowledge for long-term application. In other words, one interpretation refers to teaching job-related skills, whereas the other denotes employee development. For example, an employee may receive job-related training to improve skills in using, say, a new computer, whereas employee development may be the goal of a management course on effective leadership (the designated objective is that the trainee will develop into a more effective and, therefore, more efficient producer). What is important to note is that the HR department cannot be held responsible for the entire spectrum of training. Line management needs to play a proactive role if a meaningful, effective, and worthwhile program is to be enacted. Training has both current and future implications for job and system success. Whereas it can immediately contribute to higher productivity, fewer mistakes, greater employee job satisfaction, and lower turnover, it can, in addition, enable the employees to cope with organizational, social, and technological changes. The company, in effect, becomes responsive to fast-paced change while simultaneously exposing itself to the beneficial use of information tools. And inasmuch as training is a learning process, whether its focus is on initial job skills, developing employee potential, or retraining the organizations technologically underemployed, learning has to occur for training to be successful.

People have a greater predisposition toward learning when the material presented is important to them. They also tend to learn such information better when (1) they are fresh (at the beginning of the workday rather than at the end of it) and (2) they believe that the process benefits them personally. The latter is especially true when the information presented is meaningful to its audience. Meaningful information, as it turns out, has a greater propensity to become fully stored in a person's conscious or subconscious mind for later retrieval (howbeit molded to a person's perception).

With this foundational background, the ERP project has the need for both conceptual (education) understanding of what ERP is and the impact it will make on the future business model as well as job and ERP solution-specific (training) understanding. Because the education and training has such a long time span to execute the broad user group, timing the commencement is crucial to user community's assimilation of the material and effectiveness of the material. Conceptual education may be disseminated early (and should be throughout the project span); however, training (how to use the tool and do the job) needs to be back-scheduled from GO LIVE, so as to be as fresh as possible to users doing their job with new ERP tools. Based upon the scope of the ERP project, the education and training footprint can be enormous for large organizations, requiring 10–20 train the trainers. Mid-size and smaller organizations usually have a smaller mass of users to deal with, so they may get by with a couple of trainers. Ideally, training should be configured to each individual or job title. However, this is seldom viable, so a good middle ground is engineering training by process or other broader category that makes good sense. Appendix D.2 has a couple of good matrices that might help guide the project team in tailoring the education and training framework and deployment strategy.

6.1.2 Commitment

There are a variety of dynamics that affect the education and training framework. There is the general organizational model that refers to the training needed for employees, in general. There is skill-specific training model that assists an employee to do a better job and function within the culture of the organization. This skill-specific model also provides avenues for promotion, or career-expanding opportunities. The third model is an initiative-oriented model, which in our case is the ERP implementation project. Let's look at these briefly.

Here, organizational commitment refers to the relative strength of an individual's identification with and involvement in an organization. It can be characterized by at least three factors:

1. Strong belief in and acceptance of the organization's goals and values
2. A willingness to exert considerable effort on behalf of the organization
3. Strong desire to maintain membership in the organization

When viewed in this manner, commitment represents something beyond mere passive loyalty to an organization. Instead, it involves an active relationship (entrepreneurial spirit) with the organization in which individuals are willing to give something of themselves in order to help the organization succeed and prosper. Real commitment often evolves into an exchange relationship in which individuals attach themselves to the organization in return for certain rewards or outcomes.

The attributes of organizational commitment and skill model include the following factors:

- Personal factors where older and more tenured employees tend to be more committed (security factor), whereas younger employees tend not to be as committed (or tend to be more volatile or flexible); studies indicate that women tend to be more committed as a group than men.
- Role-related characteristics influence commitment. Employees working on enriched jobs as well as employees reporting low levels of role conflict and ambiguity tend to be more outwardly committed.
- Structural characteristics where, for example, employees in decentralized organizations and in worker-owned cooperatives report higher commitment levels.
- A series of work experiences have been found to be related to commitment when, for example, employees believe that the organization is dependable and is at least moderately interested in their welfare; when employees believe that their jobs are particularly important to the organization; when employees are highly involved socially in organizational activities; and when, in general, employees believe that their expectations have been met on the job.

Restated, the major influences on the degree of employee commitment can be found in the person, the job, and the situation or work environment. In view of this, the job of building commitment is certainly no easy task.

In like fashion, it is imperative that an organization realize the importance of showing their commitment to their employees. Many managers boast the importance of their HR as being among, if not actually being, the organization's most important assets. However, many fail to realize the need to sufficiently stimulate HR so that they will continue to grow and not stagnate. For if not, then many of the firm's employees can legitimately expect to reach a point of decline in their respective careers. Many Japanese firms, for example, provide lifetime employment as a means to demonstrate its organizational commitment to its employees.

Education and training programs are perceived to aid significantly in exhibiting organizational commitment to employees. Similarly, by helping to mold the attitudes and perceptions of each student employee, the programs also help to establish and improve employee commitment. Studies have indicated that the earlier the employee commitment is made and instilled, the higher the employee retention success ratio. This contrasts to similar organizations making or instilling either no commitment or a deferred commitment. Consequently, the organization's timely execution of an education and training program often results in a rejuvenated spirit of commitment from its employees.

Furthermore, an employee typically receives an annual performance evaluation. Frequently, this may be the only time that the supervisor provides the individual with positive reinforcement such as "A job well done." Supervisors seem to get caught up in the pressures and stresses of deadlines and tend not to spend important time patting people on the back. Quarterly performance reviews are better, but if this cannot be accomplished, rewarding good performance in a fashion that invests in the employees (e.g., an education and training program) is a reasonable substitute. A well-designed education program can provide not only positive reinforcement but also an indication that management appreciates their employees' talents. Highly charged employees really do make a difference in overall organizational performance.

Again, employees truly committed to the goals and values of an organization are more likely to participate actively in organizational causes. In addition, highly committed employees generally have a stronger desire to remain with their employer (studies have indicated that commitment has been consistently found to be inversely related to turnover). The motivated employee will continue to contribute toward the attainment of the organizational objectives with which they agree.

As employees continue to intensify their identification with their organization (and to believe in its objectives), they will, naturally, become more involved in their jobs. Entrepreneurial spirited employees are more willing to expend a greater amount of effort on behalf of the organization than those employees who are not.

Over the years, firms, realizing the potential rewards, have begun to concentrate more directly on ways to increase the commitment of their employees. Many have found that a key factor to consider entails increasing the level of employee involvement, knowledge, and feedback. The goal then becomes to make the employees an integral part of the organizational team. Education and training programs serve as one mechanism, which leads to such increased efficiency.

In the case of an ERP implementation project, the education and training is focused upon preparing the employees for the new system, the associated analytical tools, new decision process, and so on. The attributes discussed above also apply but in a different fashion. In the organizational/skill area, the time frame is not pressured, is unaffected by employee involvement mass, and may span one or more years. In the ERP project model, there is a time constraint and mass of employees, which add pressure to the process. In addition to pressure concerns, the ERP model has a host of variability in the subject matter. In the organization/skill models, the employee sees an existing structure limited to confin the subject matter (organization, culture, skill, etc.), whereas the ERP project is broad-based by redefining the business process framework, its tools, its decision process, and a variable list of other dynamics. To give a practical example, if a product breaks, there are at least two steps to get it fixed. The first step is to troubleshoot to determine the cause of the failure and then engineer a corrective action, and the second step is to follow the prescribed process and the material list to fix it. An ERP implementation project is equivalent to the troubleshooting and engineer corrective action aspect. Troubleshooting involves an array of potential cause factors (mechanical, electrical, fluid, environmental, process, etc.) and repair strategies (cost, resources, material availability, etc.), whereas the process to fix the product is typically well defined and using cookbook methods to repair.

Regardless of the model being considered, there needs to be a well-engineered framework and measurement process used to help ensure that the expected results align actual results. Engineering this framework is a performance-oriented **best practice**.

6.1.3 Setting Goals

To be successful, an education program must have clear goals. For example, employees must support the system, identifying their own success with that of the system. They must believe that the system will work and that they are in part responsible for making it work. They must be convinced that the system will provide them the means to do an outstanding job.

Consider the opposite situation: employees not dedicated to seeing the system succeed or not understanding its purpose and employees not using the system or using it improperly. This results in apathy, fear, and subterfuge.

To achieve the stated goals, an organization must provide a thorough, quality education program. This will require an education coordinator, someone who will see to it that a quality program is developed and then followed as a viable program that is implemented and used over the years.

An intrinsic program goal is that it should be tailored for each "individual" based upon their specific skills inventory as well as upon their personal job and career objectives. The more tailoring the employee perceives, the greater the value the program has to the employee, and, consequently, to the organization.

Another such program goal should be that of an ongoing "process" and not a one time event. Lasting benefits come from lasting investments; therefore, the marginal difference between investments and benefits should provide overwhelming yields.

Goals are inconsequential if the results are not rigorously measured. Therefore, a reasonable rate of return should be sought and monitored based upon the education and training goals. This practice of tracking benefits (hard cost savings) is another performance-oriented **best practice.**

The ERP project model should align to the precepts discussed earlier. Education and training should be an integral aspect of commitment (definitive expected results that are tracked) during the senior leadership collaboration workshop (see Chapter 3). In addition, the education and training deployment strategy should be aligned to the nested internal customer/service provider process measures discussed in Chapter 4.

6.1.4 Achieving Quality Education

Because education consists of providing the knowledge, skills, and attitudes necessary to produce long-term behavior changes (also termed goals and objectives), quality education can be defined as the efficient achievement of these objectives.

Efficiency requires that:

- An individual states his or her objectives; the long-term changes in behavior, knowledge, and skills that are expected of learners; and the criteria that will be used to assess learning. For example, an objective could be that All sustaining engineers will be able to follow the procedures for introducing engineering changes, and the criterion might be by describing the process manifested by entering a change into the data base.
  Another might be Planners will interpret the Supply/Demand profile by describing the contents of a Supply/Demand Report and explain what actions should be taken based upon the information presented on the report.
- An individual provide the methodology resulting in a mix of the desired changes. These methods should include workshops, education sessions, reading and video assignment, conference room pilot, and so on. Methods should be engineered after the objectives are stated.
- Certain conditions must be prevalent for learning to occur and should be considered whenever education programs are developed. One must essentially recognize how adults approach learning. This means providing education that is
  - *Motivating*—The adult student must be "inspired" to
    - Change behavior.
    - Pay attention to instruction.
    - Accept what is being taught.

- *Sequenced*—Learning must be structured to provide sufficient repetition, hands-on practice, and quick feedback on the results. The proper sequencing reinforces newly acquired knowledge and skills, and transfers them to real-world applications.
- *Understandable*—The content must be presented in context with what the adult student already knows and understands.

Of the three, motivation is probably the leading factor. If the adult student is not sufficiently motivated, the education program for that individual will be a failure. Motivational education is:

- *Stimulating*—The sessions introduce elements of excitement and humor.
- *Relevant*—The adult student understands and accepts the purpose of the education as personally relevant.
- *Endorsed by the adult student*—Fear and other inhibitors have been removed.
- *Applicable*—The subject can be tried out immediately.

The reference to overcoming fear might seem unusual. However, many employees fear plenty. Most education and training programs (hereafter referred to as "education programs") require substantially more procedural rigor and formality than most companies culturally maintain.

The introduction of any new system requires new methods, new inputs, and new levels of quality. New systems require changes in processes and, often, changes in organizational roles and functions (e.g., changing from expediting to planning within a manufacturing system). This change can be threatening toward employees and these threats cause fear.

A beginning task in education, therefore, is to overcome an individual's fears that tend to inhibit learning. By mitigating fear, it becomes feasible to develop an atmosphere that will promote support for the new system. An individual is then able to become a system champion who wants to succeed by using the new system and by using it properly.

This will happen if the proper time and effort is expended to see that all employees are aroused to understand, accept, and believe that the system will be personally beneficial.

- Be instilled an awareness that their jobs will be better, that each department (and, consequently, the entire company) will function more efficiently, and that resorting to informal methods to circumvent the new system is self-defeating and must be prevented.
- Achieve some reward and prestige by proudly showing that they have learned how to use the system correctly. (This may be in the form of a certificate program; see Chapter 5 for an expanded discussion on certification.)
- View the need for the new system as a challenging, competitive, and opportunity, not a painful problem. Every person should profit from the system. They should grow professionally, feel an increased sense of worth to the company, and an increased respect from it and earn an increased respect from it.

As discussed in subsection 6.1.3, education must result in "quality" deliverables as monitored and tracked. Chapter 4 goes into detail on the meaning of quality and the need for a standard. As the ERP project core team is developing the education and training framework, it must consider the need for technical as well as end-user requirements. As you reflect back on Chapter 1, it should be clear that as the core team engineers the education and training requirements, they must span the following environments:

- Technical environment (technical mastery of new ERP system)
- Business operating environment (process change mastery of impacts on the day-to-day business)

- User community environment
- Project environment (project team skills mastery)

Therefore, education and training materials must permeate these four environments. I choose the word "mastery" if the ERP project is to achieve stellar results; mastery is the level of expertise pursued. This is a performance-oriented **best practice**. Unfortunately, most ERP projects do not have the budget or training intensity to achieve stellar results. Consequently, they settle for nominal results instead.

In sum, support must be gained from everyone affected by the new system. Only then will people be sufficiently postured to learn. Learning mandates motivation. Motivation will occur only when each adult student

1. Knows the goals of the education program.
2. Understands and accepts these goals.
3. Believes that the new system will work and be of personal benefit.

6.1.5 Planning for New ERP System Education

Engineering a solid education plan is absolutely necessary for education and training to be effective. The key attributes of the education and training framework are as follows:

- *Be specific*—Tailor to individual job function/nested process whenever possible.
- *Realistic time frame*—Ensure that the roadmap is doable to the ERP project life cycle, as well as to the individual's "day job" commitments.
- *Has timelines*—If it is important, it needs a timeline.
- *Measureable*—Ensure that the plan is measurable and measured.
- *Achievable*—Similar to realistic time frame, this attribute implies the quality aspect of the deliverable.
- *Relevant*—Avoid conducting training that "might be" essential to ERP deployed. Ensure that it is mission relevant.
- *Under change control*—If changes are needed, they must be aligned to project plan and project life cycle deliverables.
- *Buy-in by all stakeholders*—This needs to be a committed plan, not a suggested plan.

Done properly, the plan results in a program that facilitates the right people, learning the right things, at the right time, and with confirmation that exceptional understanding was delivered. Education planning is an integral part of the detailed ERP project planning process. It also conforms to and has deliverables aligned to disciplines discussed in Chapter 2.

Planning is accomplished by identifying the basic elements of education and training, then systematically relating these elements to the company's current posture. To accomplish this, determine:

1. The *field* of who needs to be educated. Group logically based upon new system roles.
2. The *what* which needs to be learned by each group. This means defining the educational goals and objectives consistent with the rationale for pursuing them for each audience.
3. *How* education will be attended to. This involves determining how internal and external classes are to be incorporated.
4. *Timing* when each group will be educated. This requires close coordination with the implementation phases.
5. *Location* where the process will occur. Choosing the right location will enhance the adult learning process.

6.1.5.1 The Field

Between the beginning of a project and the time that the new system becomes fully implemented, as many as 90% of the firm's employees should have received some degree of education related to the system.

A proven approach is to create a list of the groups of people who will require training. In Figure 6.1, there is a sample group matrix (role) with associated topics and courses.

Topic	Course Name		Roles										
		Course#	Accounts Payable	Accounts Receivable	Configuration Management	Configuration Service	Engineer	Fixed Assets	General Ledger	HR	IT	Inventory Management	Procurement
Allocations	Fundamentals	FIN-1	X	X				X	X			X	X
Budgeting	Fundamentals	FIN-2	X	X		X			X		X		X
Configuration Management	Configuration Management	CM-1			X	X					X		
	Appraisal Method	CM-2			X								
Contracts	Fundamentals	PUR-1				X	X						X
	PPV	PUR-3	X										X
	Requisition Management	PUR-4										X	X
Financial Analysis	Fundamentals	FIN-3	X	X		X	X		X			X	X
Financial Modeling	Fundamentals	FIN-4					X		X				
Fixed Asset Management	Fundamentals	FIN-5							X			X	X
General Ledger	Fundamentals	FIN-6	X	X		X		X	X			X	X
HR	Payroll	HR-3								X			
	Human Capital Management	HR-1								X			
	Benefits	HR-2								X			
Information Workmanship Standards	Basics	IWS-1	X	X	X	X	X	X	X	X	X	X	X
	Advanced	IWS-2					X			X	X	X	X
	Nested Measurements	IWS-3	X	X	X	X	X	X	X	X	X	X	X
	Process Performance Management	IWS-4	X	X	X	X	X	X	X	X	X	X	X
Inventory Control	Warehouse Management	IC-1	X	X		X			X			X	X
	Cycle Counting Techniques	IC-2										X	
	Physical Inventory	IC-4										X	
	Record Accuracy	IC-3				X						X	X
Managing Variances	Fundamentals	FIN-7	X	X					X			X	X
Pay Cycles and Vouchering	Fundamentals	FIN-8	X						X				X
Process Improvement	Process Overview	PI-1	X	X	X		X	X	X	X	X	X	X
	Process Improvement Skills	PI-2	X	X	X		X	X	X	X	X	X	X
	Appraisal Method	PI-3					X						X

Figure 6.1 Sample group matrix (role).

In this example, we look at a matrix with various roles within an organization (the horizontal axis) with the vertical axis representing various topics (processes). It also displays various courses associated with the process. The "X" represents the recommended attendance. In lieu of an X, the number of hours that the course takes may be substituted. For example, the accounts payable team members would take a fundamentals course in the allocations business process, as well as budgeting, contracts, and financial analysis.

Many people will fit into more than one group. For example, some department managers will be members of the core project team. This and subsequent example worksheets have been synthesized from an education program centered on an ERP implementation project.

Figure 6.2 shows a worksheet to estimate the number of people in each group. There is a space within each group to list the number of managers, supervisors, administrators, and staff (user

Topic	Course Name	Course#	Roles										
			Accounts Payable	Accounts Receivable	Configuration Management	Customer Service	Engineer	Fixed Assets	General Ledger	HR	IT	Inventory Management	Procurement
Allocations	Fundamentals	FIN-1	1	1				1	1			1	1
Budgeting	Fundamentals	FIN-2	1	1		1			1		1		1
Configuration Management	Configuration Management	CM-1			1	1					1		
	Appraisal Method	CM-2			1								
Contracts	Fundamentals	PUR-1				1	3						2
	PPV	PUR-3	1										2
	Requisition Management	PUR-4										1	2
Financial Analysis	Fundamentals	FIN-3	1	1		1	1		1			1	2
Financial Modeling	Fundamentals	FIN-4					1		1				
Fixed Asset Management	Fundamentals	FIN-5							1			1	1
General Ledger	Fundamentals	FIN-6	1	1				1	1			1	1
HR	Payroll	HR-3								1			
	Human Capital Management	HR-1								1			
	Benefits	HR-2								1			
Information Workmanship Standards	Basics	IWS-1	1	1	1	1	3	1	1	1	1	4	2
	Advanced	IWS-2					3			1	1	4	2
	Nested Measurements	IWS-3	1	1	1	1	3	1	1	1	1	4	2
	Process Performance Management	IWS-4	1	1	1	1	3	1	1	1	1	4	2
Inventory Control	Warehouse Management	IC-1	1	1		1			1			4	1
	Cycle Counting Techniques	IC-2										4	
	Physical Inventory	IC-4										4	
	Record Accuracy	IC-3				1						4	1
Managing Variances	Fundamentals	FIN-7	1	1					1			1	2
Pay Cycles and Vouchering	Fundamentals	FIN-8	1						1				1
Process Improvement	Process Overview	PI-1	1	1	1		3	1	1	1	1	4	2
	Process Improvement Skills	PI-2	1	1	1		3	1	1	1	1	4	2
	Appraisal Method	PI-3					3						2

Figure 6.2 Number of employees taking each course.

associates, production, technical, etc.), as well as others. The scope of the education effort will be determined by roughing out an estimate of the number of employees to be educated.

As the planning progresses, names of the people within each group must be given. One such format may be used to list the names of those people who comprise the executive, project team, manager, and instructor groups within the organization. These are the people who must be educated first. To begin identifying the company personnel who will need education, one might proceed as follows:

1. *Executive management*: Managers with overall responsibility for the company, generally the president and his staff.
2. *Project team*: People chartered with planning and implementing the system.
3. *Department managers*: Within the manufacturing department, for example, one would address the directors of supply chain, engineering, HR, purchasing, production control, accounting, and information technology (IT).
4. *Administrators and supervisors*: Middle management, foremen, and others from the functional units engaged in the new system.
5. *Staff within each functional area*: Team members such as design engineers, associates, administrative assistants, and direct liaison employees.
6. *Instructors*: Those with responsibility for educating the company management and staff.

These initial lists of groups, processes, and courses are rather straightforward. However, depending upon the size of the organization, the number of locations (and/or time zones), the operating footprint (24 × 7), and a variety of other factors, the training matrices may become rather complex.

6.1.5.2 The What

The what addresses the concepts, knowledge, and skills that must be acquired by each of these groups. A step-by-step process of defining goals, assessing the employees and the positions relative to those goals, and determining the educational activities necessary to bridge the gap is as follows:

1. Specify the goals or the learned behaviors that each group must demonstrate. These behaviors should relate to your overall goal of maximizing new system benefits. Identify for each group the specific level of knowledge necessary in all functional areas and describe the nature and quality of behavior desired. For example, determine which groups simply need to enter data and/or which groups must create data. Determine which groups need only to read certain reports and which must use them to perform their tasks. These goals need to be aligned with the overall ERP project goals as well as goals documented by the senior leadership collaboration workshop (see Chapter 3). If goals are part of a service-level agreement (SLA) or certification path, these need to be clearly stated (see Chapter 4).
2. Next, evaluate the current knowledge, skills, and ERP mastery of each group. Try to identify where expectations regarding procedures, performance, or responsibilities, change significantly. If the ERP implementation project is dealing with a significant "process" change, this skills inventory may be especially challenging and a bit daunting with alignment within the ERP project plan.
3. Finally, analyze the difference (gaps) between goals and current performance. This analysis will help determine the education plan that will best accomplish the organization's goals. This gap analysis presumes that new system performance would be equivalent to current performance. If this presumption is not true, there may be a need for multiple tiers of groups,

each with their skills' inventory levels fairly aligned. Due to a variety of dynamics (whether involving bargaining units or not, performance review model, etc.), the tier concept needs to have close collaboration with the HR department guidelines.

Each group should have goals specific to its function at this point, which are as follows:

1. Executive goals
   For any new system to succeed, top management must provide commitment and leadership. This means that these executives must all be thoroughly familiar with and committed to the following:
   a. All system benefits—business related, return on investment related (as discussed in Chapter 3), process improvement related as well as ERP software module related.
   b. All system costs, including money, personnel, and other resources.
   c. The general features, organization, and functions of the system.
   d. Planning and financial functions that support top management activities, such as resource planning and production planning.
   e. Executive roles in the implementation and profitable use of the new system.
   Senior management must lead and be examples for the entire organization. This is a stellar performance **best practice** attribute.
2. Project team goals
   A nucleus of high-level managers will have major responsibility for planning and implementing the new system, including support functions such as education. Broad goals, some with numerous subordinate objectives, are as follows:
   a. Build a robust project team relationship by organizing projects to effectively resolve conflicts.
   b. Develop a detailed project plan by acquiring and maintaining support of top management (see Chapter 1).
   c. Identify and sequence the activities involved in each phase of implementation.
   d. Establish controls, assign personnel, set priorities, and estimate completion dates for all activities.
   e. Establish and implement cost control criteria for measuring the success of the project implementation effort.
   f. Coordinate and supervise project deliverable activities according to a detailed project plan.
   g. Understand the features of the system and the relationships of its individual modules.
   h. Define the responsibilities of functional departments in their use of the new system.
   i. Understand the detailed functions, features, and use of each module, as shown by knowing "how" and "what" to input into the system, as well as how to interpret and react to output from the system.
   As the project progresses, the project team undoubtedly will develop additional goals and specify more precisely the objectives that enable these goals to be accomplished.
3. Instructor goals
   From time to time, nearly all employees need to be educated (or reeducated). The best way to accomplish this is to use, whenever practical, the organization's own people as instructors. As mentioned earlier, for an ERP project, more than 80% of the employee population will require some education.

   It is critical that someone be assigned the responsibility for coordinating the education effort. This "education coordinator" should be a member of the ERP project core team. Although this coordinator may actually train others, the magnitude of the overall effort usually requires that several instructors do the training. Instructors should come from each functional area of the company, if possible, or at least demonstrate a mastery of those functional area precepts.

The education coordinator and the instructors from these functional areas must be trained in effective education techniques. They should know the following:

a. General features, functions, and purposes of all new system modules and their related business processes.
b. Specific functions and applications of the new system modules they will use.
c. Techniques of effective teaching can be shown by knowing the following:
   i. What methods to use?
   ii. How to plan and schedule?
   iii. How to prepare themselves, their courses, and their students?
   iv. How to motivate?
   v. How to assess learning?

Note: If significant "process change" is expected, these instructors would benefit from being certified in black belt competencies.

4. User group goals

   Each major functional group of new system users within the company should be identified. Within the area of manufacturing, for example, these might include design engineering, manufacturing engineering, stockroom, production control, master scheduling, order entry, accounting, manufacturing, information systems, and purchasing.

   Most people in each group must learn

   a. The concepts upon which the formal system is based (generic).
   b. The general features, functions, benefits, and use of the system as a whole.
   c. The concepts or reasons behind their application area system functions (e.g., design engineers would have to know, say, how to define bills of materials that work in the system, and they would understand the concepts and mechanics for engineering change activity).
   d. How to use the system in their particular jobs.

   The critical goal for each user group is to understand the functions of each system module and be able to describe the specific activities of each group and must perform relatively to those functions. For example, under the auspices of the implementation of an ERP system, engineering change control is a function that must be performed cooperatively by both sustaining engineering and production control. Within the bill of materials module, several functions relate to engineering change control. Engineers, in such a case, would therefore be expected to

   a. Describe what reports or on-line inquiries they would use to evaluate the potential effects of an engineering change.
      i. Use those reports or inquiries for decisions.
   b. Enter the appropriate transactions that enact engineering changes.
   c. Be expert on retrieval data as well as on data.
   d. Verify that the changes have actually been made upon the database.

   By relating system features to the specific job functions, each functional area instructor will have a comprehensive list of concepts that must be learned in order for the system to function as expected.

   As the education needs of each group are analyzed, goals should be specified. These goals may be categorized as follows:

   a. Conceptual education relating to generic system fundamentals and techniques.
   b. Education regarding the specific functions, features, and uses.

c. Module education on system interfaces.
d. Training on new (and revised) procedures and forms required by the system.
e. Training on screen usage and report interpretation.

Once the issue of what education has been resolved, the next step is to define the methods to be used. With the methods identified, a detailed education plan can be developed.

6.1.5.3 How

A central educational goal is to properly educate the employees who will operate the ERP system on a day-to-day basis. These efforts should focus upon providing everyone with a good understanding of their job requirements as it relates to the new module, on-the-job training, a conference room pilot, and practice exercises. Various instructional methods may be used: lectures, group discussions, books, and workshops.

Some or all of the needed materials may be developed by themselves or may be selected from external sources, including the following:

- Commercially available video programs that cover the fundamentals and techniques of various subjects, including, for example, modern production and inventory management
- Commercially available public seminars and workshops
- Society (e.g., American Production and Inventory Control Society, American Management Association, etc.), workshops and meetings, availability being contingent upon membership
- Conferences
- In-house custom seminars

In addition to using these sources, an organization will have to develop some education programs on its own, especially for procedures and forms unique to the company. Such customized programs are very effective because the organization itself and firm-specific products are used as examples, reflecting the company's own policies, configuration, and environment. Custom development of all courses, however, can be time consuming, complex, and costly.

At times, education is most cost effective if provided by outside sources, especially if courses are customized to fit specific needs. The education needed may only be available through outside sources to meet time schedule and provide expertise.

1. Generalized course plan
   An overall plan should include the following:
   a. Education on the new system fundamentals for top management, the project team, operations management, and key users be provided through outside seminars and video courses. When using packaged education courses, adapting the discussion portions to the company specifically should be used for maximum learning value.
   b. Executive education be ongoing, obtained through group-oriented seminars.
   c. The project team receives fundamental project management skills training.
   d. Generic system education be provided early in the implementation phase for management, key users, and the project team. Workshops can be designed to meet this objective.
   e. Procedures training must be developed and presented by the company itself and given close to the cutover date. The use of a conference room pilot may be especially helpful for this (for further information, see Chapter 5).
   f. The education program must encompass all levels of personnel engineered specifically for varying levels of detail and varying scope of coverage. If engineered properly, the

education program will have touched over 90% of all people in the company by the time the new system is in full operation.

2. Workshops and seminars

   Workshop courses are designed for use by the project team, instructors, and key users representing each functional area within the company. Workshops should include case studies, discussion sessions, and exercises that provide practical experience with the new system. These courses educate the people responsible for the implementation of the system and should be given early in the project. Their brief descriptions are as follows:

   a. *Executive-oriented seminar*—Introduces management to the objectives and elements of the integrated business planning process (ERP), and it defines the role of each executive in the installation and profitable use of the system. Its goal should be to secure top management's commitment and leadership, and thus ensure successful implementation of the system and achievement of operating objectives.
   b. *Project team leadership workshop*—Teaches the project team how to organize and manage the system implementation project effort. Using case studies and selected course materials, the project team should learn how to plan their project, evaluate techniques to manage it, and solve problems it will face, which are unique to the organization.
   c. *Instructor training courses*—These are designed to provide "train the trainer" lessons. It is intended to help instructors learn how to plan and develop educational courses, and develop effective teaching skills. The course should combine education on instructional methods with actual practice in teaching.

   In addition, there may likely be software-specific courses available. These courses, in addition to those designed for key users of the system, detail the purpose, function, operation, and use of the relevant modules. They may be taken by some or all of the project team prior to a module definition review. They are also designed to further educate instructors and key users so they can better understand and be prepared to use the system themselves. In many cases, they will also be better able to teach the system users in their respective departments.

3. Planning

   The education issues of what and how are now known. The task of planning the course must begin. Samples of several forms that will aid in this process are given in Appendix D.2. They are filled in as examples, although blank copies for your use have been reproduced and placed at the end. The forms are used for planning

   a. Education on the generic concepts of formal systems.
   b. System-specific, module-oriented, and user-oriented courses.
   c. Methods and procedures training.

   Each form provides space for identifying the groups who must be educated, the courses that can be taken, the numbers of people enrolled, the duration of courses, the total number of student education hours, and an estimate of course costs.

4. Conceptual education

   Conceptual education is available from various sources. Space in the boxes at the top of each column indicates specific courses needed (see form, Appendix D.1.2). Each column has two sections: one for detailed education and the other for overview or general education.

   The Participants column lists the names of persons or groups of individuals. The three rows at the bottom of the form list the course duration in hours. By multiplying the total number of students by the course hours, the total number of student hours is determinable. Finally, the estimated cost for each course can then be computed. This allows the education coordinator to size the course time commitments for scheduling purposes.

The worksheets in Appendix D.2 can be used to estimate the groups and number of people who will attend the courses. The education coordinator may choose to list names of groups or may use one form per group and list the names of the employees and the courses they are expected to attend.

5. Procedures and methods training

   Implementing an ERP system tends to change people's work activities, responsibilities, and (hopefully) attitudes. New reports, screens, and forms as well as different data management methods and timing affect nearly everyone. Better quality and timely information allow people to be more thorough in their work performed.

   The ERP project should be an operation to probably realign responsibilities as well as elevate management's expectations. New and revised procedures must be documented and ingrained in all those affected.

The procedures training worksheet (see Appendix D.2) can be used to help and develop the training program. The procedure courses needed are listed in the columns. Space allocated for either organizational groups or individual names in the same manner as on the education matrix forms are listed on the left side of the rows. Line supervisors are expected to rigorously teach their employees how to use the new procedures. One of the education coordinators should be available to assist them in making them effective.

6.1.5.4 Timing

The next task on the docket is to define the timing when the education must occur. Course schedules will be developed following the principle that "effective learning occurs when that which is to be learned has some immediate use." Hands-on training courses, especially, should be scheduled close to the time when it is likely to be applied. Education is an essential ingredient of the project plan, and scheduling is an integral component of that plan. The following section explains the relationships between education scheduling and major implementation milestones, and the need to interweave with the detailed project plan in a balanced manner. Do-ability is key to results realization and timing is key to do-ability.

If the subject was not addressed, during the senior leadership collaboration (see Chapter 3), then this is the time to determine what portion of the elephant will be consumed by this ERP project. Contemporary ERP software solutions may be complex and the organization must time-box the ERP project duration. To that end, there may be phases of the ERP project defined and each phase is bounded by resource, budget, deliverable, and business driver constraints. At times, a competitive window of opportunity becomes a driver in time-boxing the project deployment. Regardless of the variables influencing the ERP implementation planning horizon, a stake in the ground needs to be defined on the breadth of the deliverables and their respective time-box. We will discuss this topic in more detail in Chapter 6.

Initial system implementation tends to occur in four stages. These stages include a preliminary planning stage, an implementation preparation stage (six months to a year), a pilot implementation stage (a few month's duration), and, finally, the cutover stage where the full system is implemented. The major milestones in each stage are listed below.

Stage 1: Preliminary planning

- *Project initiation*—This stage is for getting organized, gaining management commitment, and forming the project team. The hardware, software, and support service contracts are formalized.

- *Project implementation planning*—The project team develops the initial project implementation plan and the framework for the education plan.
- *Installation of software system*—The software system is installed. This occurs soon after project initiation. The system will be used for technical orientation and to facilitate the module definition review process.
- *Module definition review*—The project team determines which system features and functions are appropriate for initial cutover deployment. It also identifies what modifications may be necessary to tailor the system to the company's needs. In Chapter 3, we discussed the senior leadership collaboration workshop where ROI commitments were made. Based upon these commitments and their realization timetable, the system feature deployment will likely be prioritized accordingly.
- *Process definition review*—If concurrent business process reengineering is an outcome of the ERP project, then a process definition and gap analysis may be developed and managed. However, in smaller organizations, this will likely be combined in the module definition review process.
- The *policies and procedures* that support the use of the system should be identified at this time. Issues affecting implementation are defined and their resolve forms a key component for the detailed project plan.

Note: Depending upon the magnitude of expected process change and the size of organization, these process deliverables may be decoupled from the ERP project and managed independently and led by a black belt competent resource. However, to align with the project ROI goals, this subproject should remain synchronized to the ERP project plan implementation horizon. In addition, during the integration testing phase, by definition, process testing is an integral aspect of the integrity of the integration test results.

Stage 2: Implementation preparation

During the second stage of the project, as a result of feature/function deployment and their respective process changers, the team will develop procedures, establish or convert the databases, and conduct a conference room pilot (see Chapter 5). At the completion of this phase, a functional specification is developed. In Chapter 2, we discussed requirements generation, representing the engineered "what" or design stage of the ERP project. The conference room pilot allowed us to test drive the software functionality and the confirming end result of what we will be using is the functional specification. Engineering a robust functional specification is a documentation **best practice**.

The data are loaded and the pilot phase begins. Most of the education and end-user training occurs during this stage. It includes both overview education and specific training for job functions. The project team will conduct a conference room pilot at this point to validate the processes, new procedures, transactions, and interfaces within the system. Using a small database, the pilot produces a valuable tool for testing the software, educating employees, and testing the team's use of the system.

Stage 3: Production pilot

During this stage, a full integration test (see Chapter 7 for more details) will be conducted consisting of hardware, software, customizations, interfaces, processes, policies, and procedures. The integration test consists of representative users performing a typical "day-in-the-life-of" use of the end-to-end processes scenario. It will also validate month-end and year-end functionality. In addition, the technical team will perform a stress test and document system performance capability. As a result, the new processes and procedures will be modified as necessary. This is the time to further resolve issues as well as evaluate the adequacy of the education program and general planning. As the result of that evaluation, the project team makes necessary modifications and conducts any corrective training. This is also a perfect opportunity to perform a trial cutover (dry run) whereby the team tests cutover procedures.

Stage 4: Full Cutover

This is the final implementation stage. The modules of the system that have been implemented are used by all departments across all lines. The company's employees now use the new system as the routine method of doing business. Initial education and training is completed, and the educational coordinator now begins to conduct advanced and continuing educational programs. The project may continue into phase II, which may include implementing other modules or adding additional capabilities from those modules implemented.

As part of the overall project plan, the project team will define, by actual name and date, who will receive what education, and when and where it will occur.

Detailed education planning is imperative if expectations are to convert the plan into action. To complete scheduling properly, the educational coordinator must keep track of who will be attending which courses and when. In addition, the education coordinator is charged with keeping track of grades, certification status, and so on. The education planning worksheet (see Appendix D.2) will help in the development of the detailed plan and in monitoring it as it is implemented.

The form may be used in three ways: (1) track participants, by class; (2) track classes, by participants; and (3) track classes/participants by departments.

Months included in the plan should be written above month numbers (e.g., 01 might be April, 02 might be May, 03 might be June, etc.). Class participants and departments should be specified. If the date of a class is known, it should be entered in the appropriate month column. If an exact date has not been set, an "X" should be placed in the appropriate month column.

Courses are conducted during normal work hours. It should also occur early rather than late in the workday. This emphasizes the importance the company places on education, training, and its employees. This also increases employee motivation to learn.

Finally, sessions should be limited to about two to three hours per day. Learning takes time and everyone has a limited ability to absorb new information. Limiting training time per day reduces the chances of information overload. Spreading training over time has several advantages. It more readily facilitates the new knowledge to be integrated into what an employee already knows before any attempt is made to present additional information, and it also allows employees to continue to perform their present jobs.

6.1.5.5 Location

Depending upon the size of the organization and the breadth of scope of the project, education and training may take place in one or many locations. Facilities will need to be provided. Depending upon the size of the company, it may require classrooms large enough to accommodate groups up to 30 people and perhaps smaller rooms for groups of up to 12 or less. When planning education sites, the goal is to develop classrooms that facilitate rather than hinder teaching and learning. While this may seem obvious, poor facilities often have the effect of compromising, in other respects, effective education programs.

Two factors must be considered when planning for education facilities: student factors and teaching factors. Although the lists that follow are not all-inclusive, they should be used as a checklist to guide the planning of appropriate and learning conducive facilities.

Depending upon the software used, there is likely an opportunity to learn via self-study courseware. The ideal training program will consist of a balance of both self-study and group study curriculum.

Student factors affect student learning and include the general classroom environment, seating, writing surfaces, workstations, and video monitors.

- *General classroom environment.* The following factors influence the general environment:
  - Temperature—The room should be air-conditioned and maintained at a comfortable temperature. Excess heat or cold will cause learners to be less attentive.
  - Ventilation—Air should be constantly circulating, exhausting stale air, and bringing in fresh.
  - Light—Light should be adequate for reading and close paperwork. It should not be harsh. Facilities would optimally have separate light controls for student and instructor areas. If possible, all lights should be dimmable.
  - Noise—The room should be free of exterior noises. Internal noise control can be accomplished through acoustic ceilings and carpets.
  - Space—The room should be large enough so students do not feel cramped. Adequate space should be allowed in front of, behind, and beside each student area.
  - Color—Pleasant neutral colors are best.
- *Seating.* It is important to consider seating if you expect students to pay attention for any period of time.
  - Comfort—Chairs should be firm, noiseless, freestanding, and able to both turn and recline slightly.
  - Flexibility—The seating should be movable. Depending on the type of instruction, instructors may sometimes want the chairs arranged in rows, auditorium style. On other occasions, they may want to set up the room classroom style or for discussion groups.
  - Position—Most education will involve some form of visual aids, workshop, video, whiteboards, or flipcharts. Seating should be arranged so that these are visible to each student with no obstructions (including the instructor).
- *Writing surfaces.* Each student should have some surface on which to write. Writing surfaces may be individual desks, small group tables, or bench-type tables. All should provide adequate space to take notes and room for student workbooks and other reading material.
  - Flexibility—Just as seating should be flexible, so should writing surfaces. Tables capable of seating four to six people may be used. These seem to provide the optimum balance between auditorium seating arrangements and those used for small group interaction.
- *Workstations.* A workstation for each user is essential. To accommodate group exercises, workstations should be accessible to groups (certainly not larger than four to six people).
- *Video monitors.* Classes may use video programs. Ideally, the workstation may be used for video playback; however, if this is not possible, video monitors may be necessary. There should be enough monitors so that all students can see. For a 20-inch monitor, the farthest student should be no more than 15 feet away. Two monitors are usually adequate for groups of up to 50 unless students are sitting at large or widely spaced tables.

Video-based classes may be taught to small groups or even taken by individuals themselves. Students must therefore be able to use the video playback units. It is probably desirable to have a room separate from the major classroom where individual students or small groups can view the video components.

Teacher factors, however, affect teaching. Teachers must be able to perform with minimum interference from audiovisual devices. The classroom should contain the following:

1. *Table, podium, or lectern for the instructor's notes*: It should be lighted so that notes can be seen when classroom lights are dimmed.
2. *Whiteboards*: These are convenient discussion aids. These with washable markers are better than chalkboards.
3. *Overhead projector*: The projector should have a built-in spare bulb feature.
4. *Projection screen or whiteboard*: A large screen is needed for use with overhead projection.
5. *Classroom-ready workstations, video playback unit, and other necessary support equipment and cabling*: These should be staged ready for each class.
6. *Dimmable lights with controls*: These are easily available to the instructor (optional).

6.1.6 Achieving Overall Program Efficiency

Thus far, education has been stressed as being crucial to the success of a new integrated system. Some characteristics of quality education have been described and some guidance in planning for education has been provided. Next, guidelines are presented, which are designed to help to achieve those plans. Effective education has several important characteristics, which are as follows:

- An essential ingredient is the support of executive management. Education and training should be recognized as a "priority." The necessary resources, money, personnel, and time are firmly committed, and the educational endeavor is not compromised. Management is concerned about progress made and roadblocks imposed.
- Education is well planned, educational needs are defined, and the methods for meeting these needs are spelled out.
- Education is administered well and carried out efficiently. All courses are monitored and tightly controlled so that the desired results are obtained.
- All participants are well prepared, and instructors know how to teach and are prepared to teach their courses. Students are incented and prepared to receive instruction.
- Courses are delivered effectively. Instructors see that the environment is conditioned for learning. Instructional materials are designed and administered properly. All educational goals are strived toward.
- Follow-up and ongoing education are provided; users practice what they have learned (validated by their supervisor through performance measurements). New employees are properly trained to use the new system.

Programs can be developed, which have winning characteristics. The most important characteristics are as follows:

- Selecting a qualified education coordinator
- Selecting and properly preparing instructors
- Preparing and motivating students
- Allowing students to feel comfortable with education
- Debriefing students and evaluating their progress
- Conducting intensive follow-up and ongoing programs
- Engineering effective courses

The primary criterion to judge the success of the program is employees performing their jobs as expected with the new tools. For optimal success, emphasis must be placed on developing courses where learning can be demonstrated by successful performance. The education "process" is as important as course content. Exposure to instruction, be it in a classroom or in front of a video monitor, does not necessarily result in learning. The focus cannot be on the teacher or the materials; it must be on the student and the process. Each student must be "special." It is easy to retrain the high achievers, but frequently more difficult are the critical mass. "All" must be equally prepared.

In all education activities, therefore, one must first define why education is occurring, what the objectives are, and what knowledge, skills, or attitudes learners are expected to exhibit as a result of the education. Second, some means should be provided for continual evaluation to ensure that individual and course goals are being achieved. Finally, throughout the duration of the plan, a constant review and revision of the plan may be necessary to meet changing needs and changing conditions.

6.1.6.1 The Education Coordinator

The education effort necessary to support successful new systems implementation requires significant amounts of time and labor. A hundred or more people may have to be trained. One may have to present hundreds of training sessions, and these sessions may occur over an extended time horizon. This is why it is frequently necessary to have a full-time education coordinator.

The major activities performed by the education coordinator are as follows:

- *Planning*—With help from other members of the project team, the coordinator determines the education needs and how they will be met.
- *Administration*—The coordinator sees that education plans are carried out and the desired results obtained. To do so, the coordinator selects courses; selects instructors and sees that they are trained; ensures that appropriate facilities are available; oversees any internally developed training classes; arranges for the education of executives, the project team, and key users; and monitors and controls the education of employees at all levels.
- *Education and training*—The education coordinator will very likely be the leading instructor as well. As an instructor, the coordinator must be a role model for those he or she teaches; know how to teach, lead discussions, lecture, and facilitate learning; know the new system thoroughly; and be able to relate to how the new system will be used in the company.

The person selected should

- Be a leader with demonstrated competence in accomplishing tasks.
- Be respected by executive management and the other members of the project team.
- Be knowledgeable to understand the company, its organization, procedures, and products. He or she needs to understand what the new system is and does.
- Be able to teach effectively and understand the importance of being a role model, realizing that students will likely adopt similar attitudes toward the new system.
- Be able to administer the complex education function.
- Be a motivator capable of working effectively with and through others to accomplish education goals.

A highly capable individual is required. One can seldom afford to delegate this job to someone on a part-time basis. The overall success of the education program is highly influenced by this person.

6.1.6.2 Selecting Instructors

The education coordinator seldom can teach all of the courses. Several instructors may be necessary. Selection of instructors must be based on an understanding of their roles. These include the following:

- *Role model, motivator, and leader*—Students adopt the attitudes of their teachers. Instructors must be positive about the system and its benefits. The most effective teachers are the known leaders, such as department managers. An instructor should be in a position of authority relative to the functional group he or she is educating and training. In the role of instructor, the individual will be in a "helping relationship."
- *Planner*—Instructors must be able to plan the courses they will teach. They are frequently responsible for seeing that students are informed about the course and prepared for it, that necessary facilities and course materials are available, and so on.
- *Teacher*—Instructors must be effective at teaching and administering the classes. They must deliver the instruction and lead discussions, relate new concepts and techniques to old problems, and effectively handle questions and objections.
- *Expert*—The instructor must know his or her subject. Students must believe that the instructor is an expert in whatever system-related topic they are teaching. Knowing that internal expertise is available helps convince students that the system can and will work in the company. The instructor must therefore know more than the material covered in the course being taught.

The proper selection of instructors, therefore, limits itself to people who

- Have a positive attitude toward the company and the new system.
- Are respected by peers and subordinates.
- Can teach, have had experience in teaching, or are capable of quickly learning how to teach (i.e., are able to communicate clearly and effectively, and can control groups).
- Are (or can become) an expert on the system as it relates to their functional area.

An important by-product of key managers functioning as instructors is the "system ownership" issue. A manager interested enough to learn the new system sufficient to teach it certainly will champion its use and success.

6.1.6.3 Preparing Instructors

Effective instructors must know their subject matter as well as how to teach. Those who are selected as instructors, however, may be weak in these areas at the time of selection. Although individuals with the best potential as instructors are selected, instructor training is usually necessary to develop these specialized skills.

New instructors need special preparation. They must recognize that they will have to learn in order to teach the material. This will motivate them to learn not only the content but also the process of teaching the course. Instructors must be impressed with both the prestige that will accompany their role and the seriousness of their assignment. Instructors must be prepared for assuming the responsibility of being a motivator and a role model.

A basic requirement for teaching any course is that the instructor be thoroughly familiar with course objectives, materials, methods, and content. Before anyone tries to teach these courses, he or she must first go through them as a student. After taking the courses as students, the prospective instructors should study the course administrator materials. Then, before teaching the class, they should practice teaching the course perhaps to a panel or a group of peers. Team teaching with an experienced instructor is also a good way to get started.

6.1.6.4 Preparing Students

Just as it is necessary to prepare instructors to teach, students must also be prepared to learn. The goal is to be sure that students have satisfied course prerequisites, are motivated to attend the class, and have arranged their schedules accordingly. The goal remains the same whether the student is a corporate executive or a stockroom associate.

Specifically, before employees attend any class, someone (preferable their immediate supervisor) should be sure that they understand the following: The purpose of the course, that is, its goals and objectives. One should try to relate the purpose to the overall implementation plan.

- *Why they are taking the course*—Discuss how learning the course material will benefit both student and company.
- *What they are expected to learn*—Employees must clearly understand what is expected of them once they have concluded the course: performance and preparation. Specific statements are necessary, describing the performance that is expected and how that performance will be evaluated. Providing this information will motivate the student to expend the effort to pay attention, study, and learn.
- *What the prerequisites are*—Many courses will likely presuppose some knowledge of the new system. Students should know what the prerequisites are and how to meet them.
- *What the administrative details are*—Students should know when and where the class meets, what materials to bring, and what travel and living arrangements to make if the class is off-site.

Every class and student should be scheduled well in advance. With proper notification and preparation, students will not have to cancel out of courses because of other commitments. The importance given to education and training must be stressed. Students must make attendance their first priority if all goals related to education are to be met.

Note: A reluctant student typically causes himself or herself to be so valuable to daily operations that he or she can't possibly carve out education time. A typical pattern is that he or she arranges to attend but seldom shows up. This is the real "test" of management commitment how important is this project "really?"

Following this simple advice on student preparation will make education efforts more effective. It will help ensure that students are motivated, know what is expected of them, and are better able to learn.

How employees are informed of the plans to implement the new system will affect the future success of the project. As mentioned earlier, people tend to fear what they don't know, and change is naturally resisted. Therefore, project kickoff and initial education are extremely important. Students should be prepared well before the education program begins.

It is recommended that the new system implementation begin by having the highest executive officer introduce the project to all of the company's employees. He or she should explain the importance of the project, its benefits to both the company and the individual employees, and the necessity of support from every employee. Following a general kickoff meeting, each department head should meet with his or her workers to answer questions, handle objections, and describe the general implementation plan. At this time, the overview course should be explained and plans made for everyone to attend it.

6.1.6.5 Debriefing and Student Evaluation

Upon course completion, it is desirable that students be debriefed. This helps to identify problems with the course and to evaluate students' success in achieving its objectives. Consistent debriefing will reinforce for the employee the management's commitment to learning. If employees know that they will be tested through questions or on-the-job performance, they will be more highly motivated to learn.

Sometimes, after finishing a course, students will have unanswered questions or lingering doubts about the system and how it and they will work. By debriefing students immediately, one can resolve these issues and thus reduce the incidence of worker criticism that can be so detrimental to the education efforts. Debriefing can also be used to reemphasize interest in each employee and his or her value to the company.

Specific debriefing and evaluation activities should include the following:

- *Supervisory debriefing*—Especially interested supervisors may choose to spend quality time understanding the extent that their employees have grasped the subject. As a "coach," a supervisor may be able to shed light upon gray areas and remove student frustrations because they can't relate.
- *Course evaluation*—Students should be asked for an evaluation of the course. The primary reason is to identify weaknesses that can be eliminated, thus making the course more effective. Specific questions should be asked concerning organization, delivery, practice exercises, attention to individual problems, and the ability of the student to learn what was expected.

  One may try to ask question eliciting constructive criticism. For example, how would the student suggest improving the courses? Should there be more prerequisites? Should something be done to better prepare students? If students do make suggestions, follow up and let them know the outcome.
- *Student evaluation*—Some methods of assessing the level of student learning should be used. Specific questions about course content may be asked. If the anticipated outcome of the course includes using the system, then students should demonstrate that they can, indeed, do what is required. Software certification is a good tool to assess students and give recognition to diligent efforts (see Chapter 4 for more details).

A word of caution: Learning takes time. Employees should not be expected to become experts when they have completed a course. One must allow time for practice and have some means of evaluating it. Full competence and efficient use of the new system will come about gradually. Typically, the systems champions (super users) seem to have the aptitude to "master" the material quickly. They are also the ones who reach out to other associates that are struggling and help them succeed.

It is important to remember that the success of the new system depends on dedicated, enthusiastic, and supportive employees. Each manager should consistently evaluate employee acceptance and enthusiasm. Nothing kills such enthusiasm more quickly than frustration resulting from the inability to use the system as management intends. Therefore, it is necessary to be supportive of employees as they grow in their knowledge and ability to succeed. There are a few key points to note at this time:

1. It is difficult for some students to divorce themselves from their jobs. Some frequently believe that "spending time in class" is merely putting them further behind, consequently, adding stress and frustration to them personally. Others tend to use the class to "mentally solve" problems they will go back to. Therefore, each student must be evaluated fairly to help ensure that they are "up with" the class.

2. Other system success factors, such as performance measurements, are also important if the project effort is to be balanced. For example, if management fails to express their "system expectations" to the organization, then members of the organization may
   a. Believe that management is not really committed to the system.
   b. Attempt to anticipate, blindly, what management expects.
   c. Not take it seriously themselves until they are goaded.

Note: Performance management has been covered in various areas already and will be the central theme in Chapter 8.

3. The composite student's time off the job is the real expense of system education. It is extremely important that this time be quality time. It must be perceived as quality by the student and engineered as quality by the instructor. It is everyone's responsibility to focus this quality in a timely and effective fashion. This may mean rewarding top student achievers through promotions or certificates/awards. But the quality issue is central to the overall effort.
4. As so frequently happens, it is the highly charged few who make a difference between success and failure. Like the old saying goes, "if you want a job done right and on schedule, give it to a busy person." These are the movers and shakers within the organization. These are also the folks who must champion the new system. Without their support, any system effort is slow to fruition.
5. Management expectations should be in line with the organization's ability to absorb the impending change as well as personnel, priorities, and so on. An out-of-balance condition of expectations versus capability will inflict extreme levels of frustration upon the entire organization. This is another important reason for an effective executive education program. An educated executive tends to align expectations more reasonably. The objective should be aggressive, but yet doable.

If students do not learn what is expected of them, then some form of a contingency plan is needed for remedial instruction. It does no good to equate hours of exposure to teaching with ability to perform afterward. Just because time has been spent in class does not mean that objectives have been achieved.

Employee enthusiasm and success, so essential to the success of a new system, will be seriously undermined unless the necessary level of understanding and skill is achieved.

6.1.6.6 Course Preparation

The need to develop and teach some of the organization's own courses will undoubtedly arise, and these will need to be adequately planned, designed, and implemented. Obviously, this guide cannot provide definitive instruction in course preparation and delivery. What follows is a brief overview of course development, content, and teaching methods, with some recommendations to consider.

Courses are developed and taught out of necessity. Students must acquire some new skills, knowledge, and attitudes in order to perform their jobs properly using the new system tools. Courses must be developed around these needs by setting goals.

The first step in course development is to identify what has to be done, that is, what the goals are. Each goal should have clearly defined reasons associated with it. Once these have been defined, specific objectives should be developed, which clearly indicate what students are expected to learn.

By definition, goals are broad statements of instructional intent. They characteristically contain verbs such as "to know" or "to understand." Objectives, however, are definitive statements. They contain verbs of measurable intent such as "to demonstrate" or "to operate."

1. *Sequencing objectives*: Having defined the objectives, the next step is to then sequence them to foster learning. Generally, the sequence should be a logical progression from simple to complex, concrete to abstract, the known to the unknown.
2. *Outlining content*: Next, outlining specific content should be conducted to support each objective. This will include all the material students that are expected to learn. Rough drawings might be included in this outline, as might figures or charts, if appropriate.
3. *Choosing strategies*: A strategy for the structure and delivery of the course is now ready to be developed. It should reflect an understanding of how people learn and should use instructional techniques appropriate to the content. The strategy and objectives should result in a course that is structured, sequenced, and presented to cause student learning, not to satisfy a teacher's need to lecture. In developing a strategy, factors needed to be decided upon include the following:
   a. What the teaching activities are and how the content will be presented
   b. For example, how students will be involved, that is, what methods of practice will be used -quizzes,-case-study analysis, discussions, simulation exercises, and so on
   c. How feedback on performance will be given. Until students can demonstrate the expected level of competence as defined in the objectives, they must be provided with adequate opportunity to practice and receive feedback
4. *Developing materials*: The actual instructional materials are now ready to be developed. These might include student reading materials such as workbooks, exercise sheets, or reference aids; lecture notes and course outlines for instructors; case studies and discussion questions; and media such as overheads, slides, or videotapes. Generally, these are first developed in a rough draft of prototype stage. Care should be taken to keep the level and complexity of the material appropriate to the entering ability of the students. The content should be organized to be motivating. Overviews, summaries, and transitions between topics should be carefully planned. Practice quizzes and exercises should be developed to reflect both what has been taught and what is expected to be learned. Overall, students should have learning experiences which are positive and satisfying rather than negative and frustrating.
5. Once the prototype course has been developed, it is recommended that it be tried out on all groups of students. This allows the course developer to see how well it works and to determine if any changes are necessary for students to meet the objectives.

After refinements have been made, production of the print and media components will be readied, and the course be ready to begin.

Adhering to all these steps can be time consuming. Attention to these details, however, will result in courses more likely to accomplish its objectives.

6.1.7 Instructional Methods

It is easy to be overwhelmed by the variety of instructional methods available. All methods, however, can be classified into five broad groups: lecture, discussion, demonstration, independent study, and a combination of these referred to as the lesson. Subsections 6.1.7.1 through 6.1.7.5 discuss the uses, advantages, and disadvantages of each type, respectively.

6.1.7.1 Lecture

Lecture is probably the most widely used form of instruction. It consists of an oral presentation by an instructor to an audience, most commonly to impart a body of knowledge.

Although lectures are widely used, they are normally less effective than other instructional techniques. Research shows that even the most motivated students have difficulty absorbing for more than 15 minutes. Retention studies show that students forget up to 80% of what is presented by lecture within a day.

As a method of instruction, however, the lecture has several advantages as well as disadvantages, and there are ways to enhance its effectiveness.

The advantages of the lecture method include the following:

- A lecture covers a large amount of material in a short time.
- Lectures are suitable for any size group, as long as the speaker can be seen and heard.
- Lectures can be suited to a wide range of audiences, taking into account such things as previous knowledge, education levels, and intelligence.
- An instructor can easily maintain control as he or she sets the pace, organizes the sequence, and controls the degree and timing of student interruption.
- Lectures usually require less preparation than other forms of instruction.

The disadvantages to consider, however, include the following factors:

- A lecture is typically one-way communication. The instructor speaks and the students listen. It is teacher not student oriented. There is little or no check on student learning.
- Students are largely passive. There is little participation and it is difficult to remain attentive.
- Effective lecturing is not an easy skill to develop. Maintaining student attention over long periods of time is a difficult task.
- Lecturing is usually inappropriate when the objective is for the student to develop new skills.

There will be times when lecturing will be used, whether it is the most appropriate method or not. Lectures can be made more effective by adequate preparation, using good delivery techniques, limiting the time and amount of content presented, and, if possible, providing students with a written outline or synopsis. Lectures should be limited to no more than one hour, and student involvement through questions and exercises should be developed.

As with any instructional technique, students should be adequately prepared to receive a lecture; they should know why they are there and what is expected of them when they leave. Organization is a help here. The lecture should include an introduction, a body, and a conclusion.

- *Introduction*—This should clearly delineate the topics to be presented, their sequence, relationship, and importance. It should state the objectives, that is, the expected learning outcomes for the students, and the relevance of topics and objectives.
- *Body*—An outline should be developed and used related to the introduction. The instructor needs to be sensitive to the needs of the audience; stress key points and make good use of repetition; clarify the message; summarize frequently; provide transitions between topics so that relationships are understood; and realize that students will be bombarded with a great deal of new information (unless the instructor enables the students to understand the content and integrate it into what they already know, there is little chance that learning will occur).

- *Conclusion*—A conclusion is critical to influencing which portions of the lecture students will recall, especially because low retention rates are inherent in the lecture method. The conclusion should relate back to the introductory statements. Any questions that were raised should be answered. If the lecture is part of a series, it is useful to place it in the context of what came before it and what is to follow.

In sum, the key steps in planning a lecture are as follows:

- Tell the audience what is going to be told to them and why (introduction)
- Tell them (body)
- Tell them what was told to them (conclusion)

Systematic repetition will increase the likelihood of the audience retaining the message.

6.1.7.2 Demonstration

Demonstrations share many characteristics with lectures. Both use telling to impart information. But, in addition, the demonstration shows how to do something and is therefore a highly visual form of instruction.

Demonstrations are frequently used to teach procedures. For example, demonstrate how to use a terminal to enter data. Learning to perform skills composed of steps carried out in sequence is often made easier by demonstration.

The advantages of demonstrations include the following:

- They relate classroom concepts to actions in the real world.
- They keep the attention of students and, when performed properly, are both challenging and thought provoking.
- They teach skill. They can be paced to fit the needs of individual students, and repeated and practiced until the skill is learned.

The disadvantages of demonstrations include the following:

- They must be given to small groups to be effective. If the action cannot be easily seen, the procedure won't be learned.
- They require careful and intensive preparation and organization. If they go wrong, the effect may be lost. Repeated demonstration of some procedures can be expensive and time consuming.

Many disadvantages of demonstrations can be overcome through careful planning and execution. As with any instruction, students should be prepared: preface a demonstration with an explanation of what the student should look for as the procedure is performed, and draw attention to the key points as the demonstration progresses. The instructor may want to demonstrate how to do something several times, first slowly in isolated steps, then at a normal speed as an integrated procedure.

Students should be allowed to practice the skill immediately (and as frequently as needed) after the demonstration. Instructors should observe each student, detect any errors, and then show them how to eliminate these errors. Practice and feedback are essential to demonstrations.

6.1.7.3 Discussions

Discussions involve active two-way interaction between students and the instructor. The method is student oriented and participatory, rather than teacher oriented and autocratic. Discussion can take many forms: question-and-answer periods in lectures, formal debates, seminars, role plays, and case studies. These are all good variations of discussion. To detail these methods is beyond the scope of this guide.

Discussions have many uses such as solving problems, actively involving students in their own instruction, exploring issues, and making decisions. Properly used, discussion can be an excellent method for bringing about attitude changes.

The advantages of discussions include the following:

- They transform a lecture into a participatory activity.
- They permit everyone the opportunity to be actively involved in learning.
- They allow the knowledge and experience of each participant to be shared with the entire group.
- They are highly motivating when properly planned and led.
- When used as problem-solving sessions, they often result in better decisions than individuals would make.

The disadvantages to consider, however, include the following:

- They may be difficult to control. Unless properly planned, they can degenerate into useless debates.
- They are time consuming. For many instructional goals, other methods are more efficient.
- They can be dominated by highly verbal or aggressive participants, or individuals with more rank.

The various forms of the discussion method can be effective educational techniques. To be successful, discussion sessions must be well planned. Both the instructor and the participants must be prepared to discuss the topic intelligently. The atmosphere of the group is important; it should be relaxed, yet organized.

Each discussion should have a beginning, a middle, and an end. Conversation should be kept flowing and relevant. Although diversions may occur, the instructor should keep control of the group and always redirect members toward the major purpose of the meeting. All sessions should have a definite ending; no discussion should be allowed to just gradually die. A good technique is to have someone summarize the key points at the end.

6.1.7.4 Independent Study

Employees will not always have the luxury to learn in group situations. Direct supervision by an instructor is not always necessary; sometimes participants can learn by themselves.

Independent study requires instructional materials designed to be used without supervision or group interaction. Books and other reading materials, some video programs, conference room pilot, and other forms of media may be used. Practice assignments using a terminal and a test database provide good independent study activities. (This saves the embarrassment of slow achievers.) The advantages of the independent study method include the following:

- Adult learners can progress at their own rate. They can repeat material as often as necessary until mastery is achieved. Self-checks or tests must be built into the material so that students can assess their progress.

- Adult learners can receive feedback through self-tests. They feel accomplishment as they progress successfully on their own.
- Students must accept responsibility for their own learning. This tends to solidify ownership.
- Instruction focuses on the student and mastery learning, rather than on the instructor and content delivery.

The disadvantages to consider, however, include the following:

- If the employees are not highly self-motivated, are not able to use reading or other methods, or do not have necessary prerequisite knowledge, then independent study may not work.
- Independent study is very solitary and will not work for learners who require the presence of others.
- Independent study requires special instructional materials that may be time consuming and difficult to produce or hard to obtain.
- Allowing students to progress at their own pace can cause logistical problems. Some may progress too quickly, others not fast enough.
- Due to its nature, third-party monitoring of progress is difficult. This may introduce a degree of frustration by the students. After a few frustrating sessions, the individual may simply "give up."

Independent study requires students to be given specific assignments. Each assignment should have detailed deliverables that can be measured. This is accomplished by prefacing the assignment with a clear statement of the objectives the learners should achieve and how their performance will be assessed. Instructors should be sure to follow through with this assessment.

Some individuals react badly to too much independent study. An individual may prefer the structure and discipline of the training room to the solitude of working by themselves. Independent study, therefore, should probably be limited to no more than an hour and a half, and be alternated with other instructional methods. One procedure is to follow an independent session with a short lecture that summarizes the session's content, then with a group discussion or problem-solving session.

6.1.7.5 Lesson

The lesson incorporates features of all the methods discussed in Subsections 6.1.7.1 through 6.1.7.4. Typically, it begins with some form of instructor-led presentation and ends with an independent or small group assignment. The main feature may be a lecture that includes a great deal of discussion, debate, and questions, or it may be a video presentation followed by discussion.

The lesson is one of the most versatile and useful instructional methods. It can be used with advanced or beginning students to teach both knowledge and skills or to change attitudes.

In effect, the lesson presents material in such a way as to gain maximum group activity. As we have stressed, learning involves the student doing something it does not happen passively. The lesson provides for participation, calls for active student involvement, and readily maintains student attention, a key to motivation.

The advantages of the lesson technique include the following:

- It is flexible. It incorporates most instructional methods and is easily adapted to most teaching situations.
- It encourages, demands, and sustains student activity and participation.

- It ensures that both the instructor and the students cooperate as members of a team.
- It can be fluid in nature by highlighting a recent event and using it as a small group discussion or case study.

The disadvantages to consider, however, include the following:

- It does not lend itself to the subject matter so detailed that group activity interferes with the sequencing of instruction (not normally a drawback in the kind of education in-house programs typically present, however).
- It is complex. It requires more care, thought, and time to prepare than other methods. A lecture or videotape tells and shows; a lesson requires asking and doing. It also requires a very versatile and quick thinking instructor.

A lesson has three components, which are as follows:

- *Introduction or preliminary explanation stage*—Usually a short lecture or presentation; this should take no more than 10%–15% of the total lesson time.
- *Middle or development stage*—This should take 50%–60% of the lesson time. It is here that the real group activity occurs. Questions should be used continuously both to sequence the material and to enable the instructor to assess student learning. There is usually a great deal of flexibility. This alone requires a knowledgeable, prepared instructor who can keep the session moving toward its predetermined goal.
- *End or consolidation stage*—This is the final stage and should take about 20%–25% of the total lesson time. Consolidation allows for summary, practice, feedback, and direction so that students can demonstrate actual learning.

The intent of this short section on instructional techniques was for review of the major teaching methods. It was not designed to make potential instructors able to use them. Specific related education can be received from commercially available instructor training courses.

6.1.8 Ongoing Education

The education program cannot stop at cutover. It cannot stop when employees have received their initial training. Several factors make it necessary for organizations to plan and administer an ongoing education program:

- The system will continue to evolve and new procedures will be developed. Personnel will need to be educated in their use.
- The software supplier may provide a product maintenance program, which continues to improve the software feature richness. These new features require education and training before they are absorbed into computer operations.
- New employees hired in the future will need to be trained to use the system.
- Some employees will require a reeducation periodically.
- Personnel will be transferred or promoted, and this will require new or more intensive training.
- Advanced education becomes desirable as employees become more knowledgeable about the theory and operation of the new system on a holistic scale, and thus more valuable to the company.

Don't underestimate the importance of the ongoing education program. It is interesting to note that in many initial education programs that the "information overload" is so massive that the student is totally disoriented. It is very valuable to repeat a course after users have used the system for two to three months. The next layer of absorption can greatly yield benefits through added capability and lessening the frustration load.

In sum, the system shall be growing and the company's personnel need to grow with it. This guide outlines many issues related to education and training. If a company is prepared to use it to plan and execute its education program, then there is no doubt that it will succeed in the use of its new system.

6.2 Implementation Framework

6.2.1 Overview

The implementation plan is an essential ingredient for successfully implementing a project with the vast business implications of ERP. Virtually every department within the organization will be impacted to some degree. It is important that representation from each department be provided with the intensity dictated by the stage within the project life cycle. For example, dedicated engineering assistance is essential early in the project as bills of material are validated and loaded. Later on in the project, engineering involvement becomes rather passive. Project representation will require multiple resource levels of involvement. Representation is needed by executive level personnel in the form of steering committee (guide group) by operating management level as core project team members for day-to-day procedural involvement. Within each software module, a "systems champion" from the working level should evolve, who becomes the expert of that software capability. The next few pages are essentially a high-level recap of precepts covered earlier, but essential to a robust implementation plan.

6.2.1.1 Project Planning and Control

6.2.1.1.1 Physical Project Plan Tracking

Many project teams will want to use computer-based tracking software such as Microsoft Project. Other companies have found Excel to be suitable for updating and producing project plans. The tool used to track the project will typically rely upon the company tool standards and resource expertise. The size of organization, culture, and budget also influence this decision.

6.2.1.1.2 Detailed Project Plans

The project plan should consist of an overall project plan that spans the life of the project and the short interval schedule that details project steps covering the next—two to four weeks. A detailed project plan is presented in Section 6.2.1.6.15.

6.2.1.1.3 Project Control

Regardless of what project tracking technique is selected, successfully controlling the project involves much more than just developing an initial project plan. Feedback on task status to the project team and Steering Committee is communicated through the project status report. Upon completion of a project step, a document representing the deliverable must be forwarded to the project core team to ensure that perception conforms to reality of task completion. Project replanning is important if a task falls behind schedule.

6.2.1.2 Project Budget

1. Developing a project budget
   Some companies choose to not establish a separate account for the project. Others have a very formal budgeting procedure with a great deal of detail. Whichever technique is used, a current budget, projecting as accurately as possible, is essential to avoid "cost surprises" later. Ideally, the actual project costs should be within ±5% of the budget.
2. Project accounts
   A partial list of accounts that might be used in the project budget is as follows:
   a. Facilities
   b. Hardware
   c. Software
   d. Customization and testing
   e. Education and training
   f. Project team
   g. Data collection and organization
   h. Data accuracy improvement
   i. Documentation
   j. Product consulting
   k. Management consulting

6.2.1.3 Education Plan

It must be integrated with the project plan and the detailed project steps.

6.2.1.3.1 Project File

Place project artifacts in an electronic location accessible by all project team members (project file).

1. Value of the project file
   a. New project team members can gain better understanding of the project.
   b. Useful reference for all members of the project team and Steering Committee as a review of the decisions made to date and as a baseline for other plant implementation projects.
2. Contents
   a. Project charter.
   b. Project status reports.
   c. Task documentation.
   d. Correspondence with management consultant.
   e. Communication to and from the Steering Committee.
   f. All internal memos, statements, minutes, and so on.
   g. Written explanation of key problem resolution.

6.2.1.4 Implementation Audits

Implementation audits are necessary to keep the project on track. Audits should be conducted to compare project results, business objectives, systems objectives, and project objectives. As discussed in Chapter 3, the committed ROI realization is central to the project success.

1. Business objectives
    a. The Steering Committee should regularly compare company performance to established business objectives (inventory investment, service levels, etc.).
    b. If changed, realign project activities/deliverables accordingly adhering to change management practices.
2. Project audit
    a. Predetermined review points should be established to objectively assess progress being made. The old adage, "You can't see the forest for the trees," tends to be a reality in project implementations such as ERP. It is worth the time investment to provide management with a periodic review as to budget and progress to plan.
    b. As discussed in Section 1.1.14, toll gate reviews may be used for ongoing project audit.
3. Project objectives
    a. Project objectives (such as on-time completion of tasks) are audited by the project manager, Steering Committee, and independent consultant. Project audits should include behavioral as well as technical issues. Attitudes of managers, project team members, users, and systems people are essential to a successful implementation.

6.2.1.5 Implementation Success Factors

From its earliest days, an APICS/University of Minnesota and Clemson study* found that unsuccessful users spent just as much money as the successful users in their implementation efforts. The differences between the successful and unsuccessful implementations can be attributed to a limited number of factors. The following list of success factors was condensed from the above study and an in-depth survey from various articles from production inventory management, information systems, decision support systems, and organizational behavior journals and magazines:

1. Top management leadership and involvement in the implementation process
2. Top management's view of the system as a means of executing the business plan, the marketing plan, and the production plan rather than simply a data processing system or production control system
3. Clearly defined business objectives, including objectives for the system
4. Extensive education and training at all levels in the organization in order to insure that people understand the concepts and procedures
5. Procedures, disciplines, training, and accountability needed to develop and maintain an accurate database
6. A well-managed project team using a formal approach to project planning and control
7. Involvement of key users early in the project
8. Willingness and ability to overcome resistance to perceived and real changes in system formality, job descriptions, power structures, authority relationships communications patterns, and performance measures
9. Assignment of accountability (responsibility) for making each component of the system successful
10. Availability of good software

---

* John Anderson and Roger Schroeder, "A survey of MRP implementation and practice," *Proceedings of the Material Requirements Planning Implementation Conference*, September 1978, APICS and University of Minnesota, Minneapolis, MN, pp. 6–42; http://www.sciencedirect.com/science/article/pii/0272696381900310.

6.2.1.6 Implementation Plan

The following implementation plan outline provides a guideline for a project manager. Smaller organizations may be able to combine organizational levels (e.g., Steering Committee and project team) and streamline the implementation time frame significantly. Each task listed on the implementation plan should have a due date and responsibility listed. In addition, each task should be detailed further into working documents (see 15 below), which define tasks into nominal 40-hour, or other reasonable deliverable period, segments. Each segment should have a due date and responsibility assigned, which functions as the "short interval schedule." See Appendix D.6 for a modular guide and checklist to develop a detailed implementation plan. The following nested guideline may be used as a checklist of ingredients that should be included in the implementation plan:

1. Define and establish project implementation team
   a. *Select project manager*—This individual should have excellent project planning and control skills and good communication skills; will ideally have in-depth understanding of current manufacturing and material control practices; and must be full time to the project.
   b. *Select project team members*—Every functional area of the company should supply a full- or part-time member to the project team. This should include but is not limited to the following areas:
      i. Material planning
      ii. Product engineering
      iii. Manufacturing engineering
      iv. Purchasing
      v. Cost accounting
      vi. Finance
      vii. HR
      viii. Manufacturing operations
      ix. Quality control
      x. Master scheduling
      xi. Field service
      xii. Facilities
      xiii. Customer service/sales/marketing
      xiv. IT

   Depending upon the size of organization, resource constraints, and budget, ideally, IT should supply a full-time project team member to head-up and coordinate the IT technical effort. In companies where there are a significant number of employees requiring training, a full-time training and education coordinator should also be assigned to the project team. Depending on a needs evaluation, each of these project team members may be assigned additional full-time and/or part-time resources.
2. Define and establish the Project Steering Committee
   a. Composed of top-level management from each functional area of the company.
   b. Chaired by the senior executive or functional area vice president most affected by the project.
3. Prepare and execute project team education plan
   a. Generic education.
      i. Outside seminars
      ii. Video
      iii. Training

    iv. APICS, ISM, and other educational forums
    v. Recommended reading
    vi. Onsite seminars

  This was discussed in detail in Section 6.1.5 and is only mentioned here to emphasize the importance of this activity for integrity of the nested implementation plan outline.

4. Develop project objectives
   a. Benefit goals/priorities.
      i. Inventory reduction
      ii. Increased inventory turns
      iii. Reduced purchase part costs
      iv. Increased productivity
      v. Better customer service levels
      vi. Reduced scrap and obsolescence
      vii. Capital expenditure deferral/avoidance
   b. Establish measurements (see Appendix D.3.1).
      i. Responsibilities
      ii. Methods of measurement
      iii. Scope
      iv. Frequency
      v. Monitoring techniques
      vi. Reporting
   c. Explain and establish the Excellence Awards program
5. Develop project milestones
   a. *Module priorities*—In developing module priorities, the following should be considered:
      i. Company objectives
      ii. Benefits schedule
      iii. Resource requirements
         A. People
         B. Budget
         C. Process changes
         D. IT
   b. *Phase definition*—It is typically not recommended that all ERP software modules be implemented at once (at least not ALL the functionality in all the modules). Instead a phased approach should be taken. It is important to define how many phases will comprise the project and how many software modules will be part of each phase.
   c. *Site sequence*—If more than one plant/site is involved.
   d. *Customizations and interfaces* (integrating third-party software into the core ERP software)—These, if needed, should be identified as much as possible, as early as possible. It is important to identify the magnitude of effort required for interfaces to facilitate proper planning and resource definition and acquisition.
      i. Order entry
      ii. Cost accounting
      iii. Forecasting
      iv. Distribution
      v. Payroll
      vi. Trading partners
      vii. Banks

   e. *Project timeline critical path*—For projects that have significant complexity, numerous customizations or interfaces, in-house legacy integrations, or other high-impact variables, it is essential to have a clearly defined critical path from which to manage deliverables and to quickly be able to assess variance impacts. Companies interested in both schedule and cost may want to consider using the earned value model to track project progress.
6. Develop and receive approval for the project charter
   a. The project charter is a written document containing the following elements:
      i. Project name
      ii. Objectives
      iii. Scope
      iv. Organization
      v. Authorities
      vi. Accountabilities
      vii. Responsibilities
      viii. Milestones
      ix. Review procedures
      x. Resource identification
      xi. Adjudication guidelines
   b. The project charter is approved and signed by the Steering/Guidance Committee.
7. Communication plan (Comm13)
   The purpose of the communication plan is to contribute to the successful implementation of the ERP project with the right communication delivered to the right audiences at the right time. This was discussed in Chapter 1 in more detail.
   a. Spread knowledge and status about the key deliverable and upcoming process changes.
   b. Facilitate the "ownership" process by end users.
   c. Provide information and ideas for greater productivity in the future.
8. Risk management plan (RiskMgmt4)
   Managing risk is a foundational precept for a successful ERP project. Risk management is the process used to identify, quantify, and rectify issues that can adversely affect the success of a project. To that end, the project needs a risk management plan to identify and mitigate known risks. This topic was discussed in Chapter 1.
9. Project health (ProjHealth10)
   Managing, tracking, and reporting project health (report card) is a key precept in helping ensure that the project is on schedule and progressing within cost boundaries. This was discussed in detail in Chapter 1.
10. Present executive seminar
    The executive seminar is a one-day program for the executive staff. The seminar describes the top management's activities, responsibilities, and use of formal ERP system components to support accomplishment of the company's objectives.
11. Define existing in-house systems
    In-house, or legacy, systems may need to be incorporated into the new ERP software. This is a typical requirement when the new ERP solution does not have a specific functionality and it is too costly, or not desired, to modify the new software. Because the inherent architecture of the new ERP solution and the legacy are so different, data mapping will require special focus by the technical team members. One of the key objectives will be to rationalize the two disparate source data to eliminate potential duplication and

present user data in a consistent format. A representative starter list of attributes that need to be rationalized is as follows:

   a. Inputs-samples
   b. Outputs-samples
   c. Paper flows
   d. Use/purpose
   e. System interdependencies
   f. Maintenance responsibilities
   g. File layouts
   h. Database element definition

12. Install software
   a. Conduct on-site installation seminar
   b. Perform preinstall tasks
   c. Execute system install verification
13. Develop detailed education and training plan (TrainingPln8)
   a. Who
   b. What
   c. When
   d. How
      i. Software module video (if available)
      ii. Software module series workshops (if available)
      iii. Video-based generic education programs
      iv. Conference room (training) pilot (see Chapter 6 for details)
      v. Hands-on role play
      vi. APICS, ISM, and other educational forums
      vii. Outside and on-site seminars
   e. Education room planning
   f. IT requirements

   This was discussed in detail in Section 6.1.5 and is only mentioned here to emphasize the importance of this activity for integrity of the nested implementation plan outline.
14. Conduct module definition review for phase I modules
   a. Compare software to existing system
      i. Functions
      ii. Features
      iii. Data elements
      iv. Inputs and outputs
   b. Determine sources for new software data
   c. Record outstanding issues
   d. Conduct gap analysis to determine missing functionality and associated corrective action plan
15. Develop detailed implementation plan
   a. Assign responsibilities
   b. Determine start/stop dates
   c. Estimate task man days
   d. Determine resource/skill requirements
   e. Synchronize with detailed education and training plan

16. Execute detailed education and training plan (TrainingPln8)
17. Develop functional specification and test interfaces
18. Develop functional specification and test conversion programs
19. Plan and execute conference room pilot
    a. Establish database
    b. Determine scenarios
    c. Review inputs/outputs against expected results
    d. Develop procedure outlines
    e. Determine additional system edits

    This was covered in detail in Chapter 5 and is only mentioned here to emphasize the importance of this activity for integrity of the nested implementation plan outline.
20. Develop test plans (Testing16)
    a. Data validation
    b. Develop/modify test plans
    c. Unit testing
    d. End-to-end integrated test plan
    e. Process testing
    f. Regression testing
    g. User acceptance testing
21. Environment and performance testing
    a. Design database environment
    b. Performance benchmarking
    c. Production database tuning
    d. Production database sizing
22. Develop final production system definition
    a. Audit operating instructions
    b. Review job streams
    c. Review backup/recovery/reorganization procedures
    d. Review security parameters
    e. Review hardware resources
    f. Review disaster recovery procedures
    g. SLA validation
23. Develop user manual (UserDoc11)
    a. Input requirements
    b. Reporting requirements and distribution
    c. Screen usage
    d. Policies
    e. Procedures
    f. Process changes
    g. Transaction matrix
    h. Reason and activity code descriptions
24. Develop and execute production pilot

    (Some companies may decide to skip directly to 27.)

    The production pilot is a partial cutover to the software system usually limited to one or two product lines. Although, at times, this approach to "going live" may be difficult to implement, it is highly recommended.

   a. Determine pilot content
   b. Determine IT resource requirements
      i. Database files (disk space)
      ii. Run times
      iii. Response times (throughput)
      iv. Report distribution
   c. Review final system control parameter settings
   d. Develop data conversion method
   e. Develop measurement and audit criteria
   f. Review cutover checklist (see Appendix D.4)
   g. Review procedures for common item handling
      i. This is very important because most production pilots will involve parts, which are common to many product lines.
25. Review results of the production pilot
   a. Review policy and procedure effectiveness
   b. Review measurements
      i. Record accuracy
      ii. Exception statistics
      iii. Process and procedural awareness
   c. Reassess detailed plan
   d. Review final production system definition
   e. Determine any additional education and training needs
26. Conduct IT postproduction pilot audit
   a. Determine full production IT requirements
   b. Review all IT policies and procedures
   c. Review all operating instructions
   d. Review all interface testing
   e. Review all input and data entry instructions
   f. Review all output distribution
   g. Review all conversion testing
27. Develop and execute production conversion plan
   a. Determine cutover requirements
   b. Review cutover schedules and sequence
   c. Review post-cutover team assignments
   d. Develop rollback plan
   e. Review measurement and audit criteria
   f. Review start-up final checklist (see Appendix D.5)
28. Conduct post-implementation audit
   a. Review measurements
   b. Determine fine-tuning requirements
   c. Determine additional education and training needs
   d. Determine need for additional policies and procedures or rewrites

For all additional project phases, begin at step 14 of this outline.

WRAP-UP

This concludes Section II where we examined the practical deployment framework essential for success, and includes a variety of tools that position the company for success.

We took a close look at **The Information Workmanship Standard** and the importance of specifying a standard, rather than allowing the team members to guess at the leadership expectations regarding the quality of information. This standard spanned the minimum acceptable quality level for transactions, job functions, work processes, and ultimately the resulting information. Without a standard there is confusion regarding work expectations. The Information Workmanship Standard (IWS), such as financial, quality, and a variety of other standards, clearly defines the expectations associated with information. In addition, we fully developed the nested internal customer and supplier of a service framework to define process-based performance metrics that reflect in-the-trenches end-to-end business process activities. We recognize that engineering "process-based" metrics allows players from the entire organization to understand their specific contribution to profitability, which is lacking in traditional hierarchical metrics.

We then discussed the importance of **The Conference Room Pilot**. Test-driving the blending of functionality reflected in the design (features, functions, and capabilities) as well as processes (how the design is configured) and the environmental structure (policies, procedures, and performance metrics within the players culture). This chapter pursued not only project team piloting but also senior leadership piloting, customer/supplier piloting, and other business partnership piloting.

Finally, we examined the **Education, Training and Implementation Framework**, the backbone to system success. This encompasses the entire user community, demonstrating the need for the competence and mastery of the new system. We examined the building blocks essential to educate and train team members to achieve system exceptional success. We also examined the implementation framework. The project requires these elements if comprehensive results are to be achieved. The appendix D included a variety of checklists to guide the project core members to take a thorough look at the preparedness. Like all competency processes, this must be achieved through proper design and fulfillment.

As discussed in Section I recap, obtaining improved business results will align with doing a good job implementing the enterprise resource planning (ERP) solution. However, to obtain stellar results (even order of magnitude improvements) will only result from managing critical success factors and adhering to best practices, these include the following:

- Creating a comprehensive education, training, and implementation framework
- Setting goals and monitoring to help ensure their achievement
- Mastery of the ERP software and process solution impacts system performance results
- Senior management must lead and be examples for the entire organization
- The engineering of robust functional specification is a documentation best practice

Now we'll begin discussing the topics of Section III.

PROJECT MONITORING AND DEPLOYMENT

Overview

This section addresses the practical deployment framework essential for success and includes a variety of tools that position the company for success.

- *Chapter 7*—Ensuring the fulfillment of project tasks and commitments, reporting the status, and invoking Steering Committee guidance and day-to-day issue/decision management are the tried-and-true practices of good project management. However, merely doing these things is inadequate for a best practice deployment. This chapter discusses these essential tasks but also exploits the practice that transforms a good project into a best practice project. Things such as behind-the-scenes salesmanship, removing risk barriers, executive ownership process practices, monitoring rules of engagement, and other differentiating elements are discussed.
- *Chapter 8*—As introduced in Chapter 4, this chapter peels back the layers of opportunity and explores realigning measurements to an end-to-end process basis, which allows the entire organization to understand the importance of every job function and gives the working-level tier of the organization the ability to measure their individual contribution to profits.
- *Chapter 9*—Experience has shown that many enterprise resource planning projects just are not successful. This chapter addresses how to convert potential failure attributes into critical success factors. It explores such topics as follows:
  - GO/NO GO voting decision—looking at the technical review and recommendations, functional review and recommendations, open issues, cutover plan, transition to production strategy, and other criteria for successful cutover
  - How to tell when the project is going off the rails
  - How to decide and prioritize what aspects of the system needs tuning
- *Conclusion*—Keeping sanity yet achieving exponential results ... On Time and On Budget.

Chapter 7

Project Management

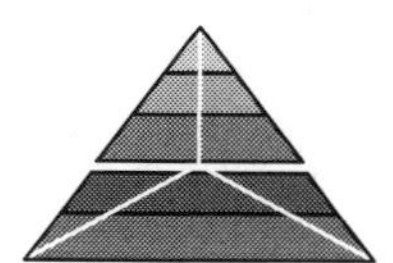

Ensuring the fulfillment of project tasks and commitments, reporting the status, and invoking Steering Committee guidance and day-to-day issue/decision management are the tried and true practices of good project management. However, merely doing these things is inadequate for a best practice deployment. This chapter discusses these essential tasks but also exploits the practices that transform a good project into a best practice project. Things such as behind the scenes salesmanship, removing risk barriers, executive ownership process practices, monitoring rules of engagement, and other differentiating elements are discussed.

This chapter deals with how to manage the day-to-day and, sometimes, mundane tasks essential for a successful enterprise resource planning (ERP) implementation. However, there is another side to this topic that is exciting, living, and represents the heart and soul of the ERP implementation effort. Let's begin here with an overview.

The best managed projects are ones where innovation is king. Yes, the tried and true ticky-tickies are necessary for good organization (and a project such as ERP implementation mandates exceptional organization and adherence to quality standards). Yet, what sets apart a nominal ERP implementation from one yielding exceptional (order of magnitude) results are characteristics such as the following:

- Visionary
- Innovative
- Flexible
- Ingenuity
- Agile
- Exceptional throughput
- Nimble

We discussed some of these characteristics in Section I, dealing with planning, as well as Section II, foundational, but it is more important in the execution of the project.

> I recall an incident while managing eight concurrent projects, including three complex earned value-oriented projects, which were in my $30 million project portfolio I ran every day; needless to say, I was busy. The large organization I worked for had a plethora of boards, rigorous standards, adhered to Capability Maturity Model Integration (CMMI) Level 3 development criteria, and without a project manager, nothing got done. The complex project environments tended to be supercharged with overwhelming stress. Some of the project teams had dozens and sometimes hundreds of resources engaged. Due to the nature of the "complex" earned value orientation, these projects were in the crosshairs of leadership because of the amount of contract value (material and labor content value), as well as the magnitude of risk and mission criticality associated with each project. Consequently, there was a continuous battle to preserve budget (middle management constantly prowled to strip off budget for political reasons). To cut to the chase, there was little room for variances or deviation from the project plan. To be successful, a variety of focus group collaborations on a daily basis were required. Many of these focus groups were ongoing contentious. To make it worse, the leadership team leaders were seldom available to have ad hoc collaborations. Because many of the resources were in three different time zones and operated on a 24×7 schedule, it was already a stretch to get the needed resources together at any given time. Therefore, creativity had to be used when scheduling meetings. To get on top of the situation, a three-tiered rules of engagement amendment approach was required. I was West Coast based, so at an inconvenience to me, if the meeting were expected to be extremely contentious, I began the meeting at 3:00 a.m. (not liked by many participants). The second tier was a mandate if the meeting attendee was unable to make it; the principal had to appoint a Delegation of Authority (DOA) to attend in their place. The third tier was that if one of the attendees could not make it, there was an immediate escalation with no appeal, meaning that the decision of those present was binding. This resulted frequently in tasks being assigned the missing resource area without their input. The result is that it did not take long before calendars became available during normal working hours.

7.1 Visionary

To the best of their abilities, ERP project core team members attempt to frame the project in a way that facilitates the essential tasks while minimizing the risks. Looking at an ERP project in a *visionary* manner, this would include such things as follows:

- What are the ASSUMPTIONS we are taking for granted? Detailing assumptions frequently uncover potential risk gremlins that would not normally surface.
- Using Pareto, what are the critical success factors that facilitate success—the 80% that help enable success—ensure that these are leveraged through the entire project to steady the momentum of project success.
- Using Pareto, what are the rocks in the road that might potentially derail the project? Many of these surface as risk agents as risk analysis is conducted. However, in addition to risk issues, there may be a plethora of project distracters that tend to sap the project vitality over time. In many of my ERP projects, there tended to be a particular "political" resource that

sat on the outskirts of the projects sapping the energy of key resources on the project and wearing out the needed energy. Playing "politics," in projects, adds no value and is always detrimental to the project goals. Besides politics, there may be a number of other "cultural" distracters to project success. Uncovering these gremlins early in the project will likely mitigate derailing salvos.
- Leveraging lessons learned from other large projects. Any advanced insights into the ERP journey may help offset delays.
- Leveraging the momentum gained from the senior leadership collaboration workshop. This process of leadership commitment to results may be the first time ever exercised within the company. Those leadership teams taking the task seriously tend to be exhausted after the ordeal (I know I was after leading a variety of these sessions). Squeezing the productivity juice from stakeholders is a salve that may be reapplied at various critical times within the project timeline.

Using visionary tactics is a project management **best practice.**

7.2 Innovative

Many engineering firms view themselves as innovative; by virtue of the products they create and build. However, this company strength may also become a liability, over time, inasmuch as it is taken for granted and not pushed to its limits on a **daily** basis. I look at this precept as directly correlated to an ERP project. As a common practice within the firm, a company may go about engineering the ERP project on a one-up basis, according to its project protocol standards. The importance and nature of the project drumbeat (monitoring deliverables in a cost/schedule rhythm) tends to create a robotic wave of tasks. The best practice-oriented projects need to engineer innovative thinking into not only the implementation framework but also the project deliverable framework. One way to effectively accomplish this is by leveraging the "process improvement" aspect of deliverables (Chapter 8 discusses this in detail). As the team commits to each deliverable, they may receive the "minimum" benefit by performing a minimum effort. A much better approach is to enjoy a maximum benefit by engineering a process that uses a fully new engineered minimum effort. This practice is not common within organizations inasmuch as they tend to be swept away with "just do the minimum" mentality. In fact, I have found that companies that lean toward reengineering each project task, with every new project, become more efficient with each project. Innovative thinking, then, would expect leveraged results (sooner/better) with less effort. To accomplish this,

- Exploit the delegation of authority model to its limits and give more latitude and freedom of team members to be the driving leader of the deliverable.
- Rather than working tasks serially, work them concurrently.
- Take advantage of cross-functional "team think" to exploit interweaved ingenuity.
- Regularly schedule a "take a deep breath" deliverable that permits the project resources to pause, refresh their thoughts, and look at the forest rather than the day-to-day trees. Infuse periodic creative mental breaks into the project drumbeat rhythm.
- Leverage insights from team members as we progress through the timeline and elicit productivity enhancements from everyone (the day-to-day doers frequently have great suggestions on improvements that may be deployed; let's listen and incorporate with vigor).

Using innovative tactics is a project management **best practice.**

7.3 Flexible

There are typically various roadmaps that may be followed to bring a project deliverable to fulfillment. It is common for the deliverable lead to take the most direct path to accomplish the task conclusion. A wise project team anticipates the need to deploy various paths within the roadmap as it seems fit. To that end, as the deliverable solution is being engineered, take time to also engineer alternate routes. Creating alternate routings is a common practice with manufacturing shop floor activities because of the importance to customer fulfillment goals. These may be precipitated because of a key resource crisis, equipment failure, quality barrier, interaction conflicts, and a host of other variables. Whatever the cause, having invested the time up-front to engineer alternate roadmaps, pays dividends when they need to be executed. In the crisis mode, to triage a failure, design an alternate and get back on track are less efficient than having already designed alternate paths, although clear thinking is prominent and, then, merely making a path decision.

Flexibility then needs to be part of the project design framework. However, flexibility is also essential when bad things occur while executing the project. This requires reliance upon heuristic characteristics of the project team core. These heuristic characteristics come with experience of the team. The wise project team expects these events to occur and be mentally prepared to conduct triage under controlled circumstances.

Using flexible tactics is a project management **best practice**.

7.4 Ingenuity

There is a difference between waiting for something to go bad and then fix it, compared to anticipating that something will go bad, and preventing a bad thing from occurring. Unfortunately, project core team leaders are seldom given the opportunity to dream; therefore, dreaming ingenious solutions is seldom practiced. A good middle ground is to constantly be on the lookout for signs of things that might begin to go bad. Part of this awareness is exposed through the project rhythm, part from experience, part from the radar webs planted by the project core team as early warnings, part from management by walking around, and so on. Regardless of the trigger, anticipating potential derailing events is keen perception. It is almost like listening for a vibration. There have been times when I would get a "feeling" that a particular deliverable was wobbling. The detailed status indicated on track (I was pinging daily) and the resource lead was not aware of any issues, but that little birdie whispered in my ear and I went directly to the developer assigned. He was experiencing program conduct that he was not expecting. The vibration proved to be real. We immediately called in a technical specialist who was able to triage the concern and it was fixed before it was actually recognized as broke. Another ingenuity perspective might be called reading between the lines. How does one affix a label to an early warning radar web. The only response is that you are on the lookout for something and you search potential anxiety areas until you determine it was a false alarm, or you validate that it was real. A third element of ingenuity might be recognizing an inflection in an individual's response to a pointed question. I have found that detailed short-interval task planning certainly helps, however, monitoring the health of these tasks via critical drill-down questioning, which frequently uncovers potential "hairline cracks" or early warning awareness.

7.5 Agile

Agile is the ability to do "quickly." This is certainly a defining quality within a best practice project manager's arsenal of abilities. It is especially acute when weaved with flexible ... how quickly can flexibility be affected. If the project manager is working with a good project plan, then executing that plan with agility helps ensure cost and schedule integrity. As a manner of practice, project managers who regularly execute as a matter of agile style typically lead the project in a rhythm of proficient achievement of deliverables. High-output deliverable realization sets the project management field apart from merely lethargic task completion exercises. Therefore, monitoring and executing project deliverables in an agile fashion tends to permeate deliverable realization.

7.6 Exceptional Throughput

From an overall business perspective, I view throughput as the conversion of a booked order into collected revenue. With that baseline, if you were to view throughput from a "maximum" potential, it would result in "instantaneous" conversion of a booked order into collected revenue. Although difficult and maybe even impossible to attain, the difference between what is realized in operational practice and "maximum" is the "opportunity." Now dissecting this precept into a project perspective, we may view each project deliverable as the equivalent process of fulfillment ... converting a booked order into collected revenue ... the maximum opportunity is instantaneous realization. As discussed earlier, the better the project team is at "engineering the deliverable process," the more efficient the realization of that deliverable. I have yet to personally experience "maximum" throughput of a project deliverable; therefore, I believe the goal should be "exceptional" throughput. Looking at the project deliverables as a series of concurrent project rhythms, exceptional throughput realization would be completion of the deliverable as follows:

- On time
- On budget
- With an exceptional customer experience
- With an exceptional resource utilization experience (resources used want to always be on your project)
- It was the best performing project within the project portfolio

Using exceptional throughput tactics is a project management **best practice**.

7.7 Nimble

The ability to do "lightly." Similar to agile, there is a "light touch" or lean way to perform the realization of deliverables. Using a "minimum" footprint to achieve exceptional value is another best practice attribute of project management. Relating back to the project rhythm, fostering a nimble style to project leadership allows the project core team members the ability to "shine" through exhibiting their abilities and potential to the maximum. Other attributes of the nimble style include such things as follows:

- *Touch only once*—Be efficient in work effort (prevent rework).
- *High quality*—Every effort leads to the highest quality result.
- *Exceptional communications*—Ensure that everyone is always on the same page.
- *Lead by example*—Team members want to emulate your work practice.

Using nimble tactics is a project management **best practice.**

I like to view the project management of an ERP implementation on multiple planes or dimensions:

- Cost and schedule integrity
- Managing detailed work packages in a detailed manner (about a week's worth), with frequent options of daily deliverables if behind schedule
- All deliverable completions come with a document (allowing the project core team the ability to evaluate the quality of the work product, not just a percentage complete on the project schedule)
- Weaving leadership, technical, process, and functional deliverables into a cohesive whole
- Having fun performing

When a project is approached in this manner, and taking into account the need to be innovative, agile, ingenious, nimble, flexible, and visionary with exceptional throughput, it is like the project manager who is similar to an orchestra director (maestro), seeking harmony, balance, and symmetry across all tasks, resources, and constraints on a continuous hour-by-hour basis.

Note: The Project Management Institute (http://www.pmi.org/) has a Project Manager Certification—Project Manager Professional. This, and its Project Management Body of Knowledge, is an excellent resource for project managers. Rather than replicating this and other credible resources, any redundancy of concepts or terms will have a different slant or connotation if included in this book.

7.8 Project Framework

There are a variety of ingredients in an ERP implementation project as discussed in subsections 7.8.1 through 7.8.3.

7.8.1 Project Plan

This is the capstone document that describes the project operating environment structure. It describes the projects' latitudes, stakeholders, customer deliverables, scope, goals and objectives, essential project operational guidelines, resource management strategy, and so on. In Chapter 1, we discussed this topic in detail, however, as a review and with a slant toward the project entity itself.

The project plan structure is as follows:

- Customer interweaved strategy
  - Project plan purpose
  - Overview
  - Scope/high-level schedule/overall estimate
  - High-level acceptance criteria
  - End-point system expected results
  - Rules of engagement
  - Requirements management strategy
  - Description of nonlabor resource needs
  - Changes to project estimate
  - Place of performance and delivery
  - Customer/buyer furnished items
  - Project health reporting
  - Customer reporting
  - Communications plan
  - Implementation strategy
  - Test management strategy
- Project management
  - Methodology
  - Process tailoring
  - Schedule development
  - Quality assurance (QA) plan
  - Performance tracking and oversight
  - Risk management approach and plan
  - Project team
  - Project team training plan
  - User training plan
  - Collaboration coordination plan
  - Requirements tracking management
  - Change and configuration management approach and plan
  - Data management plan
  - User/system documentation plan
  - Knowledge transfer plan
  - Plan for reviews (toll gates) and walkthroughs
  - Contractor agreement management plan
  - Business information assurance plans
  - Software implementation strategy
  - Alternatives analysis

7.8.2 Documenting AS IS

The baseline or beginning point from where the ERP project commences. There are varying opinions on whether to document the legacy system, whether it is cost justified and the value proposition. However, without a baseline, how will the project evaluate performance and progress? Flowcharting existing processes may be a value-streaming roadmap to recommend the needed change.

7.8.3 Project Schedule

The project schedule is a subset of the project plan. It represents the tasks, deliverables, milestones, and pay points of the project. The schedule needs to be defined in sufficient detail to enable successful completions on time. As discussed earlier, one might, for example, define a schedule task that represents a week's accomplishment. In this case, there might be 52 tasks for each resource, if each resource had deliverables that lasted 52 weeks. Commonly, the project schedule breaks the tasks into initialization, technical, customer, software, functional, process, and training nested bundles. The style used to engineer the schedule is not as important as ensuring that the schedule is comprehensive (includes all needed deliverables), is as accurate as possible (do ability), and may be tied to stakeholder requirements.

Most projects core team's use Microsoft Project (MS Project) as their project schedule tool. MS Project is typically adequate unless there are

- Significant modifications to the ERP software requiring rigorous development.
- Significant number and size of integrations requiring rigorous development.
- Significant amount of customizations requiring substantial timer and labor commitment.
- Substantial amount of finite resource management (load vs. capacity).
  - Extensive customer nested integration
  - Extensive supplier nested integration
  - Multiple supplier tier nested finite integrations

MS Project may still be adequate; however, it is limited in its ability to finite schedule. An alternative project schedule tool is Oracle's Primavera. Please note that Primavera is an exceptionally good finite scheduling and resource management tool; however, it tends to be costly and the learning curve to competent use may be lengthy. If MS Project is all you have, the following may be helpful. MS Project handles a punch list of project deliverables without resource constraints acceptably. If there are resource constraints, you will need to be somewhat creative in resource management. One of the best, and simplest, is to create an on the side SUPER PROJECT. This would result in aggregating all projects that include elements of competing critical resource. The resultant resource plan would reflect a ROUGH CUT view. The downside, it needs to be created as a one-up regularly and requires brute force rather than integration fidelity (as Primavera would yield). Yet another creative solution I've seen deployed is to use a labor estimating tool that integrates critical resources for all OPEN PROJECTS. The dollar impact will show out of capacity "overruns" in their appropriate periods. Unfortunately, there will be the need to use brute force to disaggregate the overruns into which competing projects are impacted. Again, it is not an ideal solution.

A typical ERP project life cycle is shown in Figure 7.1. In this diagram, you see three distinct authorization events, with each event having broad deliverables. Each of the broad deliverables has a tool gate review allowing leadership to review progress and participate in GO/NO GO decisions. It also shows the phase of the project.

Other attributes of a project schedule include a variety of variables that influence the success or failure of a project schedule. One of the variables might be project constraints. Figure 7.2 gives an example of project constraints. In this diagram, there are four significant constraints listed: cost, schedule, functionality, and resources. Depending upon the company, its size, budget, and a plethora of other factors, there may be only schedule and functionality constraints, or the constraints list may be increased. The list will be uniquely defined for each ERP project.

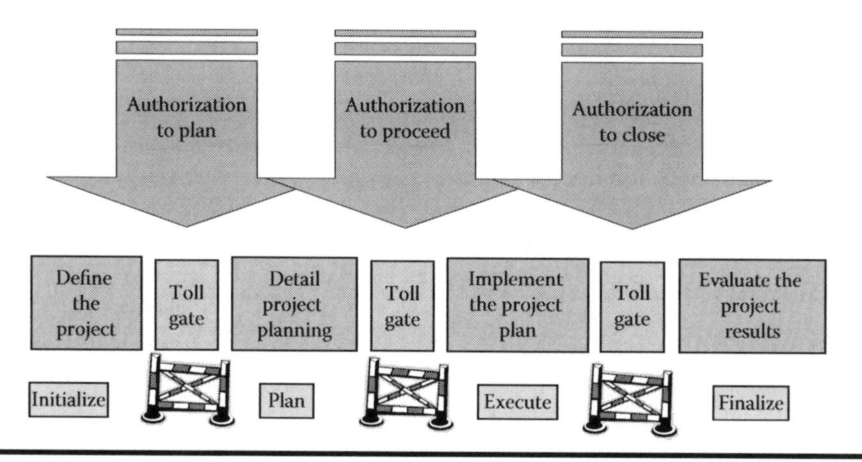

Figure 7.1 Project life cycle.

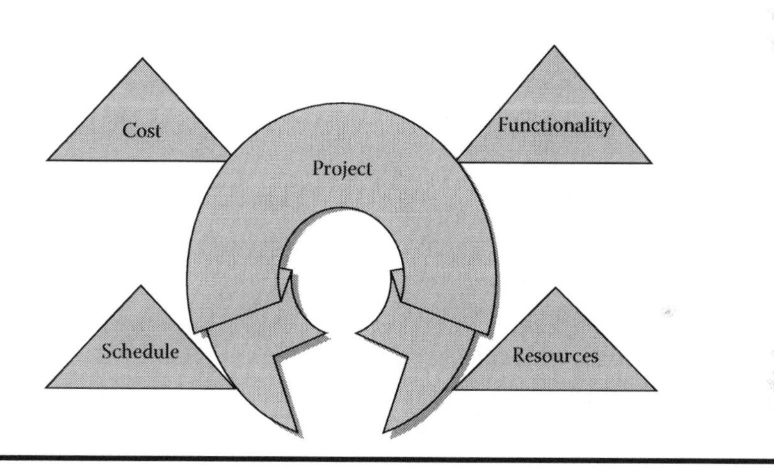

Figure 7.2 Project constraints.

Yet other variables that influence project schedule integrity are as follows:

- Rules of engagement deviation
- Risk factors
- Deviating from acceptance criteria
- Inadequate testing
- Deviation from change management rigors
- Missing deliverable assumptions (inadequate documentation)
- Missing requirements
- Significant change in requirements
- Overspecifying requirements
- Not specifying requirement tolerances
- Inadequate performance tracking metrics

An ERP project is typically a very large project, regardless of the size of company, with tentacles that cross the entire organization. As such, an ERP project life cycle seldom repeats itself within an organization. To that end, there are numerous variabilities, which might surface, that require special handling. To mitigate their impact, it is essential that the above list of variables should be incorporated (this is an incomplete list at best; best practice projects expand this list considerably based on the known anomalies within their company operating practice and culture); the project core team needs to aggressively engineer mitigation plans whenever a variable surfaces.

During the project initiation phase, resources are evaluated, budgets are being formulated, the project plan is emerging, requirements generation is postured, stakeholders and sponsors are emerging, project core team members are being evaluated, and the Steering Committee is being chartered. The project schedule is an outline of tasks, software modules, and preparatory deliverables. A toll gate review might be used as the leadership approval point.

Once the authorization to proceed is given, budgets are approved, the project plan is created, software suppliers are being sought, stakeholders and sponsors are finalized, the project core team is finalized, requirements generation is launched, and the project administration activities begin. During the project administration, the following are typical tasks:

- *Project chartered*—This section will be scaled to fit the company size, project budget, and company culture. However, compromise in this area may result in excessive amount of project-level crisis and reduced end result project value.
  - Rules of engagement finalized (RulesofEngage3)
  - Executive collaboration members assigned to project oversight
  - Change management strategy approved (CM7)
  - Approval of project plan
  - Adjudication and escalation strategy approved
  - Communication plan approved (Comm13)
  - Risk management strategy approved (RiskMgmt4)
  - Resource management strategy approved
  - Acceptance criteria approved (Accept1)
  - Testing strategy approved (Testing16)
  - Training strategy approved (TrainingPln8)
  - Project health approach approved (ProjHealth10)
  - Project performance metrics approved
- *Project sized*—It defines what is to be delivered based on the budget.
- *Resources negotiated and committed*—Both internal and external.
  - Depending upon the level of development/customization/integration, a finite scheduling strategy may need approval
- Software RFP created (prerequisite is finalizing requirements) and qualified ERP software suppliers contacted.
  - Software selection strategy and methodology approved
  - Software evaluation scorecard and weighting method approved
  - Software supplier negotiating position straw man approved
- *Project schedule is cast*—Deliverables negotiated and approved.
  - Technical environment deliverables and tasks
  - Business operating environment deliverables and tasks
  - Functional user community deliverables and tasks
  - Project environment deliverables and tasks

- For each deliverable and task, owner will engineer a roadmap that will help ensure cost and schedule realization. Roadmaps will be submitted to the project core team and project manager.
- *Rough-cut capacity plan*—Depending on the size of the project, the level of customizations and modifications, and the number of customer and/or supplier integrations, this resource management activity may be sizable.

A toll gate review (TollGate14) might be used as the leadership approval point.

Adhering to this project initiation process is a project management **best practice**.

The next phase of the ERP project management process is execution of the project and progress monitoring. This requires fervent focus upon ensuring that the project plan is followed diligently, as well as the project schedule deliverables and tasks are completed as planned. Let's drill-down a little on this topic.

For example, within the project schedule discussion above, there was the following guideline: "For each deliverable and task, owner to engineer a roadmap that will help ensure cost and schedule realization. Roadmaps to be submitted to the Project Core Team and Project Manager." On the surface this guideline sounds practical and as an owner, it seems like it would be "important" to help ensure successful completion of the activity. In reality, this may be one of the project manager's most difficult challenges. However, getting the owner to commit to creating a "roadmap" tends to be almost impossible. There are a variety of contributors to this aloofness as discussed below:

- A common response is, "I commit to doing this on schedule, that is sufficient, I don't want to spend time creating a roadmap." In reality, unless the owner spends time engineering the solution in substantial levels of detail, they are "guessing" at best, and likely do NOT have a clear picture of a roadmap for the solution. Consequently, this deliverable is vulnerable for missing its cost/schedule commitment.
- Another common response is, "My team is the subject matter expert on this deliverable ... we do it all the time ... there is no reason to waste my time developing a roadmap for our area of expertise." In reality, unless this deliverable is precisely the same as "routine" tasks done by the team, there is likely a large degree of variability between deliverables. Again, the owner must spend time engineering the solution in substantial levels of detail; otherwise, the team is "guessing" at best and likely do NOT have a clear picture of a roadmap for the solution. Consequently, this deliverable is vulnerable for missing its cost/schedule commitment.
- Typically, a resource area is committed to multitude of projects simultaneously. This ERP project is usually one of many other projects that need completion by the resource area. Inasmuch as the ERP project core team does NOT typically have any say on what the resources spend their time working on, then the project core must rely upon the resource area leadership to ensure completion on time and within budget. Some resource areas are pretty good at meeting their commitments. However, if they have ever missed before, then they are vulnerable for a miss again. One of the values of creating a roadmap (engineering a solution) is that if one of the resources working on the ERP deliverable happens to be sick or takes vacation, or has some other reason for absence, at least an important piece of the solution (roadmap) exists, which may be passed along to an alternate resource. Again, the roadmap is an important element in controlling variability in deliverable integrity.

- Leadership has a tendency, over time, to modify project priorities. It may have been with good intentions that the resource area is committed to the cost/dates at the onset; however, with a change of priority by leadership, the ERP deliverable may now be given less emphasis and less resource to devote to completion on time. Changing priorities is a common event for an ERP project core team (a good reason to have an already defined adjudication and escalation approach). The solution is typically outside the control of the resource area and project leadership, requiring escalation and resolution from management. Having a detailed roadmap, it defines milestone completions and helps the escalation process stakeholders evaluate the deliverable vulnerability. Without a detailed roadmap, the reliability of solution integrity becomes a matter of opinion rather than specific facts.

The ERP deliverable roadmap is a project management **best practice**.

There are a host of other variables that may arise to negatively influence the ERP project deliverable commitment. The important takeaway is to ensure that the deliverable owner engineers the roadmap as a means to help ensure committed deliverable realization. As the project timeline progresses, there are times when the span of the roadmap is still too large to help ensure committed deliverable realization. This requires further refining the roadmap and/or statusing their completion more frequently. I have had to get down to statusing a deliverable completion every two hours on a 24/7 calendar clock.

As introduced earlier, an important aspect of monitoring progress is helping the deliverable owner be successful in achieving their commitments. One of the key roles of an ERP project core team is to help remove barriers to the successful completion of the deliverables. Depending upon the size of the project, this can become a monumental endeavor, however, a very important critical success factor. As described above in the project initiation phase, there was mention of engineering a risk management strategy. When taken seriously, one of the purposes of risk assessment is to identify these deliverable barriers, and engineer a mitigation strategy that takes the sting out of the risk negative impact. An astute ERP project core team could use the project chartered elements as a checklist to help head off barriers early, all for the purpose of helping the deliverable owner increase their propensity for successful realization of the deliverable. Yet another owner helping strategy is to frequently review their deliverable resource load versus capacity. If the resource teams do not have a project-oriented capacity management tool to help them manage their commitments, then this becomes another vulnerability to ERP deliverable commitments on time and on budget. Let's take a closer look at yet other "helping the owner be successful" strategies that might be pursued:

- *Behind the scenes salesmanship*—As an ERP project core team member, the organization leadership has vested confidence in their team to do what is necessary to achieve the goals, objectives, and deliverables according to plan. Frequently, this requires creative insights into the company culture and operating practices, including company politics. To accomplish ERP project success typically requires an astute early awareness of waning leadership support. Each project core team should have devised the means to engineer a set of early warning triggers to assist in taking the needed corrective action before hardship surfaces. One such technique is to create a behind the scenes salesmanship strategy and work with the deliverable owner in shoring up support as time progresses. Leadership commitments need constant care and feeding across the ERP implementation timeline. It is one thing for leadership to commit to deliverables at the onset of the project, but quite another thing to ensure things happen in the heat of the day-to-day battles with budget, resource, and priority constraints. Assisting the deliverable owner in conducting these sessions is a good use of ERP project core time over the longevity of the project. This is a project management **best practice**.

- *Politics*—Unfortunately, due to the cost of the ERP implementation, the project tends to be in the crosshairs of the project investment portfolio. Depending on the size of the project and company deploying ERP, this high visibility typically puts the sizable budget into continual "rejustification" mode. Budget battles and other political nuances tend to attract company politics at its highest level. I personally believe that politics is a significant ERP success factor distracter and a nonvalue-added exercise of out-of-control leadership egos and I rather not participate in. Even though it is nonproductive and annoying; if it surfaces, it must to be dealt with all the same. One of the best strategies is to convene the attendant circle of influence of involvers (executive sponsor[s], affected stakeholder[s], resource area, deliverable owner, and adjudication designees) to address the situation and create a mitigation strategy. If this does not effectively resolve the issue, take it to the ERP Steering Committee for immediate resolution. Nipping politics in the bud early, and comprehensively, facilitates the project rhythm at the optimal level.
- *Barrier removal*—There are a plethora of potential "rocks in the middle of the road" that need be dispositioned on an ERP implementation project. A key to project success is to handle each barrier as it surfaces, and ensure that it does not resurface (rework may become a time-consuming nightmare if not handled decisively). I have found that immediate barrier resolution may be a critical success factor over the span of the project. To that end, it is wise to spend some effort in engineering a barrier removal strategy up-front and create a wishbone chart that consists of various cause-and-effect paths, so that when barriers surface, they may be handled with quick response. As discussed earlier, the use of adjudication and escalation resolution quickly has its advantages as does reporting project health RED. Regardless of the strategy chosen, the key is to hit the barrier head-on with intense impact and comprehensive resolution. Try to waste little time muddling in insignificant thorns and project distracters. This is a project management **best practice**.
- *Actively use the project executive sponsor(s)*—If the ERP project has appointed an executive sponsor(s) (and I believe this is a very important critical success factor), then keep them actively engaged in the project by conferring with them daily and rely upon their judgment, especially as politics is concerned. I personally believe that a company truly committed to ERP success will have the ERP project an agenda item at all company staff meetings, from the Board of Directors through all levels of leadership. This tends to keep ERP progress a "business priority," not just another project consuming high level of costs.
- *Posture for success*—Because of the typical size of the ERP project and the span of time essential for realization of results, the ERP project status of results requires continual broadcast. Therefore, it is important that the communication plan include frequent project status through all levels of the organization. There are times when "perception" rules and the perception of the project need to foster high-impact successful results. If this is not the case, then the project is likely on a derailment path. Regardless of the level of variability, this course correction needs to be made immediately, decisively, and comprehensively. Like an octopus, there are many tentacles concurrently engaged in an ERP project, each with the potential to lead to derailment. Again, as discussed earlier, like a symphony, the ERP project needs to be balanced and harmonic through the entire life cycle. This typically requires daily, and at times, more frequent, reengineering of the plans, strategies, resources, and other key elements that lead to successful realization of results.

Continual use of these precepts is a project management **best practice**.

A word about **project reviews** and **project status**—This is central to the communication plan, which was covered earlier. The preparation for the review may be as important as the delivery of the review. Let's look at some key precepts regarding the preparation and then the presentation of the review:

- All facets of resources participating in the review presentation should be prepared to present their status. It is important for the review attendees to hear the status from principles … those accountable for delivery. Typically, there is a mixed bag of resources: some comfortable in presenting and the other not comfortable in presenting. All need coaching to some extent:
  - Keep to the salient progress status and avoid rabbit trail adventures.
  - Be crisp in what is being presented; no long winded speeches.
  - Avoid the temptation to air out grievances; there are other venues for this.
  - Don't dodge directed questions. Be truthful about the status, and if there are variances from cost/schedule plan, be prepared to discuss with a slide that shows the variance remediation and corrective action plan. Remember that this is a review, not a corrective working session (keep to the agenda).
- If using PowerPoint, ensure that all slides use a common template and are kept crisp. Avoid undue wordiness.
- Ensure that the executive sponsors, stakeholders, and other key decision makers are aware of any "nerve endings" that will be exposed during the review (they need to be prebriefed so they may support the project core's position). Ask if they would like an advanced copy of the presentation.
- Keep to the agenda and honor the participants' time commitments. Avoid overrunning the meeting times.
- Update all attendant status reporting tools (MS project, problem logs, issues list, capacity plans, etc.) before the review.

Project status should be provided based on agreed-upon schedule. Written status is an important chronology of the project progress and needs to be posted so that all interested parties may have access. The project status should

- Be crisp.
- Update the percentage complete progress toward deliverables. Once a deliverable/task is closed, remove it from the status report.
- Detail open "issues."
- Highlight progress on remediation tasks, corrective action plans, and any other potential derailment obstacles.
- Brief on what is coming up before the next status.

Both the review presentation and the project status content should be adjusted, as needed, to help ensure that the audience is adequately apprised.

Adherence to the above project review and project status guideline is a project management **best practice**.

The final aspect of the project life cycle is **project closure**. Project closure should conform to the company's cultural and project standards. However, if the cost were an important element, then a detailed accounting of cost versus budget reconciliation would have to be performed and formally packaged and presented by the project controller. Regarding the schedule, all deliverables/tasks should be completed unless approved deviation/waiver has been granted by the Steering Committee, sponsors, and stakeholders. If a waiver is granted, the yet-to-be-completed tasks need to be cast into a follow-on schedule for tracking through closure. Depending on the size of company, operating practices, and budget, this follow-on statement of work (SoW) should be either included in the project portfolio as a separate project or dispositioned as a functional organization deliverable tracked according to their work process. This is governed by the project change control process.

A word about the **project organization**—We have discussed the importance of the project core and duties of the Steering Committee. Let's drill-down a little deeper into the project organization and their role in the ERP project implementation.

- *Project Steering Committee*—This group functions as the project governing body. It consists mostly of executive-level participants and includes representatives from all aspects of the business where the ERP implementation will have an impact. This is typically the entity "where the buck stops" for adjudication and escalation paths. Although not typically a day-to-day participant in project activities, the Steering Committee might be very "project active" whenever cost/schedule variances surface, which needs remediation to avoid derailment. Depending upon the size of the company, they may be chartered differently. For example, for small- to medium-sized organizations, the group will likely be the operation operating group compared to a separate chartered entity in a larger company. Regardless of the leadership makeup, there needs to be a project oversight forum to keep focus, actively support as needed, and inspire success.
- *Project core team*—This group functions as the day-to-day project leadership entity. Depending upon the size of company and budget, it typically has representation by resource areas that have mission-critical impacts upon the outcome of the ERP implementation. In a large organization, it might consist of 10–20 full-time dedicated team members or more, depending on the phase of the implementation. In a smaller company, it may consist of one to three full-time members and a variety of part-time resources. In either case, the project core team membership may expand and contract based on the level of effort needed to stay the course and complete the deliverables. An important note at this point is as follows: If the ERP project has significant integration activity with customer and/or supplier trading partners, then the project core team needs representation from these entities as well.
- *Technical team*—The technical team membership will likely expand and contract based on the SoW being performed at any given time. Membership will also be impacted based on the size of company, budget, and risk factors. The technical team lineup may consist of the following:
  - Engineering—Various facets of engineering might include
    - Product and/or sustaining engineering
    - Software engineering (SE)
    - Systems engineering (SyE)
    - Industrial/manufacturing engineering (IE/ME)
    - Process engineering (PrE)
    - Quality engineering (QE)
    - Platform engineering (hardware) (PE)
    - Network engineering (NE)

        - Methodology
        - Logistics
    - Information technology
        - SE
        - Database administration
        - Business analysis
        - SyE
        - PE
        - NE
        - PrE
        - QE
        - Systems architecture
    - ERP software provider
        - Architecture, design, and software product support
        - SME
        - Business analysis
        - Training
    - Trading partners (customer and/or supplier)
        - Architecture, design, network, and PE
        - Business analysis

The technical team may also include third-party partners depending upon the technology framework, span of global operations, and other factors. This membership may include platform (including cloud partners), integration partners, classified network partners, and a host of other key product/server providers.

- *Business operating environment team*—To help ensure that the ERP implementation maintains alignment and functions in harmony with the business strategy and operations plans, there may be a need to have representation from this body. Representatives might include such members as strategic planning, treasury, marketing, sales, international operations, legal, property management, regulatory compliance, mergers and acquisitions, and facilities. Although these members may also be ERP solution users, their thrusts may be externally focused— thus the need for collaboration.
- *Functional user community environment team*—This is typically the primary customer of the ERP implementation and the primary return-on-investment target entity. The functional team lineup may consist of the following:
    - Engineering—Use lineup in the above technical team as a checklist for inclusion.
    - Supply chain
        - Contracts
        - Procurement
        - Master scheduling
        - Demand management
        - Stockroom and warehouse
        - Material planning
        - Production control
        - Distribution
        - Logistics

- Sales and operations planning
  - Strategic planning
  - Forecasting
  - Throughput optimization and modeling
  - Cash optimization and modeling
  - Inventory optimization and modeling
  - Cost segregation
- Finance
  - General ledger
  - Accounts payable
  - Accounts receivable
  - Fixed asset
  - Budgeting
  - Activity-based costing
- Manufacturing and production operations
  - Fabrication, assembly, and test
  - QA
  - ME
  - Maintenance planning
  - Cost accounting
  - Repair
  - Safety
  - Training
- Marketing and sales
  - Sales order management
  - Customer relationship management
  - Customer care
  - Sales management
  - Product marketing
  - Public relations
  - Sales engineering
  - Field service
- Human resources
  - Staffing
  - Payroll
  - Compensation and benefits
  - Succession planning
  - Employee relations
  - Organizational training
  - Labor relations
  - Risk management
  - Records management

The project organization is typically managed using the matrix management structure whereby nested cross-functional team members design, test, and train on the precepts and processes of the ERP deliverables. Because a typical ERP implementation has such far-reaching impact across the organization, team members may be called upon to concurrently function as customers, service providers, designers, process experts, SMEs, testers, trainers, innovators, issue resolution facilitators, and a variety of other roles that are not necessarily part of their day-to-day outside ERP implementation operating practices. Frequently, the ERP implementation stretches the organization dimensional depth capabilities to their outer limits—a territory that may be outside the team members' comfort zone. However, when managed properly, the ERP implementation becomes a career-expanding experience for every participant, adding exceptional value to the individual, team, leadership adeptness, operating practices, core values, and corporate culture. The typical day-in-the-life-of a project organizational team member fosters the need for dynamic resilience, yet it mandates extreme levels of organized discipline, persistence, and stability. This dichotomy of behavior is best demonstrated when visionary, innovative, flexible, ingenuity, agile, exceptional throughput, nimble, and fast response are characteristics of the team members' style, as discussed in the beginning of this chapter.

With the entire project dynamics happening concurrently, many times with what seems like exploding landmines surfacing from all directions, keeping focus upon the project mission and deliverables is challenging. This makes the care and feeding of the project core, stakeholders, and sponsors on a daily basis paramount. Knocking down barriers, before they become issues, and maintaining project priority are essential to the stability and harmony of the project team as well as throughput of the project deliverables. Do not underestimate the impact of potential derailment fodder. This is where good organization and structure skills, persistence, and project character fuel the momentum essential for ultimate project success.

Before we leave project management, we need to consider a few other additional precepts that will help the project function in a best practice, compared to mediocre, manner. In Chapters 1 and 2, we discussed the importance of setting the stage for project success. With good requirements, a solid design, and clear mission statement, the project may foster good velocity and momentum as it moves through project execution. It is during this phase that developing a solid test plan and staying on top of training needs and completion tasks/deliverables in a fast response manner further foster project success. Depending upon the magnitude of ERP software modification and/or customization, interface and integration activity and the degree of trading partner active involvement may become a defining moment on ERP implementation progress. These tasks may become elusive and unwieldy if allowed even a hesitation in project oversight. To that end, be especially aware of any obstacles or delays that may surface. Frequently, these activities involve large cross-functional teams and SMEs, may require intense capacity management and finite scheduling, and need continual executive ownership renewal effort to stay the course. There are likely data mapping processes, multiple iterations of testing activities, and a variety of toll gate review focus essential for project deliverable realization. Intense project core monitoring, focused work groups, and a rhythm of execution of the deliverable roadmaps become essential milestone activities. This is where true project leadership differentiates the mundane from stellar project performance. Strong Leaders, or, captains of industry, always opt to be stellar.

Adhering to the above focus is a project management **best practice**.

The final aspect of project management is ensuring that **project closure** activities occur as expected. Reconciling project budget versus actuals, closing work activity charge numbers, performing any project auditing deliverables, and preparing and delivering various project reports are key project closure activities. Do these project closure activities well.

One last point is that project management tends to be a bit stressful ... offset the stress by having fun in performing the work effort. Look for opportunities to laugh, to poke fun at yourself, and to keep a light touch on things that tend to bring you down.

Chapter 8

Process Performance Management

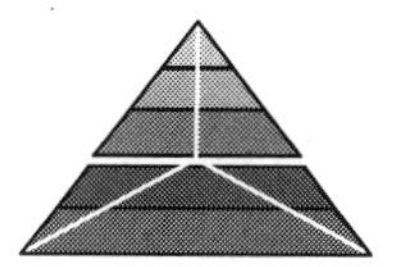

As introduced in Chapter 4, this chapter peels back the layers of opportunity and explores realigning measurements, on an end-to-end process basis, which allows the entire organization to understand the importance of every job function and gives the working-level tier of the organization the ability to measure their individual contribution to profits.

I recall an Operations Audit review I was engaged in where the client company leadership was under fire because they spent over $1 million to implement their enterprise resource planning (ERP) system and the company could show no return on investment (ROI). In reviewing the ERP project charter and other artifacts associated with the implementation, I did NOT find any ERP goals and objectives or commitments by stakeholders as to what the ERP implementation was expected to yield. After interviewing a large group of users and leaders, it was obvious that the ERP implementation merely replaced their previous legacy system. There were no engineered tasks focused upon gaining a ROI. The company used the Software Suppliers Project template as their guideline. It was also obvious that the leadership team took for granted what the ERP Software Suppliers salesman said was potential ROI. Reading between the lines, the leadership team believed by merely installing the new software, oodles of savings would immediately begin impacting the company's bottom-line performance.

As part of the Operations Audit, I was asked to put together a proposal that would identify a short-term strategy that would yield "low-hanging fruit" ROI. During the review, I found that there was nothing wrong with the software (it worked as advertised), their implementation "training" was adequate, and the project team merely replicated the policies and procedures of the legacy system. The proposal submitted essentially recommended a modified senior leadership collaboration workshop. During the collaboration workshop, I worked with the leadership team to engineer a short-term, a medium-term, and a long-term ROI strategy. One of the key ingredients in the plan was addressing the process issues plaguing the operating team. Within six months, we laid out a plan to obtain over $2 million of potential ROI.

This chapter is intended to inspire a vision of quantum leap performance improvement. I want to stroll the senior leadership team outside its warm and cozy (and maybe complacent) operating culture and insist that they apply outside the box critical thinking to imagine how they may improve their business performance, not by a mere smidgen, rather by an order-of-magnitude improvement. In Figure 8.1, transformation diagram shows the building blocks and intellectual collateral pointing the way to a new business model that permits bold stellar performance improvements.

8.1 Definition of Process Performance Management

Process performance management (*PPM*), on an end-to-end process basis, is essential to businesses that are serious about maximizing profits and employee productivity while realizing exceptional, optimized operational performance results. PPM defines the *acceptable end-to-end process performance level* necessary to achieve excellence and functions as the *internal service-level agreement* (SLA) between the internal customers and their service providers. In this chapter, the PPM will use

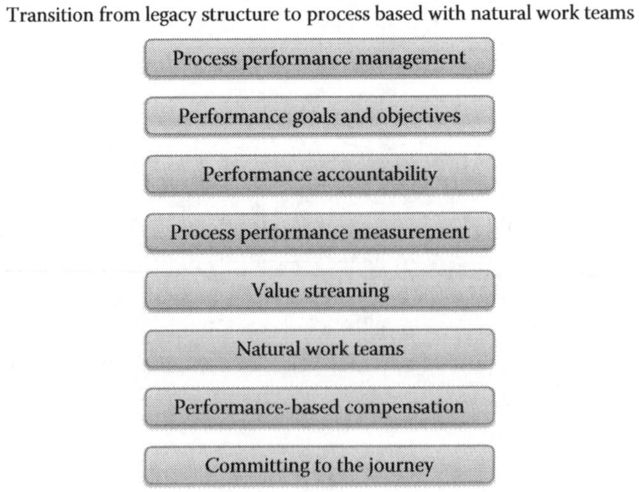

Transition from legacy structure to process based with natural work teams

Process performance management

Performance goals and objectives

Performance accountability

Process performance measurement

Value streaming

Natural work teams

Performance-based compensation

Committing to the journey

Figure 8.1 Transformation.

process performance management and **process performance measurement** interchangeably. The PPM allows the business process functionaries to know what is expected of them in their effort to manage day-to-day productivity and deliver Global Operational Excellence. The PPM is the business process **best practice.**

In short, PPM is a method to synthesize optimal business performance throughout the organization between the internal customers and their service providers on an end-to-end process basis. Today's fast paced and competitive environment requires end-to-end process as an integral asset in the day-to-day decision process. Companies that maintain high expectation and responsive business process systems leverage these **best processes** as a competitive weapon. **Flexible (easily changed), agile (do quickly), nimble (move quickly), lean (elimination of waste), quick response (elimination of unnecessary lead time), fast cycle (quick turnover), adept (highly proficient), and deft (skillful and clever)** are **best process** watchwords describing end-to-end business process characteristics. Engineering these process characteristics into processes is the tool of astute executives in their deployment of ERP systems, pricing strategies, cost variance management decisions, scheduling activities, and customer performance excellence assessments.

PPM should be a living and constantly updated leadership tool if it is to reach its potential as an integral weapon within a company's competitive arsenal. However, few companies have any semblance of an end-to-end PPM. Information that feeds the accounting system and affects the quality of book numbers is typically lacking an "in the trenches" standard. Because the sources of the accounting information lack a standard, how accurate can the accounting data be?

In lieu of PPM practices, companies must compromise. Their environment is frequently driven by crisis management, consistent profitability is suspected and affordability initiatives wane.

It is an interesting dichotomy that companies are constantly implementing or upgrading their computer-based systems. However, the performance standard that should be in place prior to implementation or upgrading of their systems is prominent by its omission. This omission is further salient by the fact that it is not unusual for a company to spend 4%–5%, or more, of its sales dollar for systems-oriented support budget. Yet frequently, as basic as the PPM is, they are not in place. Subsection 8.2 will describe the criteria for an effective PPM.

8.2 Criteria for PPM

The PPM is composed of the following ingredients:

- Each process work element (deliverable) within the nested end-to-end process will have a performance measurement associated with it (e.g., inventory accuracy, purchase order timeliness, etc.).
- Each job function within an organization shall have a uniquely defined process performance measure associated with it.
- Each nested end-to-end process *will* incorporate a composite of all the job-level measures within its business process, plus the performance measurement of the business process as an entity. In addition, the end-to-end process audit guidelines are necessary ingredients for the total enterprise-wide PPM.

One of the missing elements that many companies lack, in rolling out their corporate performance objectives, is the **vision** on how the performance is to be achieved. Another is the realization that a company consists of various business processes synchronized together to achieve common objectives. However, performance excellence does not occur by osmosis, but requires **focused and diligent effort** to realize the expected results. Agreeing upon **acceptance criteria** is the origin of the journey toward performance excellence. An acceptance criterion removes any doubt as to what is expected between the SLA team members. Once the acceptance criterion is established, **meaningful metrics** may be defined and then the **roadmap** on how to achieve these results may be contracted. Any of these critical success factors omitted will result in suboptimized performance.

An essential component of *criteria* is process validation and auditing. The auditing/validation ingredient should **not** consist of bureaucratic approvals or other methods to delay process deployment. On the contrary, auditing and validation should be engineered to facilitate more timely results.

Internal customer acceptance: By far, the best audit is not necessarily a management review at all, but the acceptance and approval of the *internal customer*, who realizes or benefits from the quality of the business process measure. *Internal customers and their service provider team members are at the heart of the PPM philosophy.* Internal customers negotiate a quality and service performance level from each of their process service providers. This negotiation agreement will be discussed further later.

Now that the scope has been defined, a review of the above ingredients in light of the scope is appropriate.

8.3 Job Functions Require a PPM

PPM for a job function should be an integral part of a job description; however, the premise of a PPM is different than the premise of a job description. *A job description typically broadly defines the scope of responsibility and tends to relate to global issues.* In comparison, a PPM is very finitely focused on specific goals and clearly defined objectives. For example, *the PPM defines a specific performance level, within a process, on a document-by-document basis as an increment of an end-to-end business process.*

Note: A document may be a form, computer process, drawing, information-triggered exception, or any other media that conveys workflow tasks needed to achieve business performance excellence.

8.3.1 Data Accountability

Data, which transforms into information, is a most valuable company asset. As such, it requires a stewardship and accountability owner responsible for its ongoing integrity. Equally important, the internal customer(s), or data user(s), work with the data owner to establish the quality standard and continuously improve the end-to-end process flow to help ensure timeliness and integrity, ultimately focusing upon a goal of perfection. At times, the internal customer may function as a surrogate for an external customer. In either case, the data owner(s) and internal customer(s) together must be diligent in their pursuit of excellence by developing the QUALITY STANDARD, defining the input edits, regularly auditing the data and end-to-end process, maintaining metrics, and ensuring that a root-cause analysis is quickly and diligently sought any time inaccuracies surface. Merely correcting inaccurate data is fine, but far superior is the diving into the root cause to determine why inaccuracies occur, then taking the necessary corrective action (tool change, process change, training, etc.) to obviate the future generation of inaccuracies.

8.3.2 Data Accuracy

Within certain job functions, it may be necessary to tailor an individualized PPM for each person depending upon the knowledge level, grade level, and/or password security level of the individual. For example, an employee who has functioned within the job for over 12 months should achieve a higher performance level than an employee with one-week experience. However, a caution is necessary here. If your objective is 99%–100% process accuracy level, even recently hired employees must function at peak performance if the goal is to be attained.

How can this be accomplished? Simple! An individual should ***not*** be allowed to perform transactions on documents against the "production database," without having achieved the specified *minimum acceptable quality level (AQL)* on the conference room pilot (training) database. This process is frequently termed *certification* or *software certification*. It is analogous to the process of obtaining a vehicle driver's license. For example, in most states you must pass both a driving test (hands on) and a written test before you can obtain a driver's license. A certification program is very similar. For certification, an individual must pass a hands-on (terminal) and written exam in the conference room pilot (training) database before he or she is allowed security access to the "production database." If the individual *cannot* pass both the hands-on and the written test, they should be reassigned to a responsibility area that does *not* involve data input (or creating any source document) or be considered for dismissal. Anything short of this process rigor is, by default, endorsement by management that data contamination is an acceptable behavior by its employees.

The certification process needs to be tailored by job (and/or employee) and requires recertification whenever an employee is hired, has a job transfer, or is promoted. Recertification is also required when new software modules are added. An annual review for recertification provides an increased insurance policy for individual proficiency.

8.3.3 Best Process Characteristics

Just as data accuracy is a key measure, timeliness is equally important. If a transaction sits around for minutes before processing, it has an impact upon upstream processes and deliverables. Introduced earlier were the ***best process characteristics***, namely, **flexible**, **agile**, **nimble**, **lean**, **quick response**, **fast cycle**, **adept**, and **deft**. Given further critical thought, this list of **best process** characteristics could likely be expanded; however, instead of an exhaustive list, it is important to recognize that well-engineered business processes, which continually improve and include these characteristics, are equal contributors to Business Process Excellence.

Let's briefly discuss why these attributes are important to a well-engineered business process.

- Standardized processes are important for consistent results, yet if the process is so structured that it is not ***flexible***, the number of daily exceptions resulting from structuring rigors may erode work effort productivity by generating numerous process rejects requiring excessive rework. Therefore, to optimize the business process, thought must be given to flexibility.
- ***Agility*** is likewise an important business process characteristic. An agile process work effort is able to navigate through thorny issues with quick resolve.
- ***Nimbleness***, like agility, allows work effort to navigate difficult barriers gracefully.
- ***Lean*** allows the process work effort to proceed with affordability in mind, with less waste.
- ***Quick response*** is also a work effort ***best process*** facilitating rapidity in deliverable closure, without compromise of quality.

- ***Fast cycle***, such as quick response, tends to reduce a given end-to-end process work effort investment ... This attribute improves SLA results while sustaining workflow throughput.
- An ***adept*** process demonstrates a resource mastery allowing process fulfillment to be accomplished in an exceptional way.
- Deftness facilitates process excellence with finesse.

There is a plethora of additional synergistic benefits that result from engineering best process attributes described above into the PPM. These benefits are synergized and leveraged when one looks at the complete end-to-end process result.

At this point, many readers are saying:

"Our company will never buy this concept ... it is too expensive, it is too time consuming, and it is too ________________" (you fill in the blank). However, I will challenge you:

What is the annual cost of obsolescence?

What is the annual cost of lost customers?

What is the annual cost of lost sales?

What is the annual cost of inaccurate product costs?

What is the annual cost of employee turnover?

Worker frustration level, leadership decisions, and operational actions of day-to-day dynamics are all influenced by PPM. Some, if not all, of the above "What is the annual cost?" items are influenced by whether a company has PPMs or not. Once a company assesses the *cost of quality* (cost of rework, poor decisions, etc.) resulting from inadequate PPM, it will be clear that it is massive compared to the cost of providing a *certification program.*

The companies surviving in the next decade will be those companies, who have and use information as a competitive weapon. The difference of a 1%–2% accuracy level may be that edge which elevates one competitor over another. Consequently, defining these performance measures on a job function and/or individual level will be essential in the future.

8.4 Performance Goals and Objectives

By nature, most individuals are more comfortable being a part of a group than being alone. Therefore, it is natural to obtain consensus on the group's commitment and process performance level before acquiring any individual commitments. Research showed that *higher quality results* are obtained when goal-setting sessions are performed as a group, before individual (or job function) goal setting occurs. Also, process performance goals should be constantly changing (at least annually), consistent with company goal changes. Specific attributes of goals are as follows:

- They should focus upon constant tightening of tolerances as time goes on.
- They should be achievable and a source for "pride" within the department.
- They need to be championed by individuals if success is to be attained.

To relate this "goal-setting" methodology to the end-to-end process PPM, each business process work element should establish its PPM objective (see Section 4.7 for an example of end-to-end process and PPM objectives):

- The performance attainment should be based upon the objective within the business process and, when appropriate, stepped up (with tighter tolerances) since the last review.
- The end-to-end business process goals should be a subset of the company-wide PPM goals and related annual individual performance evaluation. Company-wide measures are not normally attainable if all business process goals do not focus upon and add up to company-wide objectives. Some companies now extend the end-to-end business process concept through the entire expanded supply chain (customers and suppliers including multiple tiers). For expansion across tiers to be successful, there needs to be a shared benefit such that it becomes a *win–win* result.
- The logical hierarchy is that there are company-wide PPM objectives that are disaggregated into end-to-end business process PPM objectives. These end-to-end process objectives represent the internal customer and service provider team's element objectives (SLAs) that are then disaggregated further into job function PPM objectives. Ultimately, the PPM objectives are personalized to the individual. When PPMs become individualized, the profit contribution is leveraged.

The end-to-end business process (internal customers and their service providers) PPMs becomes the catalyst for encouraging yet further improved performance and tighter tolerances. The business process is the lowest operational level of accountability that maintains a cohesive association among a variety of job functions. Each business process team leader should be chartered with setting the tone for the *minimum acceptable information and performance quality level.* It is normally at the business process level where operating procedures dovetail (are synchronized) with a PPM. Because business processes typically cross organization departmental lines, there may be a bit of conflict between the departmental and the business process objectives.

This conflict is most easily resolved by establishing an *independent,* senior management level sponsor for the business process itself. *Independent* implies that the senior management appointee does not have any direct influence upon any individual's performance evaluation within the end-to-end business process itself. Therefore, the senior manager is able to adjudicate any conflicts between departmental and business process *objectives.*

8.5 Performance Measurements for Optimal "In-the-Trenches" Results

The auditing of in-the-trenches process results, described above, should constrain a vast amount of faulty data, nonvalue-added practices, and faulty or untimely decisions. However, it will not totally inhibit erroneous data. Tools such as bar coding and digitized scanning improve data entry substantially. Therefore, it is necessary to establish an *acceptable quality performance level* for each transaction, document and other deliverables within the process (as described in Chapter 4). Although the acceptable quality performance level may be tempered by input tool (bar coding vs. keyboard entry), it is the opinion of the author that *data* is data and the input tool should *not* influence the expected results; no compromise should be allowed. For example, an *AQL* for the repair order may be 99.9% to cost and schedule performance. To validate achieving that AQL requires the following:

- A reasonable ongoing means to validate the process performance level (e.g., cycle review programs)
- A valid method to identify the root cause of any performance degradation (e.g., installation procedures)
- Preventive measures to head off sources for process deterioration (e.g., bar coding, scanning, and an effective ongoing education and training program)
- Management intervention through corrective action (e.g., making it an agenda item during staff meetings, including data accuracy goals in individuals' performance review criteria, etc.)
- An aggressive awareness **practice** is needed, which clearly specifies the AQLs. This awareness **practice** may include one or more of the following:
  - Conspicuously posting achievement levels outside work areas, cafeterias, etc.
  - Quality-level performance in monthly progress reports to top management
  - Quality-level performance in the company newsletter or other house organizational literature
  - Quality-level performance in the employment *job description* so new job applicants understand the standard when they apply for a job
  - Having awards for consistently achieving a preestablished level. This may include such things as a monthly catered luncheon if goals were attained, certificates, company meetings that announce attainment, and so on

In order to optimize ongoing performance improvement, time must be spent **daily** evaluating the root cause of any individual work effort performance degradation and/or end-to-end process inhibitors while continuously reengineering any broken processes. The best process experts are those doing the work effort, not individuals standing on the sidelines as observers! Therefore, the practice of daily investment in process improvement has a great payback, if fervently pursued.

Performance is only improved by continually revising established goals and measuring performance to these goal enhancements. Within baselines, inaugurating similar attention and review to the component elements that influence the improved performance level, namely, transactions and processes, influences sustainment. Focus upon the proper entry of all relevant nested work effort is necessary if achieving process performance goals at the output deliverable level (files) is to be attained (Continued sustainment of "minimum acceptable performance levels" of each process element contributes to expected results of the whole end-to-end process).

8.6 Performance Accountability

To digress a bit, those committed companies that aggressively pursue a PPM take the position that **every employee should understand their "contribution to profit" within the organization.*** Given that the concept of a workmanship standard functions as the means to assess individual performance toward attainment of profitability, the PPM serves as its integral component. *With profitability as a baseline, and using* ***information*** *as a competitive weapon in the marketplace, the*

* In not-for-profit or nonprofit organizations, adapt profit to a term that best describes best process results.

PPM becomes a mechanism to measure the individual's performance to profitability results. There can be no organizational performance without business process performance and no business process performance without individual performance. Companies consist of people working together toward the common goals of the organization. The business process organizational structure facilitates the establishment of objectives and the monitoring of performance. **The workmanship standard serves as the means to express the *composite company-wide objectives* in a small enough increment to help ensure accountability, and contribution to profitability may be established at the lowest possible organizational level.**

The PPM provides management the confidence level (quality of information and fidelity of performance) from which they may extend day-to-day decisions to the individual level, with high confidence of success. The business process level, then, becomes the hub for information and performance accountability.

At this point, it is essential that we discuss the process performance relationship more fully. As a company migrates from a parochial entity (one with tall, monolithic, silos) to an integrated process-oriented entity, there are various dynamics activated as well.

- The isolated individual begins to function as a member of a team, rather than merely autonomously.
- The team and individual performance measures need to be aligned to provide balance and harmony.
- Accountability (taking ownership through fulfillment) as well as responsibility becomes essential ingredients.

As discussed a bit earlier, the ultimate goal is to drive accountability into the heart of the organization at the process working level. Not only accountability but also decision authority need to be given. When we are serious about process-based accountability; we must empower the internal customer, and their service providers, to define the performance expectations for quality and delivery.

These expectations then become the PPM guidelines from which to assess process fulfillment. To amplify further, if the individual nested team (internal customer and service provider) is the most finite accountability entity, then their process performance measures become the building block for end-to-end business process performance. Further, departmental objectives, in harmony with end-to-end business process performance objectives, radiate the synergy that ultimately determines exceptional organizational performance. It is simple to conclude then that if we are maximizing the nested internal customer/service provider performance elements along profitability lines, we will then be capable of optimizing profitability across the organization. By defining the nested internal customer/service provider performance criteria at a level understood by those required to perform the process, these individuals can begin to clearly perceive how they are contributing to the profitability of the organization. This is a process performance **best practice**.

Process-based measures are far superior to mere financial measures, because everyone in the organization can understand them and, individually, relate to how to achieve their success. A financial measure, such as *return on net assets* (*RONA*), may be understood by an elite sophisticated segment of the organization. However, the vast majority of the organization (in the trenches) can neither understand RONA nor relate to how **they personally** are able to contribute to RONA achievement. Whereas when teams in the trenches agree upon internal customer/service provider measures that are synchronized to maximize profitability objectives, then the broader financial measures, such as RONA, will also be maximized. The PPM then serves as the performance measurement bridge between the working-level components of a process and the intertwined end-to-end process itself.

The recognition that either information quality or process measures have been and/or are being compromised serves as the genesis of a truly effective PPM. When a company does not admit that there has been compromise, it is likened to a person who is sick and either refuses to see the doctor or refuses to heed the advice of the doctor.

Without acknowledging the *process integrity illness*, corrective action and future prevention are unlikely. Once we are able to get to this baseline, we may begin the process that will enable companies to finally realize their potential *ROI* from their business systems.

8.7 Managing Performance Expectations

Managing expectations is one of the senior management tasks as part of their commitment to implementing this PPM process. Companies that are seriously committed to the PPM process ensure that the internal customers have a significant say within the service provider's performance evaluation process. Inasmuch as the internal customers view their nested partner performance daily, they must be considered capable of evaluation of their service provider's performance. However, any given individual may likely be both an internal customer of some team and a service provider to others. Being consistent across organizational boundaries is an essential senior management theme if the PPM process is to have the balanced impact it can across the organization.

The agreement between the internal customer and the service provider nested team must clearly specify the quality and delivery performance expectations. Then the internal customer must work closely with their nested service provider partner(s) to ensure that the measurements are doable and regularly evaluated, that both parties agree on an approach to measuring progress, and that the actual measurement is consistent. Consistency must thrive between the nested internal customers/service providers team as well as across the intertwined end-to-end process work teams. Senior management must actively participate in the removal of any barriers that would inhibit the successful discharge of these performance execution tasks. Senior management must also remove the *politics* from interfering with the trench-level execution (keeping politics lofty ... in the ivory tower, but keeping maximum performance flowing). Yet another senior management commitment driver involves an ongoing task, namely, ensuring that continuous improvement is encouraged, that the PPM process becomes the second nature to every individual (acculturated), and that the process teams are continually motivated and inspired with strong leadership and commitment to the change process. This is a process performance **best practice.**

A reflective pause is appropriate here. To properly establish natural internal customer and service provider work team measurements, the internal customers and their service providers must frequently spend quality time assessing performance fulfillment, identifying ways to improve the process, and creatively thinking of better approaches (daily investment in process engineering delivers the best results). On at least a biweekly basis, the nested team should meet with their process team leaders and senior manager sponsors to discuss the merits of the proposed changes, assign a relative priority, and determine the impact changes will have across the entire end-to-end business process. Consideration (harmony and impact) must also be given across end-to-end intersections with other business processes. Then they must plan to incorporate these improvements by detailing a change strategy and developing workplans to ensure that these proposed changes become a reality, with a sense of urgency. An adjudication forum needs to be operational to arbitrate any end-to-end intersection disputes.

Establishing a standard in the form of process certification accomplishes the following:

- It reflects management's seriousness about information quality and PPM.
- It instills a pride of accomplishment in the individual having achieved the standards of certification.
- The pride of accomplishment develops into a personal source of ownership. Establishing a "personal commitment" is so powerful that it actually has unlimited potential toward goal attainment. Those companies attaining goals are the ones making money and achieving operational excellence.

Employees want and expect their boss, and the leadership organization, to specify the level of performance expected of them. The absence of "expected" performance tends to

1. Put the emphasis of establishing expectations upon an absentee leadership entity.
2. Introduce a looseness, which is perceived by the individual as "okay," but is counter to the natural need for vitality within the organization as established by a person's mind-set.
3. Become the essence and point of propagation for frustration.
4. Lead to other behavioral bad habits such as procrastination, laziness, and insensitivity. It sets the tone for the organization's culture and priority for achievement.

Establishing a performance standard, helping employees achieve that standard, measuring the performance to the standard, and taking the necessary timely corrective action are the lifeblood of a healthy, vibrant organization. **The companies found on the leading edge of performance to profitability are the ones who expect and get excellence from their employees.** These are the ones that have established doable goals and objectives. They are the ones that have clearly defined performance workmanship standards. The act of proactively expressing the desire to achieve a minimum *AQL* in the form of a certification program is the capstone of excellence. This is a process performance **best practice.**

The ever-increasing concern associated with employee loyalty can be reckoned with handily if certification standards are instilled. *The lack of aggressive standards setting has allowed atrophy to infest individual performance expectations.* Consequently, the result has manifested itself in employee turnover and a trend toward decreasing productivity. *Frustration and problem identification need to be replaced with a dose of pride, a performance standard, and distilling the perception by employees that the organization cares about them.*

There is no better expression of "organizational caring" than the establishment of performance workmanship standards and acknowledgment of individual (and group) attainment of these standards. Continuing to treat the symptoms (such as high turnover, low productivity, and low morale) may bring short-term results but lacks the endurance necessary for long-term results. Reestablishing accountability and specifying performance expectations is the formula for long-term success. This process begins by implementing a certification program. Certification reestablishes an individual's self-worth (basis for motivation). As individuals, we would demand that the organization invest that much in us. As organizations, we owe that to our employees.

Change, agility, and flexibility are the themes for thriving businesses over the next decade. Many American companies have become *complacent* in their *attitude toward change* (at least over the proper management of change). Their financial results are okay and they may even be doing better than prior periods; however, they may not be doing the things that will enable them to be acutely competitive in the future. The emphasis upon gaining short-term results frequently compromises the investment in "intellectual energy" needed to affect change, which allow a company to become a top performer in a global economy. There have been *few companies* that have clearly demonstrated their ability to be flexible and change direction and focus, with agility, at will. This is a process performance **best practice.**

"Fast cycle"-oriented change must become the second nature to every worker if the company is to survive global competition in the long run. Changing a company culture to realign their performance attributes along ***process*** rather than traditional measures is challenging, but essential to high-output performance realization and is a process performance **best practice.**

8.8 Process Performance Measurement

So far, we have focused mostly on PPM, the conceptual baseline for engineering high-impact performance measurements. Establishing proper performance measurements is essential to any stellar performance organization. The "financially driven" measurements of the past have prevented the "majority of the organization" from focusing on the top management goals, simply because the critical mass cannot relate to how they contribute toward such lofty goal attainment. A much better approach is to transition to "defining goals and objectives in the form of nested internal customer and service provider built around business processes." We will now switch our PPM term to ***process performance measure* or *process performance measurement.***

If this business process-centered goals, objectives, procedures, and job outlines are properly engineered, every employee may begin to relate to how they personally are contributing to the profitability of the organization. Once every individual is focused upon their contribution to profitability, the organization begins reaping the benefits of synergy by all individuals.

Imagine for a minute, as discussed earlier, how a shop floor machine operator would respond to a question such as, "How are you contributing to the RONA for the corporation?" The operator would feel either that management was crazy or that they deemed the machine operator position as menial. A much better approach is to align performance measurement along natural business processes, which have been engineered to eliminate waste.

PPMs must be developed with agreement from the internal customers and their respective suppliers of services. An overall guideline may be beneficial to define the breadth of tolerances and tie to the pertinent policy guideline. However, the measurement itself must be hammered out by those in the trenches who have to perform the duties on a day-to-day basis. Then leadership and human resources must incorporate these into the company performance review practices.

Accountability and empowerment must be taken and cannot be given. Responsibility can be assigned, but accountability must be championed by every individual. Therefore, to more easily gain ownership of the process and its measurement, each individual team member must participate in the design of the process and its respective measurements. On the surface, this may seem easily done. However, there is likely strong apprehension and adversity festered in the organizational bureaucracy. Therefore, this process change must be championed by senior management, and its timely resolve incorporated into the executive compensation package.

Peer review should be an integral aspect of the PPM. If peers are involved in the design and measurement criteria and peers are the process owners, then the standard will result as meaningful and embraceable by those *living with the results.*

These process-oriented measurements must be developed by all facets of the business. If designed properly in the trenches, it will result in higher level process improvements. If continuous improvement guidelines are championed, then these process measurements will be challenged continually. Waste will be removed, bureaucracy being compressed, and time needed for approvals decreased.

8.9 Retooling Information Resource Management

Information technology (IT) must become a catalyst that assists serious management teams in the process of becoming "fast cycle" oriented. We are inundated by massive amounts of data but frequently have **little to no useful information**. Rather than wasting time looking for the culprit that caused data overload, we must quickly overcome its distracting effect by shoring up the *information generation tools*.

To overcome the "lack of information" requires key users to function as the chief designer of the tools that allow them to run their segment of the business more successfully. Making the transition to chief designer requires a new mind-set in the company-wide IT approach. In essence, information resources, IT, accounting, engineering, production control, and support functions must get out of their centralized offices and into the trenches ... living and breathing with *their* customers.

The organizational change that is needed to properly transition to **information-championed hubs** is somewhat revolutionary to the stodgy structures of many organizations.

In essence, the new organization becomes a series of teams, focused upon business processes, rather than the hierarchical structure defined in the traditional organization chart. These information-championed hubs function as natural, essentially self-contained, and self-directed work teams. When properly functioning, these work teams become the information system architects specifying the information requirements needed to facilitate *their business needs*. Whether these teams are P&L empowered or not is a matter of style. Whether these teams are "gain sharers" and commissioned like the sales staff is again a matter of policy and management style. However, the spirit of this structure is the same as being P&L oriented. There is a company called Valve Corporation, Bellevue, WA, that has this concept as its guiding principle (download *Valve Handbook**).

The information tools, resulting from this new organization, should be responsive to the needs of every member of the team. For example, if real-time MRP is needed, then the team should be allowed to have that tool without an endless amount of bureaucratic delay. If team members want triggered exceptions, rather than laboriously rifling through reports, then they should be given this quickly. If team members want cost of change impacts, they should be given it. Restructuring the organization into information-championed hubs facilitates quick response results compared to legacy process of exhaustive board reviews and time delays of the hierarchical structure.

The tools being defined must migrate from a historical oriented baseline to a *projected future* baseline. For example, instead of telling me I have an unfavorable variance from an event that occurred in the past, the system should tell me how to prevent unfavorable variances as we execute the plan defined for the future.

The tools should be *cost of change* oriented. For example, when an exception comes out of the system, it would display "costed alternatives" rather than a one liner telling the recipient that it is time to commence a "research" project to determine a strategy to fix the problem. In a "cost of

* http://www.valvesoftware.com/company/Valve_Handbook_LowRes.pdf.

change"-oriented environment, the recipient could quickly select the most cost-effective strategy to pursue and then rely upon the system to generate the necessary transactions to conform to the strategy decision. (This is similar to an inventory transaction processing technique called backflushing. In the past it was necessary to process a transaction to issue every item on the picklist, which is analogous effort to a research project. With backflushing logic, I only need to process a single transaction and the system does all the picklist inventory transactions for me.)

The information-championed nested hub tools should be designed to ensure that *internal diagnostics* are performed continually to assist users in the continuous performance improvement process. For example, if I process a transaction for a completed routing step, which results in consuming more time to build than it took in previous lots (outside the allotted tolerance range) or longer than the standard, and/or expected end-to-end process results will be greater than allotted, then a message would trigger to the responsible industrial engineer (IE)/manufacturing engineer (ME) information-championed hub team member workstation to signal the need for immediate assistance. The diagnostic process would involve (1) the real-time synthesizing of recent production reporting history, or, (2) synthesizing standards data and comparing the current transaction result to the recent transactions as they occurred. This makes it possible to enact corrective action before profit margins are too seriously compromised. Continuous surveillance and monitoring controls assist in early diagnosis. The result of the surveillance can then continually use simulation and/or artificial intelligence tools to calculate continual end-to-end impacts of any point-in-time versus expected results of the end-result state.

The tools should be designed with individual *operator productivity* as well as total team and full company skills optimization in mind. Leveraging individual, team, cell, and end-to-end nested process optimization will permit the information to assist the resources attain continuous excellent results. Consequently, information tool design specifications must come from every key team members. Specifications should be designed to optimize the time and energy investment of every nested team process. These "people centered information tools" would be engineered to be consistent with every individual's IWS and PPM, and tracked via diagnostic monitoring. In this day of global economy, any productivity distracters can mean the difference between a going concern and a lifeless company awaiting death.

The IT team member, as part of the information-championed hub nested team, should function as an educator, technocrat, and facilitator. *As an educator*, the IT nested team member is constantly keeping his or her internal customers up to date and competitive by providing insight into new tools, techniques, and approaches toward information handling. *As a technocrat*, the IT nested team member maintains currency on the value of new productivity tools and sifts through the software offerings to identify those tools that will cost-effectively nudge their internal customers to use information as a competitive weapon. *As a facilitator*, the IT nested team member ensures that continuity and rationality across the organization is maintained to prevent other teams from reinventing the wheel. Another key role is for the IT nested resource to serve as the information configuration control representative on the IT change review board.

The new organizational information-championed hub nested team is focused upon the business processes and the responsiveness of the designed tools used to prevent serious departure from expected results, and likely attain stellar high-impact results. Continual surveillance and monitoring diagnostic results help ensure that continuous performance improvements are realized. The focus on monitoring should have as a goal the removal of department-oriented parochialism and truly integrate the business along natural business nested end-to-end process lines.

Management must be the champion of the change process. A *change revolution* is needed to pave the way for a viable entrepreneurial spirit being breathed into the organization. Without management instigation and daily involvement in the change process, the results will be lukewarm at best.

- Management leadership must demonstrate its vision to all segments of the organization. They must help ensure that goals and objectives are clearly defined in terms understood by every employee, and then ensure that charters are developed and executed in a timely manner. Management must be the torch bearer and cheerleader to remove any barriers to success if this change process is to occur within a reasonable time period to ward off global predators.
- Change must be inspired and leadership focused. Implementing change involves a financial investment; therefore, determining the cost of this cultural change is essential to prudent management and the realization of significant productivity enhancements.

A visionary company's quest, for the remainder of the twenty-first century, must demonstrate the following:

- A passion for quality
- A responsiveness to the customer (customer-centered vision)
- Agility to instantly respond to market changes
- An inherent flexibility in tools and processes
- A continuous improvement nature
- Nimble and lean practices
- A fast cycle adeptness with an ability to change direction at will

Managing change velocity is central to increasing productivity. However, change velocity cannot compromise quality of deliverables. In essence, a well-engineered process change model results in the following:

- Leaner thinking
- Quicker response
- Increased agility
- Improved throughput
- Expected or known impact results
- Minimized risks
- Overall improved performance on the sum total of the end-to-end intertwined business processes

Management leadership may be challenged and the company's viability placed at risk if change does not occur quickly. Thriving on change will require the management team to demonstrate unparalleled mastery of the following:

- Leadership
- Acute information tools
- Productivity improvements
- PPMs
- Reduction in time to market
- Quickness of the line

Success and long-term survival may well depend upon a passion and commitment to overcome the "complacency disease" demonstrated in the past and replacing it by breathing the above concepts into the inherent nature of the business.

8.10 Organizational Perspective

With continuing rapid technological advancements, the **potential** for information productivity improvements continues to increase. Technology has provided users with graphical user interfaces where data may be presented in an easy-to-interpret mode providing the user the means to analyze data faster and draw conclusions easier. However, we have barely scratched the surface when looking at the potential that true information may deliver. Technology also facilitates rapid decision support tools whereby the user may quickly discern possible alternatives and select the most cost-effective choice (dashboards). As technology progresses, these capabilities will continue to provide the potential for improved ***individual*** user productivity, as well as increased nested end-to-end process improvement.

However, many of our organizations are still hierarchical monoliths that tend to inhibit "process flow-oriented" productivity. These hierarchies were designed to affect command and control methods consistent with "batch-oriented" or brute force-oriented processes. Approvals were needed because information utility was limited. Data was guarded on a "need-to-know" basis because the information tools of past years prevented rapid exchange and shared utility of real-time events. Leadership was also wary of delegating authority across a broader decision membership.

The legacy hierarchical organization injected "approvals" to improve decision quality, but this resulted in process delays. Data were not well integrated, so the organization relied upon a functional expert, within the specialty hierarchy, to apply his or her knowledge, experience, and filtered down communication to further constipate to the decision-making process.

The legacy functional expert is backed by the political structure within the organization and relies upon time and experience to achieve the rank of hierarchical decision maker. The purpose of command and control structures was to reduce the risk of poor decisions while relying upon experts to consistently deploy their knowledge and past experience. This needs to change.

In the past, pressure was not exerted to make rapid decisions. In addition, decisions were expected to "improve with time" giving the decision maker a window of opportunity to contemplate and reflect upon the decision's merits. However, the business world has changed with increased global competition. Competitive pressure has brought with it the requirement to make decisions quicker (24/7) and improve the quality of the decisions along the way.

The last few decades have dawned with personal computers, and more recently iPhones and iPads are becoming accessible to an increasing number of workers. The time allotted to make decisions is continually being compressed, yet expectations for faster decision throughput being accelerated. However, organizations have structurally remained essentially the same resulting in conflicts and increased personal stress while missing the mark of potential individual productivity improvements.

Let's take a look at some of the key deficiencies of hierarchical organization structure:

1. The hierarchical organization inherently adds time to the decision process.
2. The hierarchical organization tends to promote parochial perspectives within the decision process.
3. The hierarchical organization becomes an inhibitor to fast cycle throughput.
4. The hierarchical organization's performance measures are parochial rather than process-oriented.

We must look closer at each of these deficiencies to really appreciate their distracting impact upon a finely honed management information delivery system.

1. *The hierarchical organization inherently adds time to the decision process.*
   Consider for a moment reasons why a typical employee has difficulty performing at a high rate of efficiency, or rather optimum energy to output deliverable performance. Examples I have personally observed during the past 20 years include things such as the following:
   a. Interruptions
      i. Resulting from a telephone call.
      ii. By a fellow worker asking advice, clarification, or exercising a monotony break.
      iii. From the boss who needs something immediately.
      iv. When an impromptu meeting is called.

   The hierarchical command and control structure cause the organization to interrupt business process flows, by requiring approvals and measures to be localized in order to achieve the hierarchy's goals and objectives. For example, the boss believes that it is their right and privilege to focus the energy of the group upon parochial departmental objectives. Interruptions are not only expected but encouraged to promote "better collaborative communications" within the department. Other process distracters include the following:
   a. Approvals
      i. By definition, hierarchies require next-level approvals.
      ii. Approvals require vertical activity interrupt before process continues.
      iii. Approvals may not be affected immediately.

      Obtaining approvals introduces an *interrupt* in the process flow and frequently requires that an employee to discontinue focus upon closure of the process. The employee must then take on another task only to reintroduce the task awaiting approval into the process flow at a later time, after approval is affected. Approvals by anyone other than the next individual (nested internal customer) within the process value chain introduces delay (time added without adding value; the expression "time is money" coined by Benjamin Franklin certainly applies). I'm not suggesting that all approvals should be eliminated and some are certainly necessary (e.g., grants of authority for procurements); the focus here is to assess the value-added impact that the approval (board) has on the nested end-to-end value chain. Approvals frequently necessitate inordinate process interrupts, tend to spiral into exhaustive analysis paralysis, and seldom result in improving the quality of the process deliverable.
   b. Procedural walls
      i. By focusing departments upon optimizing their self-contained performance, it is frequently at the exclusion of the impact upon other departments and the nested end-to-end value-chain (resulting in local optimization). Procedures frequently are intended to optimize department performance, not end-to-end organizational performance.
      ii. Hierarchies frequently restrict procedural latitudes by exercising authority within the span of control of the hierarchy, rather than enabling the broader nested end-to-end value chain process while promoting optimal throughput.

      Procedural walls tend to introduce conflict where none should exist by natural flows across functional boundaries. Procedures, within a hierarchy, tend to restrict throughput. Throughput has various definitions; however, I have coined the following definition: Throughput is the conversion of a "booked order" into collected revenue. Expanding further then, **maximum business throughput** would be the ***instantaneous*** conversion of a booked order into collected revenue. I realize that "instantaneous conversion" is likely impossible; however, the business opportunity and best practice for the "captains of industry" will be those companies that narrow the gap between booking the order and collecting the revenue at the fastest pace!

This throughput concept applies to all industries and activities as well. For example, a professional services firm (law, accounting, and consulting) has throughput by converting a booked service into collected revenue dollars or a retailer by converting merchandise commitments into revenue dollars. Throughput is a critical business measure, which will be discussed further in Section 7.6.

The key point is that hierarchies add time to the decision process due to the nature of the structure itself and consequential focus upon parochial achievements.

2. The hierarchical organization tends to promote parochial perspectives within the decision process. In a hierarchical organization, the boss frequently controls the calendars of the subordinates. Consequently, subordinates are postured to optimize the energy of one individual, the boss, at the expense of the energy of the broader subordinate group. In a highly political environment, a large proportion of time and resource is compromised to cater to the desires and qualms of the boss.

I recall a visit to a manufacturing facility where this compromise of the group throughput was demonstrated to the fullest. While attending a meeting of key operating personnel, whose objective was focused upon tangible operational productivity improvements, the CEO stuck her head into the meeting and indicated that she was having a garden party and wanted to bring a new product to the party. With no further conversation, the meeting was immediately adjourned and everyone's agenda became focused upon developing the product for the garden party.

Manufacturing lines were redirected and over 300 employees had immediate new marching orders. The hierarchical organization allowed this change to be effected quickly. The cost of this event … customer service schedules were impacted for the prior booked commitments, which were on schedule before the change, and the output from manufacturing for the day was reduced by 60%. The consequences from the garden party activity flurry did not occur until weeks after the event when the operational management team was assaulted because their customer service level took a hit, and costs were substantially greater for that period (unfavorable variance) resulting in morale dropping down a rung.

Unfortunately, this was a recurring experience at the company. The "boss" changed priorities at a whim, which resulted in the operations management team being admonished for poor performance. On one hand, I admire the agility and flexibility demonstrated to get a product developed in short order. On the other hand, the downside had dramatic and short-term unrecoverable consequences to the financials. This then points to the need to balance fast cycle actions with operational performance goals. A simple "cost of change" procedure could have been invoked to determine the impact before dispatching work effort. It is one thing to know and record contingencies associated with unplanned events compared to executing a strategy, only to be surprised by the impacts at a later date. With adequate process model and good prediction tools, a fully burdened cost of change impact may be determined within minutes. This is an organizational cultural change that is difficult to implement and consistently rely upon, but has outstanding strategic performance benefits.

Military organizations have functioned successfully in hierarchies because the parochialism fosters efficient discipline. Yet how many military organizations are chartered with optimizing profits and focused upon customer service? They are a cost center focused upon reducing risk with no profit objective whatsoever.

3. The hierarchical organization becomes an inhibitor to fast cycle throughput.
   As discussed earlier, throughput is the way to measure the conversion of a booked sales order, services, or merchandise into collected revenue. Fast cycle throughput is the ability to make quality decisions quickly with no wasted energy as the process time line progresses toward realizing collected revenue. A fast cycle mind-set is key to exploiting fast response to customers, whether the customer is *internal* or *external*.

   The typical hierarchical organization is not postured to be fast cycle responsive. Rather, the hierarchy breeds upon delays and tends to promote the concept of *heroes*. Everyone has likely been in contact with a *hero*. They are the focused performers who get things done despite the obstacles. They typically are very locally focused and disregard the impacts their actions have on downstream individuals or activities. They have been rewarded in the past for their outstanding individual performance and their ego is refreshed with each "atta boy." Their world tends to be totally consumed by "self" activities and they walk across the backs of their coworkers as they progress up the hierarchical political ladder. Although the hero tends to be "fast cycle" oriented, their fury to succeed (like the cartoon character "Pigpen" in Charlie Brown) creates havoc in the process, although the dead and wounded amass along the heroes trail. Throughput requires that the entire system work well and the heroes activities optimize a portion of the process; it typically suboptimizes other processes.

   Fast cycle throughput mandates that all players in the value chain perform with quality, expediency while avoiding wasted effort. The hierarchy conflicts with fast cycle by process interrupts such as approvals and procedures, which were discussed earlier.

   A fast cycle throughput process is a business performance **best practice.**

4. The hierarchical organization's performance measures are parochial rather than process oriented.
   The nature of the hierarchical organization promotes measurements, which are typically financially oriented. The reason for financial orientation is simple; those promoted up the hierarchy have been donned financial experts by virtue of the elevation. Converting performance expectation into financial terms makes it easy for the executives to rationalize. Unfortunately, very few of the critical mass (those expected to be the heart of productivity results) understand sophisticated financial measurements, and the hierarchy between the executives and the critical mass does not convert the financial expectations into working-level process expectations (many will believe that they are above the task, are not qualified or skilled at the task, or have not been given the priority to perform the task). Consequently, the working-level individual cannot relate to how they, personally, can contribute to profitability. Instead, working-level individuals are fed the line "work harder"; they are relegated to become task driven and transaction oriented rather than *results driven*. In the introduction, we discussed what the shop floor operators' response would be to the question, "How are you contributing to the RONA for the corporation?" The working-level individual would not relate to a RONA measure, and consequently would not be an active value driver toward this key executive goal. For performance to be leveraged, *every employee must be an optimized performer and all performers must be focused upon common goals that are in concert and synchronized across the entire business system.*

Let's look at another example of a hierarchical performance measurement. Various manufacturing companies are, even in modern day, afflicted with what used to be called the "month end shipping syndrome." A typical scenario depicts the organization shipping very little product in the early weeks of the month with 75%–80% of the shipments occurring in the last week of the month. In analyzing this phenomenon, one would conclude that the organization was out of control. However, if you peel back the layers of the onion, you would find manipulation of the schedule in order to achieve the dollar shipping objective (financially driven). The cost frequently translates into compromised customer commitments that are frequently ignored; individual employees are bored early in the month and frazzled at the end and costs escalate (a bunch of overtime, etc.).

The hierarchical organization model, as promulgated by the Harvard Business School, has outlived its usefulness.* The "times are a-changin," and global competitors of the future must rethink their structures, expectations, and measurements. Technology tools may be leveraged to provide the means to achieve optimal performance; however, the structures and management system must be postured to take full advantage of their offerings.

This section discussed some of the key shortcomings emanating from the hierarchical organizational structure. Key issues concerning the hierarchical organization include the following:

a. It adds time to the decision process.
b. It tends to promote parochial decision making rather than looking across the organization for impact (nested end-to-end process).
c. It is an inhibitor to *fast cycle* throughput.
d. Its performance measurements are parochial rather than *process*-oriented.

8.11 Parochial Performance Objectives

A typical company's management team continually strives to obtain maximum performance results from the organization, but seldom attains the level it expects. As a result of missed objectives, pressure to "work harder" becomes a driving theme. In addition, the performance measurements are seldom challenged, and when they are challenged, management focuses upon a particular department or task to take corrective action. The real culprits, measurements that optimize hierarchical functions, are frequently *not* challenged. Consequently, the effort spent to increase productivity, from within a faulty segment of the business, attracts high visibility and high impact. This localized corrective action spurs localized (knee jerk) results, which frequently actually improve the performance of the localized unit; however, overall performance (end-to-end value chain) does not achieve the desired result and a new culprit surfaces. The corrective action activity is then refocused upon the newly identified "weak link" and the cycle continues.

Corrective action may result in organizational team member changes, a cry for new tools and/or the traditional "finger pointing" at another organizational unit which is deemed the "real culprit." This revolving door, over time, may in fact identify virtually every organizational unit as the "culprit." Faces will change, investments will be made in tools, yet bottom-line productivity may increase only slightly, if at all.

During the heydays of the 1980s–1990s productivity appeared to improve as sales volumes rose. Yet, at the end of the heydays, companies critically analyzing "real results" were disappointed

* John P. Kotter, "Hierarchy and network: Two structures, one organization," *Harvard Business Review*, May 2011.

at the outcome. Let's draw upon a sports analogy at this point. Let's take, for example, a major league baseball team whose ownership invested in the "best" talent available, at a very high cost, but, at the end of the season, *the team didn't win the pennant*. The team may even have had the very *best* individual performer at each position, but the team didn't succeed at winning. What happened? Why didn't investing in the best "local" performer automatically result in the best team performance? Each of the individual performers may have even performed excellently, but their "all star" performance was isolated from the other team players' top performance. There is a "truism," which states that an *individual unit performing at high productivity typically results in localized optimization, yet the "total system" outcome may have been merely mediocre*. Even if the total system came together periodically, they may have not been consistent as a "system." Pitching may have had excellent results, yet hitting and errors may have been the weak link. When hitting was hot, errors may have been high and/or pitching weak. Consequently, what appears on the surface as a superior team misses the mark, yet a team with inferior individual statistics can become the pennant winners to the amazement of the management staff and to the joy of the fans.

Popular tools used to evaluate productivity results in the heydays were activity-based costing and cost segmentation, among others. These tools were valuable to isolate cost drivers that contributed to poor results. However, like the hierarchical structure, they tend to be parochial rather than cross-functional end-to-end process oriented and frequently miss a plethora of the nonvalue-added gaps within the organization.

As illustrated above in our sports example, other companies can have similar consequences. A company with highly paid industry talent may show mediocre results. Frequently, management stands dumbfounded and launches "just another project" to do one or more of the following:

- Embark on an effort to work harder.
- Change the players.
- Invest in technology tools.
- Change the local processes (not end-to-end oriented).
- Educate the critical mass.

Yet, seldom does management execute a "total system" approach and challenge the "team measurements" across the organization. Instead, management typically settles for a rather "parochial," or localized, perspective of an organization, which nets little "true productivity" benefits.

It becomes obvious that self-serving activities benefit one department, but it is also obvious that the organization "as a whole" receives little benefit or maybe was deterred as a result. The objective emanating from these self-serving individuals frequently has the "best interest of the organization at heart"; however, the approach and result have little total benefit.

If we were to view the parochial departmental objectives and their deployment, in most organizations, we would see a similar approach and a similar result. Therefore, these companies that were chosen as examples certainly are not unique. In fact, these examples could have another name or product/service label assigned and be viewed as common activities in most organizations. Although observed from a distance, these examples appear comical, yet, if the design intent were critically studied, each of these examples would be expected to yield "high productivity" results based on typical ROI calculations.

A critical thinker would likely ask the question: "Why?" How could it pass the scrutiny of a serious management decision process? *Why* didn't a *totally committed process champion* look after the interests of the organization as a whole? *Why* didn't the executive management ranks see the folly of the effort early in the implementation process and step in to cut the losses from further deployment?

When the layers of the onion are peeled back, we get at an underlying reason for this obvious omission, and frequently, many deployment activities had high expectations but frequently netted minor or negative productivity results. One key reason is the hero! Many businesses are built, fester, and frequently die, relying upon heroes. Let's look at the hero syndrome further.

Small emerging organizations are typically driven by the "vision," energy, and dynamics of a hero, usually the owner. Clearly, the attributes of the *hero syndrome* are as follows:

- The hero is stoked by past performance rewards.
- We can pick the hero out of the crowd of corporate bureaucrats.
- The hero is considered the cream that floats to the surface.
- The hero always comes through at the last moment (frequently violating most policy directives) in an effort to "save" the company from disastrous financial results.
- The hero gains a special reputation that is honored by all.
- The hero frequently evades the formal system, because for many years the individual has been able to come through in the clutch every time; therefore, he or she should not be required to suffer through the rigors that others have to follow.
- Frequently, as master of the informal system, the hero categorically knows the limits.
- Every time the hero comes through applause abounds.

Working and managerial levels alike are ecstatic that the hero has "bailed them out" again, thereby reinforcing the tactics employed by the hero. "Aspiring heroes" observe the tactics used and shrewdly foster means to emulate the hero and rise to the heroes' stature. Because the hero is praised with grandeur, he is treated like Prima Donna and placed on a pedestal above the standard for operating practices. Any compromises the hero caused the organization to endure along the way are merely written off as "necessities given the situation." Situational "mores," policies, and practices establish a new compromised standard, disregarding the cultural standards of the past. After all, everyone admires the hero and will look the other way when deviations occur.

With this descriptive hero baseline, we must look at the organizational attributes, which encourage heroes to continue this modus operandi. If a critical thinker dissects the typical company path to promotion, one quickly recognizes that the measurements are skewed toward becoming a hero. The promotion characteristics weigh heavily upon "What have you done for me lately?" and, of course, the hero heads the list by having avoided the most recent disaster. These performance guidelines are natural, because most individuals in high ranking management themselves participated in the "hero syndrome" and are well attuned to the sacrifices the hero must make to achieve the latest results, even at the cost of the team effort.

Periodically, a company's product/service quality had to be compromised to mitigate risks of not allowing the hero to succeed. Frequently, one or more customers are compromised for the "good of the whole" and the determination as to *which* customers were compromised usually boils down to their relative "revenue impact." Those customers with the right "revenue" contribution were *winners* and those without the higher revenue contribution *lost* (even though the high revenue contributors may be the worst at "paying on time," but we made our revenue numbers at "all cost"). Cash flow is sometimes compromised, inventory turns compromised, and customer service compromised (except for the "heavy revenue hitters" who may get product early). Quality may be compromised, "robbing Peter to pay Paul" is the norm, and usually the hero doesn't have time to document these "dynamic reallocations," so the formal system becomes compromised although data is exhaustively corrupted.

Heroes quickly learn that they are "politically correct" by obtaining *results at all costs.* Consequently, they are the "darlings" and at the core for the future of the organizations. With this in mind, let's look at the impact the hero has on *real results.*

1. Because corporate tradition has rewarded and promoted heroes for the past, their shortcomings are usually subordinated in favor of the recent success. The hero can usually only be a successful hero by breaking out of the pack and suboptimizing "team results" in favor of singular activities.

   Peers recognize that management favors heroes. Therefore, whatever *Rah Rah*'s that are spoken geared toward promoting *team results* are viewed as lip service by the laggard team members, who, no matter how hard they may promote team progress, are overshadowed by the hero every time.

   Over time, team ambition and zeal are eroded and the expectation surrounding the hero becomes self-fulfilling prophecies. The hero's ego is further fueled although the team players (the critical mass) productivity progressively wanes.
2. *Heroes must compromise to succeed.* Over time, they recognize what are acceptable compromises and what are not, and they normally comfortably maneuver on the edge of compromise at all times. Inasmuch as management applauds and rewards the hero, who frequently is the instigator of compromise, *a common employee* who looks at the recurring compromise process by the hero *must conclude that management endorses compromise.*
3. *Heroes exploit opportunities.* One of the key attributes of an entrepreneur is the degree of freedom with which an entrepreneur navigates without permission. A hero is typically an entrepreneur to the fullest. Opportunity exploitation is healthy as long as the hero operates within the confines of the formal system (adheres to formal policy, procedures, and system guidelines). In the "heat of the battle," however, the energy and drive of the entrepreneur (hero) elevates itself beyond the constraints of formal system (barriers). Consequently, compromise will again win at the expense of the formal system.
4. *The hero negotiates brilliantly within the "politics" of the company.* Leveraging political distractions, the hero can use these distractions to sap energy from foes and promote a personal agenda by "default." Promoting a personal agenda usually precludes the promotion of a "team"-oriented agenda.

On the surface, the traits of the hero seem admirable. How often have we heard about the value the hero contributes during a *crisis.* However, let's take a critical look at the hero as it relates to the "total system" impact.

- Whereas the hero excels when individual performance is needed, our heroes can't possibly be delayed by being a "team player." The team only *slows* the hero down. Therefore, a narrow perspective obviates the good of the mass.
- The hero is driven by individual incentives (and rewarded likewise) and has little incentive to perform as a team player.
- The hero is disguised by an appearance that allows him to look like a leader. However, without team promotion, leadership skills soon conflict with team fulfillment.
- The hero is totally consumed by expending excessive energy toward achievement of the localized goal at the exclusion of all other goals.
- The hero thrives on individual accolades and cannot share glory with others; therefore, self perpetuates the autonomy of individualism.

- The hero supports other aspiring heroes as well as other "departmental heroes," who can leverage the "away-from-the-crowd" activities to the betterment of the hero posture.
- The hero is motivated by and motivates others by promoting parochialism. Parochialism fuels autonomy and hero autonomy distances the hero from "overseer" or perceived overseer assault.
- Compensation programs are predominantly focused upon rewarding the hero for exceptional performance.

"Team-oriented" reward systems are not as lucrative, agile, nor as timely as the individual reward avenues. Individual reward systems focus upon individual and localized gains rather than broad-based critical mass gains.

A diversion is timely at this point. Let's consider the compensation issue for a moment. If we analyze a typical organization, it is pyramidal in structure. There are few at the top of the organization, and as an individual elevates oneself from level to level, there is typically an increase in financial "reward." This individualized financial reward aligns itself well with the hero mentality. If we analyze risk, until very recently, the higher up the organization a person attained, the less likely they were of losing the job.

At this point, an acute reader may speculate that the author were an *economic communist* who promotes the concept that the *masses get the rewards*. However, critical analysis of any economic order shows that the very "elite" always win and, then, there are the "others." Understanding that there will always be *the elite* leadership layer, how can we still leverage the skills, talents, and energy of the masses to gain a WIN/WIN structure for the organization and the elite as well? How does it relate back to our hero or *all star*?

Let's consider, for a moment, some way to promote and reward the hero *only through* the leveraging of the energy of the masses. The hero's drive emanates from the reward. If we restructured the reward, de-emphasizing the individual performance and emphasizing the "total system" performance, wouldn't the hero, his work group, and the organization, as a whole, be better off? What better incentive for our hero than to manifest their "report card" through the achievements of the subordinates, peers, and internal customers?

One may ask if rewarding the hero is possibly doable given the hierarchical organization structure today. We don't know because our only baseline is the existing hierarchical organization structure. However, we do have increasingly solid evidence that companies experimenting with "teaming" are making significant inroads, in shorter time and with less cost, when team solutions are exploited to their fullest.

Let's take a look at a historical initiative, *total quality management* (*TQM*). TQM stresses three key concepts pertinent to our discussion: (1) quality at the source, (2) elimination of waste, and (3) promoting team and lower level decision authority.

We can look at each of these keys in more detail.

1. *Quality at the source*: The intent of this key is to make every individual responsible for their own quality. Quality is too costly to "inspect in" a product or service. Therefore, the process of building it by every individual becomes essential. Manifesting this philosophy requires a quality mind-set, which must be breathed into the culture of the organization from top down.
2. *Elimination of waste*: It should be pursued in every facet of the business. Unfortunately, most of the time this objective is pursued only with visible entities, such as inventory and paper. One of the greatest potential savings from eliminating waste has barely been tapped, namely, waste of energy. One of the underlying themes of this thesis is the need for radical

exploitation of ***individual productivity*** (i.e., productivity among the masses). Wasted energy is the highest order of disease that wears down productivity. If the reader acquires nothing more than a burning desire to commit to taking an *order-of-magnitude leap* into exploiting individual productivity, then this thesis focus has been successful. A leading element of *productivity energy* relies upon individual energy. Energy is typically a company asset that is neither measured, managed, nor seriously considered.

Productivity improvements are almost categorically oriented toward *working harder* rather than exploiting the optimization of critical mass energy. Allusions touch on the edges by exposing phrases such as "work smarter, not harder," yet the root of the matter is seldom pursued. The symptoms of the disease are expressed by omissions such as the following:

When was the last time management decision makers asked the *common worker* any of the following questions:

What information tools could you use to help you improve your productivity?

What barriers could be removed from your job to allow you to double your output?

How could workflows be streamlined, procedures eliminated, approvals removed, and so on which would allow you to improve a quality of deliverable to your internal customers?

The ability for an organization to perform at its highest productivity levels is certainly influenced by a company's stock price, assets, revenue, cash flow, inventory, personnel, and so on. Yet the author believes that the greatest influence factor contributing to a company's success is leveraging the throughput of capacity of the intellectual energy level of its composite personnel base.

3. *Promoting team and individual accountability*: At this point, we should start to gain an appreciation for the devastating impact the hero or "all star" brings to the organization. The individual hero, no matter how talented, bright, or, on the surface, effective still adds up to a resource of *one*, whereas the organization resource base is substantially larger than one. The trade-off experienced becomes the value of one optimized resource *will never exceed* the value which could be contributed if the critical mass were optimized. Now, one may ask, what if the entire organization was made up of *all stars*, wouldn't this be the ideal? First, how many times has an organization started out with an idealistic objective to hire "only" all stars merely to fail in the process? Second, does an entire group of all stars ever function as an oiled, efficient team? Third, why not focus and leverage every team player's skill to obtain equivalent results?

Each organization has an untapped gold mine of talent awaiting to be unleashed given the proper focus, unrelenting motivation and determined leadership. This is where the real power of the organization can excel. However, as long as the hero reigns, the team will subordinate the vast resource contribution in favor of the one.

Our mind-set challenge at this point is, how do we rechannel the zeal and spirit of the hero to mine the untapped potential of the masses without losing momentum?

8.12 A Better Perspective—Value Streaming

We have now challenged the reader to intellectually construct a strategy to exploit the critical mass talent base without losing the zeal and visionary attributes associated with the hero or all star. To begin this monumental task requires us to "eat the elephant one bite at a time." To eat the elephant requires us to recognize that the behemoth structure is, in fact, an elephant. Therefore, we must regress to the fundamentals and define our beginning point. This mandates rallying support to develop a vision and inspiration for change among the critical mass and develop champions of the change process between the participants (this typically does not require much effort, inasmuch as the working-level folks have to deal with the stench of broken processes daily).

To define our beginning point requires a critical assessment of the business processes as they function today.

There's no better advocate for change than lighting a fire under those individuals who must *wallow in the mire of mediocrity* on a daily basis. With this understanding, let's tap the resources of our functional experts (those living day to day in the trenches with the business process). Who is more qualified to identify the weaknesses of the business process than our functional experts who are actors on a daily basis?

At this point, we are about to challenge the traditional "role" of a typical manager of the area through which the business processes flow. Let's look at tradition. Our functional area manager has been elevated to the authority with "discretionary" privilege to reign with an iron fist upon the area from which they have control. We use the term "control" liberally, when in fact, most managers neither control nor would know how to control the process if they were able to free up the time consumed by crisis management to attempt to control. Therefore, we must recognize that control is not based upon realization of activity, but rather deals in the realm of perception, at best.

Now, we're going to ask the manager, who has the charter to control, to give up their executive privilege, and to pass along decision authority to those who have not "earned" this privilege. Not only have the critical mass not earned this privilege, but the managers frequently view themselves as the most robust technical experts and absolutely most qualified to make "any and all" decisions associated with the functional area. Therefore, if managers view themselves as the best technical resource, anyone else proffered with making the decision will be perceived as never making the quality decisions needed.

A second point regarding the manager is as follows: If the manager believes that he or she charter is to be in control and, obviously, does not control the process, he or she becomes extremely insecure to relinquish the "seal" in fear that his or her inadequacies will surface and will be demoted or discharged. If the manager's company, in the past, has "purged" ranks periodically to separate the "wheat from the chaff," this activity adds fuel to the argument by the managers that any weakness in control will result in drastic, painful circumstances. Now we're on the horns of a dilemma. How do we convince the management team that control needs to be driven down (delegated) into the bowels of the organization without surfacing the fear that the manager might lose their job?

In years past, corporate management could point to the IBM, GM, and other Blue Chip organizations, and confidently show examples where the management team truly had job "security." Of recent years, however, these stalwarts of industry have shaken up middle and executive management and one can no longer point to job security. This, now, adds a little spice to the challenge. How do we value chart the "real" operating practices of the business processes? How do we get an honest baseline, with all the delays, faults, and bureaucracy that is standard operating procedure (SOP) for the working-level employee? How do we expose those areas that were traditionally icons of the "elite?"

If our departmental manager, or aspiring manager, even gets a hint that this exercise will produce disastrous personal results, the true process value will never be portrayed. The manager and working-level employees must feel secure that peeling back and exposing the layers of the onion are mandatory for organizational and personal success. They must believe that only when we are able to expose cycle time delays, exhaustive signatures, and bureaucratic cost additions, will the organization truly succeed and be able to reinforce job security.

The operational team must recognize that business operating practices must change if an organization is to succeed in the future. Operating practices that made the organization successful in the past will probably require significant modifications or replacement, if it is to be competitively successful in the future. *This means radical change*!!

A typical company is not postured to assimilate "radical change" into their operating practices. As pointed out previously, *few organizations are able to be flexible and change direction and focus at will.* With even the hint that the change process will become just another project (like all others that began and failed in the past), causes employees to resign to defeat before meaningful change can be affected. *Lethargy, then, works at eroding the sinew of the organization and any momentum that may have been amassed soon dwindles, and through default, change becomes ineffective.*

Let's regroup … the main reason most significant change is unsuccessful is because it is typically nothing more than another project or event. *Change must never be considered a project or an event, but rather must become an expected process of continual refinement over time.* Change must be *championed* by every individual, *no exception*!! Processes must be engineered to a new standard, leveraging flexibility, agility, lean, quick response, and other precepts mentioned earlier.

Every individual must believe that they have an obligation to instigate the necessary change that impacts their daily activities to achieve higher levels of productivity. Complacency toward change, by even a small element within an organization, becomes the disease that erodes potential productivity improvements. Change must be considered the *norm*, not the exception.

Process change must be ***delivered*** daily with formal expectations focused on the rate of change rather than *IF* change is to occur. Expected change must become as normal to the company culture as expected profits, revenues, or quality.

How does an organization go about establishing *change as a normality* within its culture? There are a plethora of written words surrounding change and it is not the intent of this book to expound upon principles well presented. However, it would be appropriate to highlight some key concepts and emphasize the need for commitment to change by every resource within the organization.

1. *Leadership must* not only advocate change, but must spend a goodly portion of "quality intellectual energy" upon the process of change.
2. The organization *vision*, guiding principles, policies, and procedures must be engineered to not only assimilate change rapidly but instigate change.
3. *Agility* must be a watchword on the lips of every employee. Every employee must become an advocate of change and identify methods to incorporate "healthy" change continuously.
4. *Change must* not occur merely because there is an objective to change but must be based upon the principles that *promote elimination of waste*, productivity improvement, and competitive posture improvement.
5. Change *must occur* and be synchronized with a "*total system impact*" in mind.

Let's look at each of these keys a bit closer.

Leadership—If the organization is not categorically convinced that the leadership is *committed* to change, then the willingness to instigate change will be compromised. By the way, I recall George Plossl defining commitment this way: "I had ham and eggs for breakfast, the hen participated *but the pig was committed*!!!"

Every employee in the organization must *believe* that every leader in the organization is supportive of change, and not just the change that the leaders instigate. That's a lot of "every's," but it's essential that the sum total of the intellectual energy of the organization be driven to identify opportunities for healthy change. They must rapidly be able to initiate a doable plan for change. The organization must be able to assimilate change and reap the benefit of change quickly. Without a committed leadership to facilitate change, energy to affect change will dwindle.

Vision—The vision from the leadership of an organization must also become the vision of all employees. For a vision to be converted into a drive that focuses every employee upon incremental change achievement, it must be understood and converted into every individual's vision as well. Every individual must be afforded the opportunity to influence the continually changing content of the vision. The vision must be a proper balance of customer focus, employee focus, financial focus as well as direction focus.

Agility—The continually demonstrated ability for the organization to assimilate positive (healthy) change into the company's operating practices with optimal impact. Agility should infer a *quantum leap*, not a gradual rate of change. *The proof of demonstrated agility results is "unchallenged market leadership."*

Productivity—The productivity objective should be marshaling the energy of masses to obtain order-of-magnitude improvements toward a competitive edge. Every individual in the organization must be a champion to ensure that productivity improvements occur *every day*! Any hesitation among the ranks distracts from the "collective" potential. With enough distraction, competition (whether external or internal initiated) toward business synergy is compromised. A business' productivity is only as good as its weakest link. If continual change, as demonstrated by continual improvement, is not achieved every day, then that day's opportunity is lost ... never to be recovered in the future.

Total system—The total system is not referring to a computer, but the sum total of all energy expended upon the aggregate of value chain end-to-end processes. Whether the process is decision making, performing a transaction, educating an employee, identifying ways to improve quality, and so on, the impact of every activity must be system oriented rather than parochial oriented. The total system should be continuously assessed for reengineering.

A critical thinker may ask, what does change have to do with value charting the existing processes? If the right mind-set is not breathed into attitudes of the organization and the stage is not set properly, then the potential that is possible in the value-charting activity is lost. Identifying the true posture of the organization and leveraging the intellectual energy of every employee are essential if this time is to be proficient. The objective of value charting is to identify what currently is being performed in sufficient detail so as to determine the following:

- Activities that add value and are essential to ensure timely customer fulfillment.
- Activities that do **not** add value but perform control reviews (approvals).
- Activities that do **not** add value and lengthen the process time.
- Each of these activities provides an opportunity to improve. Let's look at them in more detail in reverse order.

8.12.1 Activities That Do Not Add Value and Lengthen the Process Time

A critical thinker may ask, why has this been allowed to continue? First, it may have been initiated as a "one-time" event but became tradition and stayed. Second, during the "growth" periods of a business, "overkill" was okay, but "never allow a significant event such as a part's shortage to occur." Therefore, additional checks had a tendency to be added (to ensure that those "never to occur" events were prevented). Third, during growth times, politically attuned individuals looked for ways to amass numbers (empire building), which was their basis for continued power. The last time that a typical organization dissected what they did, in sufficient detail to identify waste, was long ago, if ever.

An example was an electronics government contractor we were engaged to help improve processes. The company had never critically assessed the value-added processes. In fact, years ago they were compensated by "cost plus" contracts. Therefore, it became fashionable to identify ways to broaden the cost base so the "plus" base could be broadened. The company had the opportunity to expand revenue and profits, not by producing more units or working smarter but by increasing costs. This broadening base was disguised by using many techniques including "change of scope" and follow-on business.

Our value engineering project involved value charting four significant processes: the engineering change process, the manufacturing process, the new business proposal process, and the cost management process. We formed a reengineering team consisting of a process engineer, a functional area team member, a cost analyst, and an information systems analyst. The resulting value charts highlighted layers of approvals, 80% of the process time was consumed by nonvalue-added activities, and what should have been parallel process, operated in a serial manner. The process improvement team was easily able to identify the improvement opportunities. Because the team consisted of a blend of process engineer, user, cost, and system perspectives, the corrective action process was affected rapidly. In a matter of a few weeks, these improvement opportunities were identified and change could have been affected quickly. However, the clock had run out for this division, and a corporate decision was made to shut the facility down. Had the process occurred earlier, the closure would most likely have been prevented.

8.12.2 Activities That Do Not Add Value but Perform Control Reviews

The nature of the hierarchical organization requires approvals to ensure that tasks are performed properly. However, seldom is anything but a "rubber stamp" accomplished by the insertion of their approvals. Either the approver is too busy to invest the energy and research necessary to qualify the approval or the approver has confidence in the initiator to approve "as is." This is not to say that a few anomalies are not caught in the approval process. However, is the investment associated with the approval process to identify anomalies sufficiently significant and corrective actions sufficient to warrant the throughput delay associated with the approval process? It has been the experience of the author that most approvals do not uncover sufficient significant findings in all but a few activities. There are certain investment events and certain contract events that must have another set of eyes to review. The expertise associated with this type of review can normally add value to the process by avoiding catastrophes downstream. However, for the activity to proceed to the approval process, before having a significant anomaly identified, is, in itself, a faulty process. Significant elements should have been tested earlier in the process by initiation guidelines, computer scan, or software configuration logic synthesis. Properly engineering the process is essential and the means to eliminate subsequent approvals. This all reverts back to process design, process ownership, and value process flow.

8.12.3 Activities That Add Value and Are Essential to Ensure Timely Customer Fulfillment

Although an activity is essential to the proper operation of the business, is the process as efficient as possible? Value-added activities can frequently be improved. They seldom have the cost savings potential that the nonvalue-added activities have, but they can contribute better value. Not only do these value-improved opportunities need to be highlighted, but they need to be pursued with every energy segment challenged for possible elimination or refinement. If the current process is not documented, it might be appropriate to value stream using flowcharts, it is still a good way to identify how the business process works.

To recap the foundation baselines for effective flowcharting include the following:

Change—It must occur rapidly and it must be the impetus from which the value chart activity relies.

Honest assessment—Covering up opportunity areas is counterproductive to the value charting intent.

Stratification—Developing importance criteria to leverage the value charting result.

The best approach to value charting was discussed earlier. However, it is recapped as follows:

Value charting team—Develop a cross-functional team, with representation from the attendant subject matter experts' technical skills area, who are in tune with identifying opportunities, such as process engineers, representation from the systems area (information management is usually at the core of the opportunity), costing area, the functional user, and other areas as appropriate.

User representatives should consist of individuals with insight and clout, yet practitioners. The teams must provide the energy to formulate the change focus. They must be empowered to highlight "sacred" or politically untouchable processes, and be comfortable recommending their elimination.

These value charting team members must be intuitive change agents practicing independent thinking and continually focused upon the customer's satisfaction. As "free thinkers," they usually are able to make decisions with wisdom and can visualize a new "lean and mean" organization, which conducts business in a revolutionary new manner. However, these individuals must recognize that the organization they are dissecting consists of some individuals who have invested their careers into finding ways to become comfortable. Any recommendation for change is an attack upon their turf.

Value chart convention—The primary purpose for value charting is to put on paper that currently occurs within the business processes. Fragmentation, resulting from functional structures of a typical organization, leads to process delays with ROI results frequently suboptimized. The objective of value charting is to identify ways to compress the time a process takes to complete while eliminating as many touch points as practical and substantially increasing process throughput, if possible. With this objective in mind, the convention used to document process fragmentations and delays is not as important as the identification process itself. The convention used should be consistent and follow standard document flow standards.

Value charting elapsed time—Placing on paper a pictorial view of "what really occurs" will be used as a baseline from which to identify opportunity areas. The time taken to perform a quality job needed to document actual processes should be completed within a reasonable period. However, compromising quality and comprehensiveness is foolhardy.

To reiterate, the reason we spend time documenting the "as is" is twofold

To provide the reengineering team the ability to identify "true opportunity,"

To accurately record what is really happening in the trenches.

The time invested to document the total process, along with every step or bureaucratic interlope, is essential to a competitive postured redesign of all processes. If a company chooses to compromise, at this juncture, it will certainly point to compromise when the "real issues" surface.

One more key throughput improving ingredient is time. This process should be more than an exercise. A company truly dedicated to a revolutionary approach to competitive posturing will invest executive management review "time" in assessing the "way it really is." Just in case someone misses the point, the CEO should be actively reviewing the results of this process and spending quality time, ensuring that this activity stays high on the priority list of survival strategies for the upcoming decades.

Value chart deliverables—The deliverables from this activity should consist of, but not limited to, the following:

An accurate perspective reflecting how business processes truly happen each day

An early identification of the low hanging (lush) fruit that can be harvested quickly, with immediate corrective action plans

A baseline from which to develop an engineered alternative process that eliminates 80%–90% of delays, approvals, and time to provide deft value to the customer

A proposed or "to be" recommendation on how the organization can take a "quantum leap" toward productivity improvement

Identification of ways to use "critical mass" talents to make local decisions

Deliverables are only as good as the decisions needed to affect the change roadmap quickly. Of all the areas where top management can take an active role, after commitment, this is at the top of the list. Management must be decisive, supportive of immediate change, and provide leadership to affect the change roadmap immediately. Any hesitation by management at this point is weakness and a "kiss of death" to quantum leap improvements. Any hesitation is perceived, by the critical masses, that change momentum is lost and leadership has reverted back to "business as usual," compared to a quantum leap resolve.

Without immediate championship among the executive ranks, for the need to change immediately, it is viewed as a compromise. Without strong leadership toward change, by executives, is an affirmation that other tasks are more important. *Nothing can be more important than marshaling the entire organizational energy to affect the revolutionary change that surfaces from this process*!

Let's reiterate the ingredients needed to be successful:

1. Committed *leadership* to actively lead the charge for *immediate productive change*
2. Definition of a *vision* that takes a *quantum leap* toward productivity improvement
3. *Agility* to assimilate radical change immediately
4. *Productivity improvements*, measured by quantum leaps, that occur as "all employees" focus upon customer needs, cost reduction, and waste elimination on an end-to-end process basis
5. *Total system* where every individual can clearly visualize how they are contributing to the profitability of the organization

8.13 Refining, Streamlining, and Reducing Cycle Time

Change must occur by reducing *crisis* urgency with organized or methodical actions. The intellectual energy exuded, while operating in crisis mode, may be rechanneled into engineered stability. Many organizations thrive on chaos and crisis. Organizations have become good at rising to the occasion of a crisis. People feel fulfilled by overcoming the impending disaster of a crisis. Therefore, capturing this enthusiasm and redirecting its energy flow toward revolutionary change is a bonus.

Because most organizations have mastered the process of rising to the occasion of a crisis "task," if we can direct the same mental focus and energy toward elimination of the source of the crisis process, we can achieve immediate benefit and an order-of-magnitude leap toward vision accomplishment. Individuals and organizations are very "task" oriented. Therefore, biting off more than the surface chores will require an organized methodology to affect the root cause process changes needed to take a quantum leap toward productivity improvement.

The process of taking a quantum leap, in productivity improvements, is delicate. Radically empowering the organization to define a new rule book is outside the comfort zone of every employee who is not a hero. Therefore, we need to leverage the hero energy and redefine the rewards such that successful achievement occurs only through engineered "team" effort.

Let's recap this point: to "soar like an eagle" and rise above mediocrity mandates a radical departure from the norm. The organization has been conditioned, over time, to accept the fact that only the "elite leadership" can possibly contribute to their success. This conditioning must be revolutionarily changed to encourage every individual to participate in the reengineering of the day-to-day process activities. The objective allows the organization to improve by 75%–90% immediately!! To accomplish this ambitious objective requires leveraging the "engineering and emotional energy" of every individual within the organization. Exploiting this broad swath fuels momentum by the "critical mass."

Using the "as is" value chart as a baseline, every employee should be encouraged to identify significant process improvements. Rewards should be based upon how quickly agility can be exercised to reap immediate benefit. Opening up these creative flood gates to the critical mass, the resultant comprehensive focused energy can become overwhelming within most organizations. This radical energy differentiates *true leadership* from the leadership drones. True leadership talent is demonstrated by those who can bury the processes of old and use a clean slate to define the processes of the future. The bureaucrats and drones are magnets to the processes of the past (the warm and comfortable), whereas the true leaders take an aggressive posture to affect changes and take that quantum leap toward productivity throughput.

How is this radical metamorphosis accomplished? The astute leader recognizes that leveraging the talents and energy of the critical mass can bring the *quantum leap* needed. If the leader is the only active instigator of radical change, then progress will be slight and limited by the intellectual energy level of the few. Refining, streamlining, and reducing cycle time can best be affected by the critical mass. Therefore, using these critical resources as both "engineer" and implementer is key to taking a quantum leap. However, how do we marshal the creative energy of the critical mass?

It is interesting to talk to working-level people and feel the enthusiasm and energy when they convey the massive attempts they have made to make management aware of "how to improve." However, it is also sad to hear how often these functional experts expend energy on a "nonreceptive" ear. Consistently, working-level personnel believe that other perspectives are always more important than are suggestions from the functional experts in the trenches. Consequently, their zeal and enthusiasm to instigate change recommendations are smothered by lack of responsiveness. Functional experts can be a wealth of inspiration and serve as an untapped resource to create a design that yields significant change.

Are working-level resources sincerely given a chance to instigate ideas for change? A truly committed management team, who thrives on change, must commit their working-level resources to become process design changers.

How much time should be committed? Should they be encouraged to redesign at the end of the day when all their work is done?

Parkinson's law states that work will expand to consume available time. Allotting "the end of the day," when mental energy is most likely exhausted, is certainly not a best effort. Then, what if someone were to propose allotting 20% of every functional individual's day, during early hours (shortly after the workday begins), toward the redesign of the function? This would be a valiant start, but have working-level folks gone to engineering school and developed independent activity redesign skills? *Highly unlikely*!! How about the cross-functional aspect of the redesign of a process ... will functional level resources consider upstream and downstream impacts in the designs?

These questions are certainly challenging: how do we tap the working-level functional expert engineering skills? Let's go back to time commitments. If management determines that it is important, management will certainly carve out time for things that are important. If it is important, 20% is probably a reasonable level of commitment. Now, the teams. Individuals, in and of themselves, are probably not readily skilled at self-initiated redesign disciplines. Let's place a logical group of internal customers together with their nested supplier of services and call in other functional resources as needed. Let's assign an IT user-oriented resource and deploy the skills of our process engineer. Our group is growing and this 20% commitment is now getting costly. If that is the management attitude and level of commitment, shut the process down now! The only outcome of a wavering commitment is disaster. Remember that the hen participates but the pig was committed. We must be committed if we are to gain the quantum leap benefits from our untapped resource potential.

Now that we have a true commitment and endorsement. Let's look at the structure of the team. Every functional expert must feel comfortable and not intimidated by other team members. Said another way, every participant must believe that they "are among equals." One way to gain confidence, by working-level employees, is to ensure that the organizational job title is left outside the door of the meeting room as the team convenes. How about leadership?

Team leadership is important; therefore, the team must designate a leader, or better yet allow every participant to serve as a leader at different times. There have been vast studies conducted on organizational dynamics, which might suggest that a "loose" structure is doomed for failure. Consequently, we're going to relegate this point more as a matter of style reflecting group dynamic personality, which has worked within the organization. But of key significance, if functional experts feel intimidated, the potential benefits wane exponentially.

Support must emanate from executive ranks without overshadowing the working-level experts. Assimilating and synergizing both executive and working-level employees is not a small challenge. The best approach, if practical, is to have executive participation as an integral member of the team. However, intimidation may distract results. An alternative is to assign one or more executives the job of cheerleading ... severed from process performance accountability. In other words, the executive provides inspiration, instigates creativity, and facilitates barrier removal, but does not participate in assuming process progress merits whatsoever.

Leadership by example is essential. I have found that companies most successful at managing change have leaders who embrace the concept of being an expert example of the change process. These leaders roll up their sleeves and lead the charge to streamline. They participate in identifying streamlining opportunities. Attributes of a "refining, streamlining and reduced cycle time leader" usually include such things as follows:

- Strategic thinker
- Team builder
- Spirited

As we look at these attributes in more detail, we recognize that they are the same attributes demonstrated by the movers and shakers within the organization.

A *strategic thinker* is someone who can visualize the organization as if the changes have already been implemented. They can relate to the potential problems that will pop up as the organization negotiates the change process. They are individuals who can think outside the "current methods" box visualizing how the processes would work with a 70%–90% less effort.

A strategic thinker recognizes the importance of, and are adept at, identifying "systems champions." The systems champions are the process owners who have a vested interest in improving the process.

A *team builder* is an individual who enjoys rallying team inertia and driving team performance to new levels of excellence. A team builder acknowledges achievement only as it relates to the total team activity. A team builder is motivated to isolate "weak links" within the team and jointly develop corrective action strategies to eliminate the weakness. A team builder ascribes to the belief that synchronized team performance relies upon its agility to maneuver around potential "rocks in the middle of the road."

A refinement leader is *spirited*. Their vitality seems endless and their drive toward success has no room for compromise and no means to fall short of achieving excellence. Spirited refers to having a "fire in the belly" with a seemingly endless drive toward immediacy. Their sense of urgency is unparalleled. A spirited leader has "perfect quality" as his or her goal and anything less than perfection is entirely unacceptable.

Now that we understand the leadership skills, let's take a closer look at the refinement, streamlining, and reduced cycle time implementation process.

8.13.1 Refinement

There are leading reengineering experts who believe that refinement should not be the goal, but rather focus upon reinvention. Yet few companies are ready to "risk it all." Therefore, a more practical approach to process improvement is to refine existing reengineered processes. Refinement can provide immediate results. Small successes are very important to teams who are not experienced at, nor comfortable with, taking the leap of faith into a blank sheet of paper process reinventions.

I'm not saying that the long-term solution shouldn't reflect the elimination of processes by radical change to the current process, rather, in the short run, unless radical change is fully embraced and actively led by executives, an acceptable short-term solution is refinement. Few organizations are willing to take quantum leaps by designing radical process changes without piloting. Therefore, an interim benefit may be attained by refinement.

8.13.2 Streamlining

This process improvement technique relies upon the baseline that it is customer focused. Therefore, if customer value is not added, by any aspect of the process, then that nonvalue-added business process should be a candidate for elimination.

Streamlining targets are activities such as inspection compared to quality at the source mindsets, auditing rather than providing means to prevent errors from occurring, and approvals rather than empowering individuals with decision authority. Specific examples that come to mind include the following:

Inspection—The practice of hiring quality inspectors whose purpose is to locate quality-deficient product. Can enough inspectors be hired to statistically ensure that 100% of the product meets 100% of the specifications 100% of the time? Empirical evidence is showing that there are insufficient funds and significant disadvantages to relegating "quality" determination to parties outside the customer focused value chain participants. There is the problem of "interpretation" of specifications (esthetics, tolerance ranges, etc.). Rather than hiring inspectors, use the nested internal customer of the value adding process to assess quality. When nested internal customers are empowered to reject product by their nested internal supplier of products and services, they serve as the best judge of quality. "Quality at the source" expects each participate, along the value-added chain to generate perfect quality. The judge is the nested process customer. Process collaboration is the meeting of the minds between the internal customer and the nested internal supplier in order to cooperatively engineer the process, measurements, and deliverables. This process allows those in the trenches to design the quality expectations and eliminates contentions and animosity derived by someone not on the firing line (e.g., inspectors).

Auditing—The practice of auditing is age-old and is a form of inspection. However, the audit frequently occurs substantially later than the time when the transaction transpires. Consequently, the audit can seldom prevent error from occurring, yet the audit can frequently distract. Auditing the process redesign to ensure it conforms to specifications is essential at the beginning. However, if an internal customer of a deliverable is not fully empowered to call the shot, then auditing becomes another inspection point and the integrity of the empowered internal customer is challenged. Human nature then takes over. The auditor must find "something" wrong so as to justify the perpetuation of their audit job in the future.

A better use of an "audit" type is to commandeer their expertise to assist in the best process engineering activity. Empower the auditor to work with the process owners to perfect the process. Instead of the auditor becoming an obstacle in the value-added chain, he or she can be an asset to identify ways to prevent defects or errors from occurring in the first place.

Approvals—We earlier discussed how approvals were distracting in the value-added chain. The excellence of the Ritz Carlton Hotel operating philosophy comes to mind. A while back Ritz Carlton empowered *every employee* with the authority to, on the spot, provide immediate guest satisfaction. Each employee had the authority to spend up to $2000 without any approval. If there were a problem, it could be immediately resolved.

Approvals may be necessary within the value-added chain. However, they should be minimized as much as possible and timing sensitive, at all times. A better alternative is to spot approve ... randomly observe decisions in process. Not only does the approver get a hands-on observation (manage by walking around), but he or she has the potential to identify value-added chain distracters and inhibitors firsthand. Rather than delaying the process by approval, their energy could be redirected to identify ways to allow the value-added process to become more successful and become a best process.

Streamlining opens up the gates to encourage task elimination whenever possible. If elimination is not possible, then removing time delay tasks becomes second priority. Reducing time improves throughput (remember that the process of converting a booked sales order/resource into collected revenue).

Defining the streamlining objective becomes a challenge! A creative approach is *to compensate every employee* as if they were a commissioned sales representative; this is a novel compensation program. If this approach were deployed, then the execution of streamlining would become an overnight miracle. When employee livelihood is on the line, waste of any kind is not tolerated. When it is "other people's money," waste is not only tolerated, but frequently expected.

Short of radically changing the compensation program, how can streamlining provide positive results to an organization?

1. Process owners must serve as chief architects to the streamlining activity.
2. Every individual must be afforded an opportunity to be creative streamline engineers. That means that a portion of *every day* must be focused upon streamlining activities.
3. Collaboration between process owners and the nested upstream and downstream process partners must be encouraged and elevated as a priority for every day's output.
4. Streamlining should be so revered that senior management invests a portion of their *every day* toward working with process owners to identify opportunities to become more successful at streamlining.

If frontline executives are not gasping at this point, I'd be surprised. Echoing their perspective, how can *anyone* afford to carve out a portion of *every day* to spend on streamlining? The old saying might come to mind, "We're spending so much effort fighting alligators, we don't have time to drain the swamp." This is certainly an appropriate response to a "we don't have time" mind-set. Herein lays a symptom of the "overworked" and "no time" disease that afflicts many companies. The executive mind-set must change. They must divorce themselves from executing "crisis duties" sufficiently to aid in affixing a cure.

8.13.3 Reduced Cycle Time

This element is the final precept. Most folks would agree with the saying, "time is money." On a personal basis, recognition of the time value of money is part of everyday life. We pay interest on credit card balances, we pay interest on mortgages and we receive appreciation on house investments over time. Companies have become more sensitive to better management of their inventory investment, cash flow impact, over time, and the increasing impact of throughput (converting booked sales orders/resources into collected revenue). One facet of throughput is cycle time. How long does it take us to convert our raw resources into revenue? In a manufacturing company, cycle time is converting raw materials and labor into a shippable product. In a service company, such as insurance, it is converting a claim into a disposition. In a law firm or doctor's office, it is converting an inventory of expertise into billable revenue.

Cycle time reduction is the process of converting these resources into revenue using less time. Inherent in the concept are the facts that (1) quality will not be compromised and (2) a by-product of the "less time" will "coincidentally" improve productivity and reduce cost.

If this is so simple a concept, why has it taken our businesses so long to recognize the value of cycle time? The answer to this question is not so simple. First, until recently, our technology tools have not been adequate to support a quantum leap in cycle time reduction. Examples of technology tools include CAD/CAM development tools, numerical control equipment, and programmable logic controllers. Second, our IT tool capabilities were limited. Although integrated systems have been around for decades, our use of these tools has been limited. Our investment versus payback has been poor. One of the main reasons for poor payback may be categorized as lack of "user friendliness." Another payback deficiency manifests itself in competitive pressure for "fast response." Yet another investment concern is that the cost of hardware processing capability continues to decrease. Examples of fashionable technology tools include ERP, product configurator, and activity-based costing. Third, the time it takes to develop software has been compressed significantly. What, in previous years, took months and years to develop are taking weeks and months today (software cycle time is improving). These technology development tools include such things as computer-aided software engineering, neural network, communications tools, fiber optics, and laser.

Cycle time reduction objectives address a variety of positive business influences, including the following:

- Productivity improvement
- Eliminating waste and frugality
- Accomplishing more with less (philosophical ideal)
- A notch toward agility
- Inventory reduction
- Quicker response to customer
- Improved potential for schedule attainment

A characteristic of companies actively achieving continuous cycle time reductions tends to be their ability to demonstrate a higher degree of management by specification, whereas companies that avoid cycle time reduction tend to be more crisis management oriented.

To synthesize key ingredients of this section, we have a tremendous opportunity, in the form of a sleeping giant, awaiting to be awakened. Our vast talent base of working-level functional experts has not been adequately challenged nor permitted to actively participate in improving our process designs. To marshal the potential, leadership must surface, which stretches the limits of the organization's talent utilization capabilities.

Not only should every individual be encouraged to devote part of their day developing process improvements, but management must champion the improvement process and actively remove barriers preventing their achievement.

Process improvements will be engineered to refine, eliminate, or streamline business processes in every aspect of the business. A particular focus should be cycle time reduction.

The resulting activity may be the difference between a management that wallows in the mire of mediocrity or one that rises to the challenge and soars toward excellence in business performance.

8.14 The "Vision" of the Business Process

The development of a "To Be" model will depend upon many different factors, not the least of which includes company strategic plan, culture, management style, vision, IT strategy, and competitive spirit.

The purpose of this section is to convey essential ingredients necessary to provide sustained results in an increasingly competitive global posture. The intent is to challenge the biases and age-old practices of many organizations. Simplicity will be the watchword of the To Be designer, yet an underlying expectation is that strong leadership be demonstrated continuously.

As a review, earlier we dealt with defining the "As Is" or current operating practices. As presented, the primary reason for investing energy in this As Is activity was to peel back the layers of the onion in order to really expose the full potential for improvement. This activity should be used as a rallying point to obtain ownership by the functional systems champions.

Having clearly depicted the As Is, the organization does not have to speculate upon where the real opportunities lie.

The red lining task is associated with an approach to develop a lean To Be. However, the emphasis was improving the current business processes by refining, streamlining, and reducing cycle time. Let's further enhance the experience by stepping outside the box and encourage maximum throughput improvement; the ***instantaneous*** conversion of booked sales order/resources into collected revenue dollars (although instantaneous conversion is improbable, let us push the envelope and achieve as close as possible results).

To more fully exploit the To Be design, we recognize that certain business processes should be reengineered so radically that the end product would reflect a mere shadow to the former process flow.

Let's take a look at a popular trend today. In an effort to increase throughput time, a company that has extensive receiving inspection time would most likely consider reducing the inspection effort. There are three dominant approaches one may take, and I will use the inspection process as an example.

- *Refinement*—This will entail identifying methods to increase sample sizes or reduce the elapsed time associated with inspection of every lot received. Refinement will result in quick response benefits; however, inspection still occurs.
- *Streamlining*—This approach will be more aggressive and will result in greater payback. Yet it may take a longer period to implement and, if not performed properly, will increase the risk of quality compromise exposure. An example of a streamlining approach would be to take all products within a category, say, bearings, seals, and precision fasteners, which will encompass multiple supplier deliveries and/or lots and sample from this broadened baseline. Then increase the "sample size," by, say, an order of magnitude, and observe quality reliability impact. The next step would be to broaden the commodities and/or the suppliers using the new "sampling" approach. This streamlining would continue until the entire sampling base (commodities/suppliers) was converted to the new sampling specification. The final stage would be to increase the sample size to the degree that a mere infrequent spot check was conducted periodically. The ultimate goal is to eliminate inspection altogether.
- *Radical reengineering*—This approach will take a completely different tact. Instead of reducing the amount of inspection, radical reengineering will eliminate the need for inspection at all. A radical reengineering strategy might result in outsourcing inspection to the supplier (or third party) by implementing what is called supplier certification and source inspection. *Taken a step further, radical reengineering might explore product redesign to eliminate every product component, which requires any supplier inspection at all.* Now, that's radical!

Understanding when to deploy various approaches associated with developing a radical To Be design is crucial. Before we detail guidelines to determine which approach to use when we need to ensure that we understand the To Be design.

A working definition of a *To Be process design is the "vision" of the business process when a significant amount of change has been affected.* This value vision might be best understood by red lining the "As Is" business flows. Highlighting where process steps are consolidated, refined, streamlined, and, where appropriate, eliminated. Examples of streamlining might be as follows: The incoming inspection will be replaced by source inspection; the receipt into system will be bar code scanned; and the packing slip reviews will be replaced by an advanced shipping notice by the supplier. Examples of radical reengineering might be as follows: (1) conduct a Pareto analysis on those parts most exposed to require inspection and redesign these parts in a manner that would eliminate 80% inspection liability, and (2) team up with multiple tier trading partners and together reengineer the design to reduce cost by 50%, inspection by 75%, and manufacturing cycle time by 50% through the nested trading partner value process. This nested reengineering process would leverage the engineering innovation and manufacturing process prowess of all included tier trading partners. Each trading partner would benefit from lower costs, improved processes, and, likely, substantial throughput improvement—a WIN/WIN for all participants.

If effectively engineered, the To Be process designs will remove 20%–90% of the time and effort associated with value-added activities to customer products and/or services. But even more important than affixing visual pronouncements of change, the To Be process design conveys a change in cultural philosophy, methodology, and measurements. *Change* to cultural baseline designs includes the following:

- Throughput improvements will be very evident.
- Functional accountability will be clearly defined, emphasizing the value-added chain, which resoundingly points toward customer focus.
- Businesses are dynamic and must continuously change; therefore, the To Be process designs must also continuously change.

Now that we have a better picture of a To Be process design, we can develop the guidelines defining **when** to deploy each approach in a manner that will optimize resource utilization, improve the customer value stream, and minimize exposure and risk.

The guidelines used for *refinement* approach might include the following:

1. Use a *Pareto distribution* and then identify significant value-wasting activities in the customer value stream. When we apply a Pareto distribution to our opportunity, we will see that approximately 20% of the activity will impact 80% of the customer value stream. As we review of the As Is, we should clearly point to areas of greatest opportunity first.
2. Focus *priority* on the significant activities that can deliver rapid return on effort invested. Plucking the low-hanging fruit not only reinforces the reason we're doing this in the first place, but motivates the team to continue to pursue these and other opportunities.
3. Form a process improvement team consisting of cross-functional talent. This team might include the following:
   a. An individual who is able to represent the customer perspective—if practical, invite a key customer to be a collaborating consultant to work with the team; if not, appoint an individual who is able to function in a "customer empathetic role"
   b. One or more functional experts
   c. Someone who can function as a visionary
   d. Someone with authority who can execute team decisions without approval
   e. A financial team member who can document a cost/benefit analysis
4. In small to mid-market companies, the CEO will kick off the process by *empowering the team* to use methods that affect *rapid change* with *maximum agility*. In larger organizations, the division general manager would likely be the top executive leading the effort.
5. The team appoints a project *leader*, develops a *charter* (statement of scope, authority, deliverables, resources budget, and measurements) and defines collaboration guidelines (such as don't limit creativity, limits of compromise, and adjudication approach).

We can see in our receiving inspection example that the refinement process would limit its activities to policy and procedure review, forms modification, and internal customer and supplier of service interactions and measurements.

The guideline for *streamlining* will embrace all five elements depicted above, but also include the following:

1. Define *latitudes* (responsibility, accountability, consultative, and inform [RACI]) to follow when considering process elimination. Streamlining may radically reduce the activity content but seldom promotes process elimination. Ensuring that executive staff defines project "boundaries" at the beginning of the process will prevent significant surprises downstream.
2. Define the IT *budget* and financial limits that may be deployed in support of streamlining. In our receiving inspection example, an IT budget may be something like $20,000 and technology may be limited to purchases of bar code readers, calibration equipment, or similar tools.
3. *Eliminate redundancy* serves as the primary focus for streamlining. Although this may uncover an opportunity to eliminate activities totally, the thrust is upon simplification.
4. Cutting back on approvals, auditing, and other nonvalue-added activities will provide a delineation perspective for highlighting higher visibility elements.

Referring back to our receiving inspection example, streamlining results will likely deploy preventative processes such as (1) on-site quality audits of suppliers and/or (2) working with supplier manufacturing process engineers to identify ways to limit quality rejection exposure.

The pinnacle of process improvement activities displays a radical reengineering approach. The guideline that impacts this approach is a bit different. It will include the guideline steps of refinement and streamlining approach previously defined, as well as the following:

1. To perform radical reengineering properly, most of the operating *guidelines* that the business currently conducts business within must be *challenged*. Radical reengineering may be deployed within any given department; however, because of the significant cross-functional impact that the business process has, it is very difficult to reengineer change in a localized manner.

   In our receiving inspection example, radical reengineering would most likely **eliminate the receiving inspection function entirely**. In order to perform radical reengineering, the change management team must be given broad latitudes to propose the elimination of any activity, which is **not** customer focused and is **not** integral to the customer value stream. Empowering the team with such latitudes must be monitored closely. If not handled properly, compromises may be made and company politics may surface, distracting from the potential with which the project could deliver substantial results.
2. The senior management team must demonstrate *agility in their thinking*; otherwise, the radical change proposals may appear threatening to the senior staff. Radical reengineering must be embraced by the entire management team. The vision for the new culture must be willingly practiced by every employee.
3. When extending this to one or more tiers of trading partners, the team broadens appropriately; however, concurrent engineering across trading partners tiers adds a bit of complexity and needs to be managed properly.

Developing a *vision* which reveals that the new "reengineered" culture is essential to focus the direction for the organization. The challenge becomes that involve every employee sufficiently so that he or she may embrace the vision and marshal vision ownership. One of the themes articulated in this chapter is "the productive use of *intellectual energy* by every employee." Streamlining operations can certainly serve as a catalyst to unveil *intellectual energy* opportunity. As the new To Be vision is realized, the nested internal customer and service provider end-to-end process team must concurrently revise their PPMs as needed.

Intellectual energy can be productive when employee focus is optimized upon customer value stream activities. When a customer focus and thrust is sincere, many of the *intellectual energy* distracters vanish. *Intellectual energy* distracters, such as continual crisis management, deplete the energy bank and redirect activities and priorities away from the core customer focus value stream. *Intellectual energy* deployment will be expanded into more detail later.

The section discussed thus far is recapped as follows:

- Significant management focus must be deployed, ensuring that every business process is customer focused.
- Removing the customer focus value stream distracters requires refining, streamlining, and reduction of process cycle times.
- The act of introducing change may range from minor adjustments to radical reengineering. Risk and related potential results are directly correlated to the degree of effort expended and the nature of change effected.
- Management commitment is central to evolving a successful process.
- Leveraging *intellectual energy* will provide the synergy to modify attitudes, exploit talents, and eliminate waste.
- The streamlining process, itself, must be continually refined and improved upon.
- The *survival of many firms* will depend upon making a *quantum leap* in *productivity enhancement*. Therefore, continuous improvement must define quantum opportunities as well as refinement opportunities.

Developing To Be models will enable an organization to become more agile. Let's look at some of the ***strait jackets*** that *limit* the *agility* within organizations. These include the following:

- Job descriptions
- Hierarchical structures (inadequate)
- Productivity incentives
- Performance measurements (meaningless)
- Lack of leadership
- Waste management (inadequate)
- Intellectual energy conservation
- Agility deployment (lack of)

One of my favorite areas from which I challenge companies is "job descriptions: frequently, the box, from which good, meaningful organizations entrap and limit their employees." Job descriptions, typically, establish barriers beyond which employees cannot display their natural talents. These descriptions were originally developed to provide guidelines and structure to individual work assignments in a hierarchical organization operating structure. Yet, in practice, these frequently serve as fences beyond which employees may not tread.

Not only does the organization limit the employee talents, but the employee limits themselves by staying within the box. The hierarchical nature of most organizations politically imprisons the employees and dares employees to step beyond the specified boundaries defined. Therefore, employees limit their potential for conflict and stay within the prescribed confines of the job description.

Wouldn't it be novel to take a different approach by defining what the employee should ***not do*** while ***encouraging full exploitation of all other abilities***?

Fencing people in by limiting job scope inhibits the broad base population of employees. Only a small number of strong leaders are able to look beyond these job description barriers in order to exploit broad-based employee talents. If an employee happens to venture outside the line prescribed by the job description, and if the organization imposes threat at all to an employee, then the employee's maximum benefit to the organization will be limited to the latitudes defined within the job description.

The job description, no matter how meaningfully designed, can become a talent delimiter. Most employees may contribute well beyond the limits of the job description.

However, management, through performance reviews, forces the employee to live within the constraints of the job description.

8.15 Dreaming a Bit

Let's be so bold as to dream a bit for a moment and remove the job description and the typical hierarchical employee evaluation process. Let's say that we defined a few *guidelines* and *constraints*, such as financial commitments or expenditures (grants of authority), *quality guidelines*, and *customer focus* priority. Then, let's broaden the responsibility expectation by specifying that an *employee report card* will be primarily influenced by their nested *internal customer* (rather than a hierarchical supervisor). Let's further broaden the latitudes by suggesting that the spirit of service reflects *high quality, quick response, improved throughput*, and *waste minimization*. Let's also suggest that *continuous improvement* is expected and that a team of nested peer customers, nested peer supplier of services, and nested executive champion are the basis from which decisions are to be deployed. Let's further say that in the absence of a constraint (a nonviolable policy), *performance rewards* will be driven by creative solutions that push the edge of latitudes. Let's also be creative and remove the clock as a constraint with the proviso that the *internal* and *external customers must be serviced at high 99+% levels*. (Remember that the internal customer will have "compensation-based influence" upon their nested internal service providers.) We must do one more thing: the hierarchical organization influence must be drastically reduced by giving the internal customers an 80+% voice in performance review matters and the hierarchy a 20% voice. Adjudication's will consist of appointing an executive board to handle compensation and performance review disputes.

Let's now adopt the power of the free enterprise system to do its work.

1. With the market (internal customer) dictating the level of service, quality, responsiveness, and compensation (via process performance review), survival of every nested service provider will become customer focused. Energy spent on hierarchical approvals, distractions, time wasters, excessive meetings, and the like will be regulated by market tolerance (internal customer). Focus will consequently be centered upon meeting customer needs rather than feeding the hierarchy. Politics will be minimized because it does not net the internal customers much return, especially when the internal customers have nested internal customers of their own from which to be accountable.
2. Each employee has been empowered to operate "his or her business" by exhibiting leadership skills (managing their service providers) and every employee clearly understands his or her contribution to the value stream (contribution to profitability).
3. Executives now have driven accountability into the business process arenas; therefore, they have vested interest in aggressively helping ensure that their process owners become successful. The organization has naturally flattened out because there is little value for bureaucrats who don't actively participate in the value stream. (Remember that the bureaucrats' influence has been reduced from 100% to 20% level.)

4. Because the internal customer is interested in energy waste elimination, the touch points connecting the business process flows become incented to simplify procedures while increasing quality of communications.
5. Business process owners (and their service providers) become natural work teams with a common objective and shared rewards. They also spend time daily to continue to improve the business process.
6. Business process owners, who have been freed from procedural constraints, are encouraged to push the limits on productivity enhancements. They are encouraged to thrive on use of creative problem solving and are consequently postured for success.

In contrast to our dream, let's look at a typical organization today. A seasoned conservative business executive might view this scenario as radical.

- The bureaucrat would be exhaustively threatened by seeing his or her empire dissolve.
- The typical employee would be convinced that he or she would never be afforded such latitudes.

Consequently, if we believed such press, we would be defeated before ever giving serious consideration to the merits of such a proposal. Are we willing to *wallow in the mire of mediocrity* imposed by a limiting job description or will we **soar like eagles and rise above** the "how we've always done it" and view the proposal as a *possible cure for complacency*? Are we fighters and winners or are we losers who will not consider challenging the practices of old?

8.15.1 Bringing Down the Job Description Walls

Let me be so bold as to suggest that there is a small contingent of fighters who are willing to challenge the constraining barriers of the job description and hierarchical structure. Let's also presume that they will muster sufficient support to, at least, pilot a change process. How would they go about breaking down the job description wall?

I'm going to exercise another "it depends" at this point and take one possible road. I'm going to propose something radical and presume that one or more process champions have sufficient "fire in the belly" and will "take on all challengers," such as human resource departments, union perceptions, deadwood bureaucrats, compliance lawyers, mentally retired executives, and non-value-added wage collectors. Let's further presume that "man really can walk on the moon," that the "Berlin wall can fall," and so on. Some of you skeptics are saying that I'm mixing apples with oranges and that these examples are sociopolitical; however, at one point in time, these monumental achievements were equal to the challenge and fear that eliminating job descriptions will bring within the typical organization.

8.16 How Can a Radical Visionary Begin the Process?

A few points to consider include the following:

1. A *vision* toward success must be embraced by the *entire executive team*. Remember that they must have vested interest in ensuring that the process owners are successful.
2. The process owner team champions must be groomed to function as "change transition" leaders.

3. The time schedule for the pilot must be *quick response* oriented with successes surfacing in short order.
4. The change process must maintain a *high visibility profile* with every employee encouraged to participate in the new design (defining the "can't do policy" as well as being creative in barrier removal ideas).
5. Every employee must *believe* that this transition *will happen* and that their future with the organization will be dependent upon how successful the transition occurs.

These five points go to the heart and become the greatest inhibitor to the success of this change process, and represent the mind-set and attitude of the employer, management team and working-level employees have toward empowerment.

Let's develop this a bit.

- *Empowerment cannot be delegated. It must be accepted and championed by individuals.* Management must provide an environment conducive to empowerment and, then, they must become cheerleaders, motivators, and "barrier to success" removers so as to support the empowering process.
- The task of transferring the decision making to working-level employees within a process is not necessarily an easy transition. Many managers (bureaucrats) believe that control and closely holding information is the essence of their power base and they are extremely insecure when it is suggested that they give this up.
- An inhibitor, in the process of transferring "power" to those at lower levels, is an insecurity by management that lower level folks will uncover "bad decisions" that were made in the past and politically damage those having previously made the decision. Poor results, from historically poor decisions, frequently occur from allowing the decision to be made in a vacuum, and with inadequate time allowed to gather facts. However, we need to recognize that many decisions only rely upon a 30% fact basis anyway. This "lack of facts" compromises results from inadequate "information" (not inadequate data), inadequate time, and the inability to gain concurrence among the team because the team is not postured with a "fast cycle" mind-set.
- In a highly political environment, insecurity and fear of past decisions are magnified. Therefore, a tiger team may be needed to facilitate the transition process and remove the "fear" from those relinquishing the power. However, fear of future reprisals for past decisions is a serious issue within all levels of the organization. Consequently, overcoming this concern must have a proactive strategy and action plan with continuous encouragement by the top brass.

In his book *Teaching the Elephant to Dance,** Dr. Jim Belasco highlights the examples of barriers to the process of empowerment. Essentially, he says that there are many reasons for empowerment not to occur; however, one of the biggest is a psychological barrier that prevents those being empowered from accepting the "ring" and authority and making it happen. His example of the elephant and its leg irons conveys this concept. In the case of the elephant, at an early age, it is shackled with a leg iron and is chained to a stake in the ground preventing the small elephant from wandering away. After the elephant grows beyond the point where the stake can no longer prevent

* James A. Belasco, *Teaching the Elephant to Dance: Empowering Change in Your Organization*, Plume Publishing, New York, 1991.

it from wandering, it has been conditioned to believe that it cannot move beyond a certain radius because it still has on its leg iron. Only when there is an emergency, like a fire, where life is threatened, will the elephant realize that there is no power in the stake preventing it from wandering.

People, frequently, have a similar mental "leg iron" imprinting, which prevents them from wandering out in the decision process. Alternatively, if they autonomously "assisted" in a decision process, without asking approval, they were blasted for doing such an evil thing by their supervisor. Therefore, overcoming this mental barrier becomes a significant challenge and one that must be thoroughly addressed if true empowerment is to occur.

Another significant empowerment challenge is as follows: *conveying to the empowered employees that management is "truly committed" to passing along the authority.* This barrier will take more than words to overcome. In practice, this can only be overcome by *management living the example,* being a fire-breathing advocate of the process and cheering on the successes in a continual, and very visible, fashion. Cheerleading, in this regard, is more than just accolades. Cheerleading is a day-to-day commitment to seek and destroy all barriers to the success of the process, to the extent that a significant part of top management's job becomes championing this transition process. The premise also assumes that "all members of the executive staff" participate in the championing process. This cannot be delegated to one executive only, unless that one executive is the chief executive, in which case, by definition, it is practiced by the CEO's subordinates.

8.17 Piloting Change

The stage has now been set to accomplish radical change.

- The executive champion(s) have been appointed.
- The rules and latitudes for change have been laid out.
- Every employee has been incented, not only to be supportive but to actively drive the change process.
- Empowerment is a go forward committed process.

The pilot (target area) needs to be determined. Even though the pilot area selected is important (it should represent the more spirited attitudes in the company and employees who don't accept defeat), every employee will participate in the change process. The pilot area will devote significant time to the change process (at least 40% of the time). However, every employee will be required to devote a minimum of 30 minutes a day, reviewing the results surfacing from the pilot activity. Then, once a week, every employee will allocate time to document and refine a plan to remove his or her job description within his or her functional area. These ideas will be shared among cross-functional teams as well as between the various business process change champions and stealth resources. The business process change champions will meet weekly with the pilot team to share ideas that have surfaced from among the general employee base.

Wow—***now that's radical.*** The naysayer's are actively mobilizing their ammunition at this point to lobby for abandonment on "loss of productivity" grounds. Those who are politically concerned are fearful of mounting a frontal attack, yet they are identifying ways to make themselves so important to the day-to-day activities that they couldn't possibly designate the 30+ minutes per day and 1+ hour per week to this change process. Yet others are huddling in secret to identify methods to sabotage any possible success. The battle lines are being drawn. The character of management commitment essence will soon be exposed (remember that the hen participates, but the pig is committed).

The elimination of the job description pilot must now proceed with the air of visualized success. Each pilot member is encouraged to be a triage leader. Embrace their job as if they were acting CEO. Their activities are both external and internal customer focused. They are interested in eliminating all delays and concluding the business process with an eye toward maximizing throughput (converting booked sales order/resources into collected revenue dollars). As triage agents, they identify the most important events and opportunities. They execute with the swiftness and agility of a surgeon's scalpel. Without violating the preestablished rules nor waiting for approvals, they execute actions, make decisions, and deliver results.

Every couple of hours, the pilot team convenes a tag-up and collaborates on barriers they have encountered, develops strategies on barrier removal approaches, and documents a debrief on success and failures encountered. Serious barriers and breakthrough successes are passed along to the team champions outside the pilot process (known as consultive champions). Consultive champions convene their collaboration teams as appropriate, but at least twice per week to discuss possible solutions to serious barriers.

Creative resolution feedback is directed to the pilot team each day. The pilot team assesses the merits of the creative resolution and executes, as they deem fit, autonomously. When the pilot team selects a consultive champion solution, and upon executing the solution successfully, visibility is given to the consultive champion's team by recognition at the weekly CEO's debrief. The CEO personally meets with the consultive champion team, at their next scheduled meeting, to personally provide accolades. This high visibility shows the entire organization that this process change activity not only continues to be important to the company but is also a means of gaining an audience with the executive staff.

In addition to providing feedback to the pilot team, the consultive champions are encouraged to independently determine how the pilot team successes may be quickly assimilated into areas outside the influence of the pilot activity. They are encouraged to execute triage leadership latitudes and affect change as long as major disruptions to the respective business processes are avoided. The consultive champion change activities are rigorously documented and roadblocks are elevated to the Executive Change Counsel staff for weekly decision activity. In their weekly meeting, the Executive Change Counsel must discharge one of the following:

- Commit to personally devoting the necessary effort to remove the barriers (maximum time to affect change is two weeks).
- Document reason that barrier removal provides undue risk to the organization (maximum time to document reasons is one week).
- Commit to rework the affected policy, organization structure, or resource constraint, which reduce the risk so that change may be immediately deployed (maximum time to document an action plan is two weeks).

All documented failures elevated to the Executive Change Counsel must be actively resolved, deploying one of the above three strategies, and an activity log of which strategy was selected. This log is reviewed weekly with the CEO so as to avoid gridlock and procrastination. Decisions by the Executive Change Counsel are immediately conveyed to the functional consultive champions. Change strategies are then executed with progress and barrier status conferences held jointly with the pilot team, functional consultive team, and Executive Change Counsel biweekly.

The objective of the quick turnaround is obvious: broadcast successes throughout the organization with speed and agility while leveraging the enthusiasm and synergy of pilot successes. **Success breeds success**. Exploiting cascading success quickly magnifies benefits with swiftness that consequently fuels continued agility. After a short time, the ***change*** culture starts becoming the norm.

The section on elimination of the job description is recapped as follows:

- Job descriptions imprison broad-based talents.
- Job descriptions can be an inhibitor to accountability empowerments and restrict the creative process engineering activities of the critical employee base (the bottom of the pyramid).
- Transitioning to a continuous change culture will require total employee involvement if it is to be successful.
- Implanting the spirit of change, leveraging "intellectual energy," and optimizing total employee productivity must become the second nature to every employee quickly.

The goal of job description elimination is to unleash and leverage the suppressed talents of every employee.

8.17.1 Hierarchical Structures

A curse to the productivity opportunity within the typical organization is the hierarchical structure. Steve Jobs has been arguably one of the leading innovators in recent days. In his book *Leading Apple with Steve Jobs: Management Lessons from a Controversial Genius,** Jay Elliot worked many years along side of Steve Jobs and was considered the voice of wisdom for the younger Jobs. The following is a quote about "Dangers of the Middle Management: Fear-of-Change Syndrome:

> In the traditional organization, there has always been a problem with the role of so-called middle managers. Do you know why Ford can't make a car as good as Volvo does? It goes back to back to the middle managers. Steve and I visited several car shows together, and we were always amazed at the so-called concept cars. The designs were magnificent, but we knew from experience that none of those cars would be turned into production models that customers could buy.
>
> Why? Because the companies weren't following the lead of their head designers, their creators, their thinkers—people who were coming up with breakthrough technologies, only to be defeated by middle management.
>
> Too often, middle managers are wage earners who see any change, however small, as a risk. Confidence lies in continuity, in continuing with what worked yesterday.
>
> This was the culture I saw at IBM. Innovation there, it seemed to me, happened largely when the company acquired an outside firm that had developed some innovative new technology, or when IBM spun off a unit that was then able to function beyond the reach of the IBM corporate-think.
>
> Within most large U.S. companies, innovation isn't impossible but usually requires such a struggle that the true innovators either leave or become plodders who have discovered that their best ideas are not likely to see the light of day and have just given up. Too often it's incredibly defeating to try to be innovative within the prison of a large company. This experience actually prompted me to write a letter to IBM Chairman Tom Watson Jr. and was one of the major reasons I left the company."†

* Jay Elliot, *Leading Apple with Steve Jobs: Management Lessons from a Controversial Genius*, John Wiley & Sons, New York, 2012.

† Ibid.

Organizations consist of a series of business processes, most of which flow across organizational lines. Yet the hierarchy (vertical silos of autonomous resource) typically controls the priorities, energy, and budget, which influence the lethargy of business processes. The care and feeding of the hierarchy is frequently not customer focused. To add insult, the typical hierarchical internal customer is the vertical boss within the hierarchy. The bosses' interests are best served by focusing on their vertical influence recipients (their internal customers) who are their respective boss. This care and feeding of the boss frequently has a distractive impact, of up to 90% productivity loss, distracting from the goals of the horizontal business processes. The magnitude of the productivity loss is affected by the level of politics and the strength of the vertical hierarchy. These productivity distracters are represented by things such as approvals, decision procrastination, and time-consuming analysis.

The hierarchical organization structure relies upon a "command and control" style of management in order to be effective. As discussed earlier, the vertical hierarchy develops goals, objectives and priorities based upon vertical hierarchy influence, whereas a business' natural functions are horizontal, going across various vertical hierarchical lines.

Vertical hierarchical performance measures are consistent with the ineffective vertical structure. This command and control structure was developed to provide a means to quickly execute within the spheres of functional specialties. The original design concept relating to functional specialization was aimed at providing maximum output within its cocoon area of expertise.

Today, however, the organization must become customer focused and its energy spent should optimize a combination of the following:

- Customer service
- Maximizing throughput (converting booked sales orders/resource into collected revenue faster)
- Streamlined structures
- Continual quality improvements
- Continued value streaming leading to reduced costs
- Innovators to achieve voice of the customer visionary products

With the pressures expected from these new critical business drivers, the optimized functional organization (hierarchical structure) actually becomes a determent. Command and control decision making is not as effective if customer-focused quick response objectives are pursued.

The contemporary *quick response* mind-set requires decisions to be made at the working level. Encouraging the working level to collaborate with their nested internal customer peers, then empowering them to make timely decisions is essential for quick response. Deferring decisions to traditional hierarchical decision-making methods causes undue delay and is a nonvalue-added process.

The working-level employees must be coached into accepting their new decision-making role. With the flattening of hierarchies, a new role for the key executives is not to only empower the working level but also to take the necessary action to ensure that the quality of empowered decisions meet minimum acceptable guidelines. In other words, the executive expects quality decisions; therefore, they must coach (lead) the working level to guide the process to ensure that decisions meet their expectations.

One additional expectation is that information, from which quality decisions will rely, must be provided in a timely manner. This presupposes that working-level employees are provided timely "information" from which to make timely decisions. Working-level employees must be provided with IT assistance. IT resources will help enable users to define the information tools they need in order to make timely decisions. To reiterate, IT service providers must collaborate with their nested internal user customers to assist them in defining the information tools needed to make timely decisions and posture the organization to use information as a competitive weapon. There should be a transition from legacy data manipulation to such capabilities as decision dashboards, triggering, cost-of-change modeling, and predictive option management. Information needs to drill-down sufficiently to facilitate quality and timely decisions.

This emphasis transforms information management focus from primarily hierarchical decision points to working-level process-oriented decision points. Working-level personnel must be armed with the information from which to make timely and quality decisions. Effective decision support will rely upon the team consisting of executive mentors, IT facilitation, and working-level functional expert.

The more progressive and radical organizations will empower the working level to cross organizational lines in an effort to make the needed decisions in record times, without compromising quality and customer focus. The working-level employee transitions from decision lackey to functional decision leaders and information stewards.

Working-level functional leaders most likely will participate in more than one business process. Therefore, their effectiveness must be closely monitored by their respective nested internal customers to help ensure that the proper balance of effort and performance to the process is achieved. When conflicts arise, the Executive Change Counsel needs to adjudicate. As these functional leaders become proficient, their cross-process experience serves to refine process measurements and optimize resources more effectively. This process-based dynamic hones the new process-based structure and fuels exceptional performance. In addition, the empowered working level soon sees how they are contributing to the profitability of the organization, not by lofty measures such as RONA but by daily seeing the improvements taking place such as improved throughput or reduced effort and ultimately harmony and deftness across business processes.

8.17.2 Productivity Incentive

Productivity improvement is defined as the ability to convert the same level of resources and capital inputs to higher levels of output. Productivity measures, which are not bottom line and customer focused, are frequently energy distracters. For example, if I were to take a couple of popular manufacturing measurements such as *efficiency* and *utilization*, we can see the energy-distracting nature from which they consist. A company's work force can be very efficient working on product which is "easy" to produce, but which has not been ordered by customers or which is not aligned to the master schedule. If a company is *efficient* at producing product that cannot be quickly converted to collected revenue, then high efficiency has little business benefit.

A similar scenario may be developed for *utilization*. If we keep a piece of equipment busy chewing up inventory not needed to quickly convert inventory to collected revenue dollars, then high utilization has little business benefit. However, if we redirect our performance measurement into something like throughput (converting booked sales orders/resources to collected revenue), the measurement becomes meaningful and very visible by the working-level resources.

Until the organization gets a grasp on their definition as to what is truly a *productivity improvement* and conveys this definition to all employees, providing productivity incentives is meaningless. True productivity must be defined in a way that the common employee understands and can relate on a day-in and day-out basis. Meaningful productivity gains can only be accomplished when each productivity producer has ownership in designing the expected measurements. With ownership comes accountability and with accountability comes best practice results. Gone are the days when profitability merely results from revenue gains (I'm not convinced they ever did; I believe we may have been hoodwinked into believing in this correlation by smoke and mirrors and creative management accounting). Competitive pressures toward continuous improvement in quality, delivery, value, and reduced cost (faster, better, cheaper) will persist in the foreseeable future. Therefore, how do we go about improving productivity?

As stated earlier, ownership is key to leveraging effort toward productivity gains. Our marketers have demonstrated that an inducing strategy for increasing volume (performance-based commission plan) is a balanced effort and becomes a powerful incentive. Could this concept possibly work in other segments of an organization?

There are various ingredients needed for performance-based compensation to become successful inside a broad-based segment of an organization.

1. *Balanced*: For incentives to work, we must understand Newton's law, whereby for every action there is an equal and opposite reaction. When we implement incentives, there is a change of focus and a cost associated with quickly marshaling that focus on other areas. Restated, we must ensure that key business drivers such as quality, delivery, reliability, and customer service are not compromised as we drive productivity increases. In fact, if incentive programs were structured properly, we may likely improve all business drivers simultaneously. Therefore, balance is essential for a healthy incentive process.
2. *Doable*: The incentives must be *within reach* fairly quickly. If the road to incentive earning is too arduous or results in bloody travelers, the incentive no longer motivates at its level of potential. Doable, to use a cliché, means many things to many people. Doable must be viewed from the working level, not the ivory tower level. If the strategy is well thought out and creative, the ROI has phenomenal potential if it can be attained by the broad-based working-level populace (critical mass).
3. *Meaningful*: For an incentive to be pursued with "gusto," the incentive reward must be meaningful. Offering a one-time $50 reward will result in minuscule effort by participants. Yet many companies have given critical thought to how they may retain "long-term" employees and have developed a vesting strategy that is progressive (increases over time). By encompassing more than one year effort, after a period of time, this progressive incentive is difficult from which to walk away. Meaningful productivity gains can only be accomplished when each productivity producer has ownership in designing the expected measurements.

   We don't have the luxury of extended time to wait for productivity improvements; therefore, time can be an attribute, but it must be significantly compressed to be effective. Consequently, "meaningful" likely means that there are short-term, intermediate, and long-term vesting aspects to be considered. Meaningful must be engineered in a WIN/WIN manner, the successful employees must get short-term and long-term benefit and the company must get the same.
4. *Measured*: One of the greatest challenges surrounding the development of incentives is determining the beginning baseline, which will be agreed upon by both management and employee. Next comes the joint performance tracking and process engineering review to ensure that continual improvement of the process occurs over time. Once the agreed-upon milestones are met, rewards should be immediate. Whether these rewards are financial, time off, gifts, promotions, or any combination, recognition and subsequent deployment must be performed swiftly and advertised broadly. A *fire storm spirit* needs to permeate the

organization with a highly magnetic attraction. Now that we have enthusiastic ownership that is channeled through aggressive leadership, which is functioning without job description barriers, we are postured for exciting change.

8.17.3 Performance Measurements

I have briefly discussed the measurement aspect of a focused process. We need to bounce out of the micro perspective of measurement and look at the macro perspective. Performance measurements can either be boring, dull, and bureaucratic or be exciting, dynamic, and agile. Most legacy performance measurements we deal with fall into the former, boring and dull category and consequently dreaded by all. However, they need not be. If the performance measurements are jointly developed (ownership), customer focused, monitored continuously, forward (rather than historical) focused, and agile, then they can be exciting in nature.

Most performance measurements we work with are historical representation and *dis-incenting* in nature (they amplify missing the mark). We can do nothing about history. However, if our measurements become future oriented, we have a chance to affect the outcome. Now that's management!! I'm sure that more than one of you are asking, "How can we possibly measure future actions?" *Now that's exciting* **and it is a radical change from the comfort zone we typically live in daily.**

As a former business architect and consultant, of all the areas I'm excited about, this is an untapped reservoir of abundance where "intellectual energy" can provide tremendous ROI. To master this forward-oriented opportunity, you must read this precept a couple of times and intellectually master the principle themes presented. These themes include the following: We must eliminate barriers such as job descriptions, we must exploit fully our talent base, we must change our assertiveness and leadership style, we must migrate from hierarchical to process orientation, we must become agile performers, we must redeploy our functional experts, and we must ensure that decision support information tools are actively used by every employee. Now that's a tall order.

Let's look at future orientation a little closer. First, we must recognize that most of us rely upon historical data to make decisions. We muddle over various things that happened "ago." We dispatch research projects investigating reasons why events occurred in the past (justifying variances). We distract and waste "intellectual energy" justifying reasons for failure in the past. Is that productive? Why do we do it? We have mastered the ability to analyze our failures to the maximum. We hire a staff of individuals whose sole job is to muddle in the past mire of mediocrity and failure. We're so good at it that we do it over-and-over month-in and month-out without fail.

8.18 Future Orientation

Let me now be so bold and so radical as to propose that this intellectual energy could be better used if we focused upon the *future* to prevent past failure from reoccurring.

Information can be past, present, or future oriented. However, our information design focus must be future oriented if we are going to capitalize on future-oriented tools. Future-oriented tools, such as decision support, rely upon systems champions to help ensure that the information tools meet the needs of the day-to-day users. We need to be dealing with the organization structure conducive to future-oriented tools. As a preview, let's consider for a moment, what it would be like if our IT team were focused upon their nested internal customers (an amazing concept!). Every employee would have an IT team member as his or her support resource. The IT team member would spend 80%–90% of his or her time in the trenches working with his or her nested internal customers identifying ways to continually improve his or her information decision process. Remember that our goal is to optimize "intellectual energy" by providing effective information from which every employee can improve the quality and timeliness of their job performance.

Focusing upon future-oriented information can best be visualized by such things as *cost of change* impacts, *projecting future variances*, and tracking current detail activity versus most recent past (such as operation by operation labor tracking for current job compared to a similar job and best performed job of the past). Companies exploring "best practices" can adapt best practice disciplines to daily information management. For example, when we report labor and material activity to a work order, we can establish milestone control points and compare the current activity to the *best ever achieved* in the past and provide the information "real time," as transactions are recorded. Tracking to the best performance and reporting real time provide visibility to the supervisor allowing them to affect corrective action along the way. Conceptually, this is adapting the concept of statistic process controls (SPCs) to performance management control (PMC). An objective of SPC is to provide an *early warning signal* before poor quality occurs. Our PMC conceptual model would become an early warning signal to notify management before poor performance occurs. Companies that seriously adhere to SPC disciplines significantly improve their quality performance. Isn't it likely that companies adhering to PMC concepts and disciplines would significantly improve their performance? Couldn't this be a way to have performance measurements come alive and transcend from "boring and dull" to "meaningful and exciting?"

8.18.1 Lack of Leadership

We have previously discussed the importance of leadership and some common ailments and by-products of poor leadership. However, to properly articulate the *strait jackets that limit the agility of the organization*, we must address leadership with an *eye toward competitive vigor*. One fundamental expectation of excellent leadership is to *marshal high quality and productive performance* from all individuals within their charge. Another expectation is to continuously look for ways to *improve the business processes* with which they are associated. The "To Be" model of excellent leadership requires the leader to be a visionary—one who *capitalizes on windows of opportunity*. This requires a "fast cycle" mind-set and continual eradication of bureaucratic paralysis. Our excellent leader is committed (remember that the hen participates, but the pig is committed) to removing the barriers that impede their team's progress. Excellence in leadership mandates an air toward TQM discipline.

A leader, according to *Webster*, is "a principal performer in a group." Therefore, leadership is giving direction to a group and marshaling optimized results. "Committed leadership" may be defined as the ability to contribute expertise, example, motivation, goals, and attitudes. The only real success measurement of the leader is his or her ability to actualize success through the achievements made by members of the group through excelling in process performance. We're not looking for heroes, rather we are looking for incisive improvements in our end-to-end business processes, improvement in throughput (converting booked orders into collected revenue), and stellar external customer service.

With a rapidly changing technology base, as well as ever-increasing global competition, leadership requirements are changing. Gone are the days where the leader was merely a nontechnical expert. Today, a good leader must exhibit multidimensional attributes. Successful leaders must

- Champion the vision of the organization.
- Remove "barriers inhibiting success" to accomplishing the vision.

- Inspire their team to overperform on every goal.
- Focus upon both short-term and long-term objectives.
- Be a torch-bearer for making change happen.
- Drive accountability to the lowest organizational levels.

Leadership, for the current lean and agile decade, mandates that companies exploit every skill residing among their team members in order to achieve goals faster, while incurring less cost and with an eye toward continually improving quality. The exploitation of skills must be performed within preestablished company guidelines or boundaries. As Igor Stravinsky, a well-known composer once said, "My freedom comes from my restrictions." Senior management must define the boundaries. The boundaries must exploit the skill pool sufficiently to incur less cost and continue improving quality. Every employee must understand that these boundaries cover (1) financial restriction, (2) resource restriction, and, (3) time restriction, so that true freedom may occur. Therefore, leadership in the current decade must exhibit an outrageous purpose to achieve success without compromising ethical, moral, and spiritual values.

Excellence in leadership expects management to execute change and champion the continual change process. A *change revolution* is needed to pave the way to allow an entrepreneurial spirit to be breathed into the organization. Without committed leadership in the change process, the results will be lukewarm at best.

Management leadership must expound its vision to all segments of the organization. Leadership infers ensuring that changes achieve preestablished goals and objectives by action, as the torch bearer, cheerleader, and "barriers to success" remover.

Change must be inspired and leadership focused. This means that management must demonstrate *a living example*, being the fire breathing advocate of the change process and cheering on successes in a very visible fashion. Cheerleading, in this regard, is more than just accolades. Cheerleading is a day-to-day commitment to seek and destroy all roadblocks to the success of the end-to-end combined business process. Reiterated, this means that a significant part of an excellent leader's day becomes championing the change process.

The role of leadership is changing. Many of our stodgy leadership practices of the past must give way to the "fast cycle"-oriented competitive spirit practices focused upon productivity improvements at the individual employee level in the future. Dexterity and finely honed skills of our new leaders will require significant departure from procrastination practices of the past. Planting, and subsequent harvesting, the seeds of leadership excellence requires an unrelenting focus upon the skills, capabilities, and potential of the leaders charge.

The To Be model will rely heavily upon the leadership excellence demonstrated. Transitioning from the As Is to the To Be quickly and in a coherent manner relies upon leadership's ability to be nimble, flexible, responsive, innovative, and adept at transferring the customer-focused competitive spirit throughout all facets of the organization. The lack of such leadership will perpetuate mediocrity and inhibit the organization's ability to transition quickly, if at all.

8.18.2 Waste Management

Most companies have participated in programs geared at such themes as profit enrichment, cost constraint, and control. However, these programs seldom demonstrate the focus needed for long-term profit improvement perpetuation. Therefore, their benefits, if any, soon vanish. One of the primary reasons that benefits are fleeting is that our mental attitudes fail to be *waste management* oriented. The operating practice and goal of *global competitors* specifies waste, which, in any form, will not be tolerated.

Frequently, we relate waste to things such as scrap or, in a services industry, excess office supplies. Seldom, however, do we relate waste to such things as "intellectual energy," energy expenditures, decision support practices, and business process throughput. The waste management challenge of the future goes well beyond the standards of the past. Companies that will become leading competitors of the future must rethink waste and redefine it in terms of management effort and total employee utility, in concert with maximizing throughput and stellar end-to-end business process performance.

Waste persists as a productivity detractor. Waste gnaws at the very fiber of productivity within the organization. Employees become creatures of habit and seldom recognize the erosion of productivity that occurs by waste. Permitting waste in any kind to occur, in any facet of the business, gives license to perpetuate waste elsewhere. A pertinent example of permitting waste is the quality term "AQL." There is only one AQL, if waste management is to become a guiding principle in the organization, and that level is 100%. There is only one AQL in data accuracy and that is 100%. The computer doesn't care whether information is accurate or not, but decisions must be based upon accurate information if they are to be reliable quality decisions. We need to recognize that frequently day-to-day compromises have been accepted as NORM and we need to redirect our energy away from compromise toward achieving a 100% quality level.

Some of you purists will say: Now hold on, how can we possibly attain 100% level (consequently zero waste) in every business process? Others may point out that there is a high cost associated with pushing the standards beyond the statistical >97 percentile. Does that mean it is beyond management's ability to achieve perfection? Does that mean that we shouldn't set our goals on total waste elimination?

Recall the story of Motorola that set a Six Sigma product quality standard and their leadership probably asked similar questions, but who were "solutions"-oriented and found a way to exceed their Six Sigma goals. Was that a management team that suggested that striving toward perfection was impossible? *Hardly*!

The business success stories we recognize today consist of leaders who have attitudes that are actually tuned toward solutions and who refuse to accept mediocrity as a viable alternative. Progressing upon a waste management journey is not easy, but significant improvement has been achieved by those daring to pursue a trait of excellence.

As indicated earlier, the degree of success may well depend upon the balance the organization can maintain along the journey, but, if we are mentally defeated (before we get started) by merely giving lip service to the concept (we like the benefits but unwilling to make the investment), will result, like so many other projects, into never really getting off the ground.

As we look at waste management opportunity, let's not forget about waste of intellectual energy, waste resulting from poor information quality, and waste resulting from hierarchical measurements. The benefits can be phenomenal if we're willing to step outside our comfort zone, exhibit an outrageous purpose, and be radical in our commitment.

8.18.3 Intellectual Energy Conservation

I've attempted to convey a deep conviction related to intellectual energy management. One of our greatest untapped resource pursuits should be optimizing the "intellectual energy" of our *entire resource team*. There's an inordinate number of activities, events, and procedures that inadvertently sap the intellectual energy from our management teams. Crises, excessive meetings, employee problems, competitive pressures, quality problems, company politics, and so on all grind down the available time and energy of our organizational resources. Hierarchical organizations naturally foster intellectual energy deprivation manifested by the buzz associated with care and feeding of the organizational unit activities (the bureaucracy). The lack of exploiting the full spectrum of people talent causes the individual to perpetuate intellectual energy drain. The inability for organizations to leverage the return on IT investments as well as the lack of future-oriented decision support, which should be a substantial enabler, have not proven to help optimize intellectual energy. Throughout this chapter I have highlighted distractions and leaches busy sucking the intellectual energy lifeblood from individuals, teams, processes, and organizational entity as a whole. What can be done to conserve this energy bank drain and limit usage to value-added tasks?

We must recognize that the discussions on job description, hierarchical structure, productivity, and so on involve corrective action opportunities, which will support an intellectual energy optimization strategy. But without recognizing and acknowledging these distracters, a cure is not possible. Companies and individuals must engage critical thought upon understanding the magnitude of day-to-day intellectual energy drain. After recognizing the causes of energy drain, a priority must be assigned to cure activities. A by-product of migrating from hierarchical to process-based measurements is a giant leap forward.

Let's clearly define *intellectual energy*. It is the time, bodily energy, focus, and influence spent in value-added decisions, design, or process execution that directly contributes to the profitability of the organization. At this point, I like to challenge executives, managers, and working-level personnel to critically assess their *intellectual energy productivity level.* Do they invest 90% of their time in intellectual energy investment for the company? 80%? 20%? The typical response ... I've never really thought in terms of intellectual energy management. Consequently, few executives can provide an adequate response. Most respondents talk about their stress level, frustrations, problems, and other symptoms of intellectual energy deprival, but few can relate to even reasonably high productivity. Without an intellectual energy optimization strategy, the intellectual energy bank will be depleted regularly and progress will not be made toward corrective action.

The potential for improving intellectual energy productivity becomes enormous in most organizations. Taming the sapping shrew is the challenge.

8.18.4 Agility Deployment

Agility is the ability for the management team to perform in a nimble, flexible, fast cycle-oriented, and maximizing intellectual energy manner. Examples of agility are continuous improvement processes such as the following:

- Reducing cycle time in half (or more)
- Reducing product development and new product introduction time by 75%
- Increasing quality and reducing cost at a rate substantially quicker than competition
- Providing substantially greater levels of service for the same cost of the competition
- Reducing setup time to "single-minute exchange of die (SMED)" swiftness
- Responding to customer needs immediately

Agility will become a means to separate the real winners of the *next decade* from those who merely participated in the competition. It requires a change in our mental attitudes. Not only do our actions need to be nimble and quick, but our problem resolution thought process needs to become swift solution oriented.

This section was devoted to defining the "to be" or better design attributes within an organization on an end-to-end process basis. We took a look at the following:

- Refinement, streamlining, and radical reengineering deployment guidelines for the end-to-end business process network
- Various strait jackets that limit the organization ability to become more agile, which included the following:
  - Job description
  - Hierarchical structure
  - Productivity incentives (lacking)
  - Performance measurements (nonprocess-oriented)
  - Lack of leadership
  - Waste management
  - Intellectual energy conservation (dissipation)
  - Agility deployment (inadequate)

The To Be design sets the stage for enhancing organizational productivity potential of the future. We can continue to *wallow in the mire of mediocrity* or we must rise above mediocrity's stench and strive for excellence for the future.

8.19 Developing Action Plans That Deliver Effective and Orderly Change

A bundle of change ingredients has been presented, but what is now needed is the creation of action plans, which are crucial for change to occur successfully. With the avalanche of downsizing activity, poor financial results, increased levels of competition, and tight money supply, both individuals and companies are eagerly postured to affect significant change. Where do we start? What are the risks?

As expressed earlier, commitment and leadership are crucial to transition from the As Is to the new To Be culture. Because many of the new cultural concepts are so radical (compared to how we operate today), focusing upon the journey ahead can look intimidating. However, developing transition plans can be meaningful if we employ the full talent skills of our entire resource base … gee, it sounds like the journey should take advantage of contributions from its entire employee base. In fact, if we cannot demonstrate our ability to exploit the working-level resources during the transition effort, why would we think it will work later?

Now we're really concerned … you mean that we should expose the "keys to the kingdom" to that great body of "unwashed." For over 20 years, I have worked with companies which had said that they were committed to making "significant" change in their business management systems. The formal systems implementation projects spanned virtually all functions of the organization. Some projects were a matter of convenience and others were a matter of survival. Yet, in every case, I would always critically assess the attitudes of the employees. What did I conclude … I'll take a positive "attitude" toward change over despondent technical skills any day. Consequently, commitment and leadership play an emphatic role in performance-based outcomes.

Transitioning from the As Is may require a bit of education. First, let's presume that the principles expressed in the As Is and the flowcharting sections will be closely adhered to. Let's further surmise that functional consultive champions have been appointed as discussed in the Section 8.12. Second, a critical review of the As Is by the functional consultive champions is in order.

Recall that one of the purposes for conducting the As Is task was to identify opportunities to remove nonvalue-added effort from the business processes. The consultive champions can become a significant resource to be used as *change catalysts.* However, they must have a mastery of the current flows, assess the nonvalue-added impacts, and assist in developing a risk assessment. Here's where *process mapping* becomes a *critical success factor.* Process mapping has two structural forms:

1. It identifies the following:
   a. Process owners
   b. Internal customer/service provider agreements
2. It transcends the following:
   a. Old to new
   b. Conference room pilot

The process map describes *entities, attributes,* and *relationships. Entities* are simply things such as people and data elements. *Attributes* describe things. For example, attributes of a person would be hair color, age, height, weight, and so on. *Relationships* represent how things interrelate. For example, people relationships would be mother/son, husband/wife, student/teacher, and so on. Developing this logical *process map* will be helpful in transitioning into the To Be design.

Don't be surprised to find that many of these supposed "logical" definitions actually turn out to be very *illogical.* Recall that an earlier dialog which explained that various of our business processes simply occurred not by design, but rather merely by circumstance. Our business processes may *not* have been developed based upon good business or using logical reason. Consequently, the essence of activity existence is without rational basis. Stated another way, what we find is that we may end up putting effort into activities that have little or no payback. The process map helps us define many of these logical flow discrepancies.

As we develop the process map, we are able to highlight the nonvalue-added tasks (entities), events that cause delays (time perspective attributes) and points which are not customer focused (entities, attributes, or relationships). Let's take a look at the billing business process and a process map. The business flow might look like this:

Prepare to ship product	Generate shipping papers
	Bill of lading
	Packing list
	Product certifications
Ship product	Process shipping transaction
	Inventory relief posted
	Shipment posted
	Sales order relief posted
	Sales history posted
	Billing advice generated
	Accounts receivable entry posted

(Continued)

Accounts receivable (A/R)	Billing
	Generates invoice/remittance advice
	Prepares billing statement
	A/R aging
	Aged and projected A/R
	Commission management
	Dunning notices
	Post receipts
	Cash application
	Bill backs
	Credit memos
	Partial/overpayment
	Nonsufficient funds
	Freight charges

Let's now generate substantially abbreviated entities, attributes, and relationships table (Figure 8.2).

We could continue to list *all* the entities associated with the billing process, each entity's associated attributes and then the related relationships. To place an order of magnitude on just the billing process, we may have hundreds of entities, hundreds/thousands of attributes, and hundreds of relationships. Therefore, a business process, which, on the surface, may appear to be simple, may have a complex network of activity associations. This is especially true if we are process mapping our company and our multiple-tier trading partners to look at the entire community opportunity, not just the local opportunity.

Once we define entities, attributes, and relationships, we are poised to conduct process mapping, which is combining the As Is physical flows with the associated entities, attributes, and relationships. Now we have sufficient detail to identify ways to eliminate nonvalue-added activities and redundancies. As we begin analyzing the process maps, we must present the following challenges:

- Is the process/task customer focused? (customer-oriented)
- Does the element/task/activity add value? (add value)
- If there is no value added, is it still necessary? (can we eliminate)
- Is it redundant? (eliminate)
- Is it needed to prevent errors from occurring? (audit)
- Does it negate an activity previously performed? (rework)
- Can we combine it with other tasks/processes? (combine)
- Can it be simplified? (simplify)

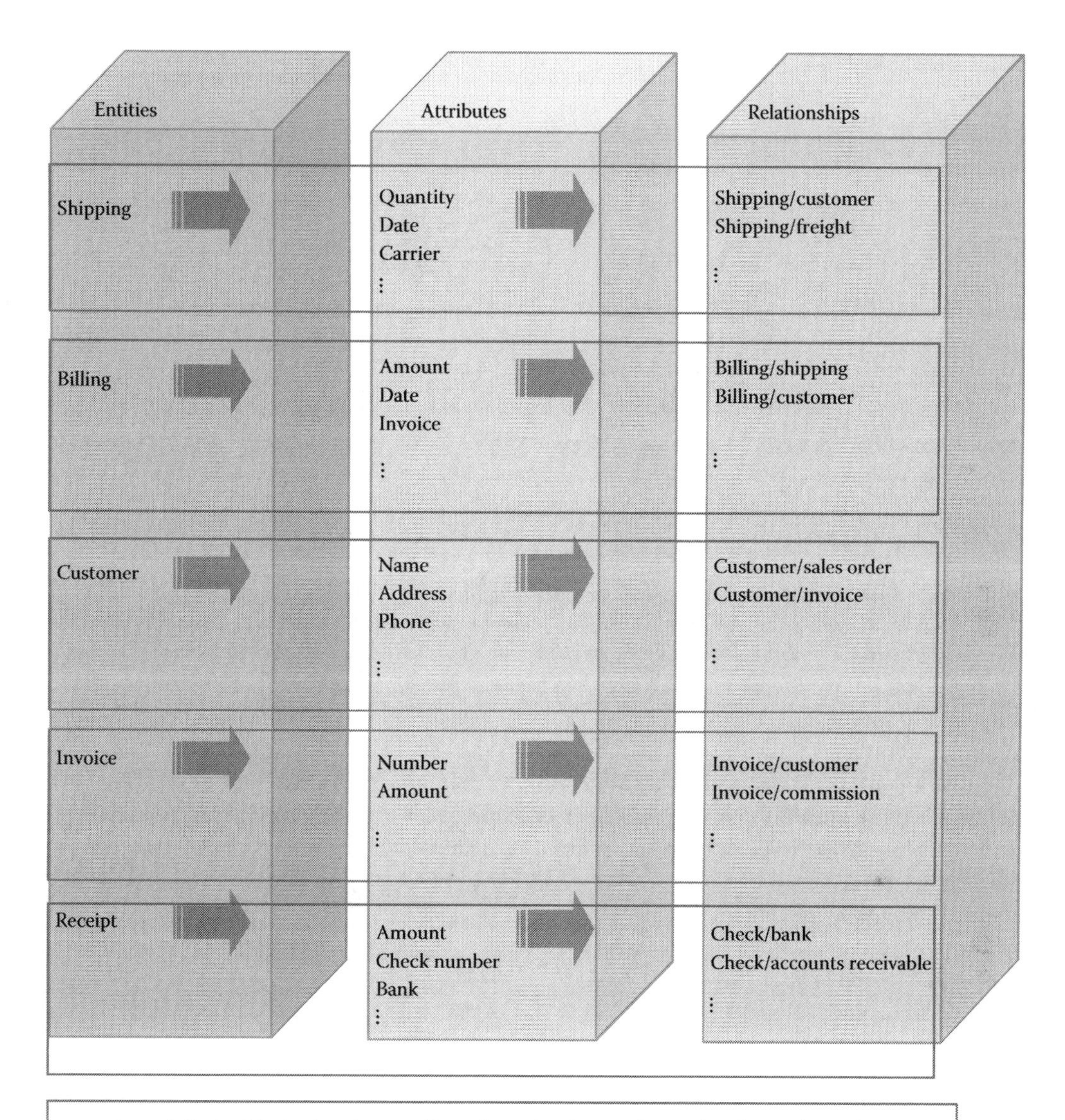

Legend: Entity shipping has three attributes (quantity, date, and carrier) and two relationships (customer and freight)

Figure 8.2 Entity, Attributes, and Relationship.

- Can it be enhanced or performed in a better manner? (improve)
- Can we reduce the time it takes to perform it? (cycle time)
- Can we contract it out at a faster, better, cheaper manner? (improve)
- Can we eliminate it from being in the constraint path? (constraint)
- Can we downgrade the skill level to perform it? (deskill)
- Can the information system do it automatically without human intervention? (deskill)

When we summarize the challenges, the result falls into five categories:

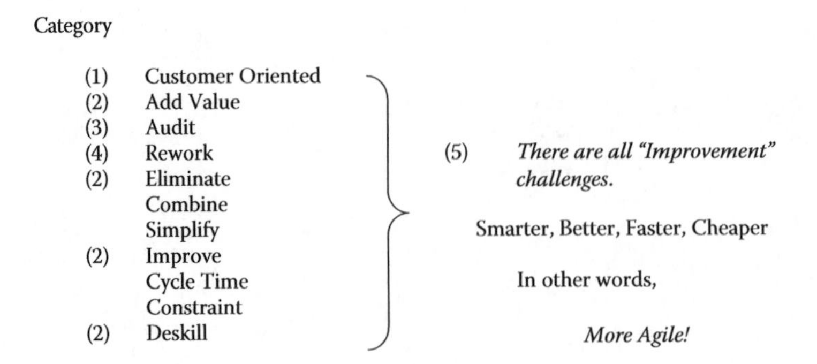

8.20 Prioritizing

The opportunities resulting from the process map must now be prioritized. Depending upon the magnitude of reengineering activities, priorities should normally be established based upon a mix of *quick hits*, *return on effort*, *profitability impact*, *risk*, and customer and *user preferences.*

Prioritizing must also be viewed along two planes: *strategic* and *tactical.* The strategic plane is interested in longer term impact, whereas the tactical plane addresses the immediate day-to-day impacts.

Other important ingredients are *attitude* and *championship.* If there is a burning desire to affect change, then the probability of success increases dramatically and the timeline compresses.

Finally, priority consideration must be given to resource impacts and the number of overlapping change implementation teams working simultaneously. Let's refresh our memory of the pilot approach previously presented.

Baseline ingredients:
- Vision defined
- Change transition leaders appointed
- Quick response orientation
- High visibility
- Will do theme

Activity ingredients:
- Executive champion
- Latitudes defined
- Employee incentive
- Empowerment permutations
- Pilot determined
- Process results assessment by every employee
- Consultive champions appointed
- Executive change counsel established

We are now postured for success. However, we must develop the concept of empowerment a bit more thoroughly if we are truly going to be very successful. The contemporary rhetoric regarding teams and empowerment has too broad a base of interpretation.

Back in the 1960s, a concept labeled "managerial grid"* discussed organizational dynamics, which developed a matrix grid showing *concern for production* along one axis and *concern for people* along the other axis. The ideal was to have a 9,9, whereby the highest level of production (9) and the greatest concern for people (9) converged. In sum, there were three distinguishing principles conveyed in this 9,9 convergence:

1. Committed people produce the highest level of quality work.
2. "Success" interdependency surfaces when there is a common stake.
3. Purpose leads to relationships of trust and respect.

These certainly sound a lot like empowerment ideals. We get "committed people" by shared ownership. We maximize cross-functional productivity by exploiting the interdependencies of the nested internal customer. Trust and respect are essential for teamwork. Here it is decades later and we're still trying to implement "empowerment" principles. Were these ideals that difficult to comprehend?

I submit that one reason that empowerment has taken so long is primarily due to the hierarchical organizational dynamics. The 1960s brought with it a growth within organizations. The 1960s–1970s was mainframe and centralized computing as well as growth in "middle management" needed to use these new tools. The 1980s–1990s brought with it the personal computer and drive for decentralized ownership of information. With global competitive pressure came the need to downsize our organizational behemoths in order to survive. With the hierarchical organization shrinking (drastic reduction in middle management) came *ownership at the source* and facilitating this ownership at the source came empowerment. Don't get me wrong, there's been a tremendous technological explosion, a resurgence in attitudes toward customers, and a tremendous broadening of products and industry. However, I am a firm believer that the hierarchical organizational structure has been the greatest barrier to empowerment.

When a company truly embraces empowerment principles, the management ego must deflate. Decision authority must be transferred from the "gridlock few" to the working-level "critical mass." Information must be designed by and shared with the working level. The reward and compensation must also be shared with the working level if "real" long-term success is to occur. The principles to be embraced include the following:

1. Empowerment begins by *exploiting (optimizing) teamwork*. Transferring decision authority into the hands where process impact occurs is much more efficient than elevating decisions and approvals up the organizational ladder. However, teams must be properly trained to accept this new responsibility. Therefore, training the workforce will become one of the greatest challenges companies will face in the next decade.
2. Teamwork excels when team members agree upon goals and objectives, and clearly understand deliverables. This means that management must translate "financial"-oriented business drivers and metrics into working-level terminology. Said another way, every employee must understand how he or she contributes to profitability of the organization. However, the essence of this translation process is to develop, as discussed, "process-oriented performance measurements." If we can maximize our performance in the trenches (cross-functional optimization), the pent-up productivity exploitation will result in maximized financial goals.

* Robert Blake and Mouton, Jane, *The Managerial Grid*, Gulf Publishing, Houston, TX, 1966.

We must ensure that our maximization process keeps our activities focused upon customer needs, wants, specifications, and service levels. If we can merely maximize the productivity of the individuals as they relate to the team, then maximize the teams as they relate to the business process, and finally maximize our business processes as they relate to other business processes and the customer. We will maximize our company-wide productivity. However, the process must begin by defining goals, objectives, and deliverables in *working-level terms*. Expecting working-level employees to relate to financial-oriented objectives is foolhardy! Management must take the effort necessary to translate these business drivers into common-level language that is championed by every employee.

3. The working mechanics of empowerment stress the horizontal intergroup activities more than the vertical. For empowerment to yield its potential:
   a. Self-directed work teams must make the day-to-day decisions needed to operate.
   b. Tactical leadership serves the role of coach more than boss along the vertical linkages.
   c. The strategic leadership role of the executive converts the knowledge gained from the coaching activities into organization and talent strengths. These strengths may be exploited into competitive advantages for the future.

   Let's restate this concept a bit. The empowered organization downplays the organizational hierarchy and promotes the interactivity among peer groups across the horizontal (business processes). Looking back at To Be discussions, we find that the leadership role of the executive is radically changed. The majority of decisions no longer are elevated to the top, but rather are made within the working-level ranks between nested teams of peers. The more radical organizations eliminate job titles and view each other as "associates," simply with different functions.
4. Once we've converted the goals, objectives, and deliverables into process-based working-level terms, we can unleash the broad-based working-level talent into developing self-directed continuous improvement goals. Who is more of an expert and better qualified to define the changes needed than those who make things happen in the trenches day in and day out. Understand that the working-level mass needs the "vision" from leadership to align with corporate goals, objectives, and priorities.

   Let's recall that a portion of every workers day should be spent designing ways to improve the business process. If we spent 100% of our workday effort executing, how will continuous improvement possibly occur? For continuous improvement to occur, we must spend time designing improvement ideals. Who, outside the business process, is more qualified to engineer improvements to champion the cause and then execute the effort needed to affect the improvements?

   Some of you purists are saying, Isn't that opening the door for the fox to enter the chicken coop? Aha, I've got you now. Do we really believe that "inspecting in quality" is superior to "designing quality in?" Never, never, never!!! Yes, we do need proactive management leadership expended in the coaching activities to ensure that the continuous improvement is truly improvement and that the solutions are total system sensitive rather than localized and parochial. We may even need to monitor progress to goals a bit to ensure that we're meeting our strategic goals and objectives. However, isn't that what leadership and process management is really about?

   What role of management is more important than working with the organization to coach and remove the barrier preventing working-level employees the ability to achieve success? Now, looking at today, how much of your time is spent coaching and removing barriers so that your charges achieve 100%+ of their goals and objectives? The answer probably reflects the level of your organization's commitment to empowerment. Hmmmm!

5. To restate a paragraph earlier discussed, "empowerment cannot be delegated but must be accepted and championed by individuals ... management must provide an environment conducive for empowerment ... " What this concept conveys is actually the opposite end of the continuum of the way many business management teams think today. Let's refer back to our hero! Remember that this is the individual who "saves the day" and who is so important that he or she couldn't possibly function under the constraints of the formal system. We continuously reward these heroes and elevate their "win at all costs" performance attributes as being the "way we all should do it." How does continuously rewarding heroes contribute to the success of empowerment? A management team can say whatever words they would like about the importance of empowerment but "walking the talk" of empowerment discourages all forms of hero practice. Companies must decide which road to take. Do we support the superstar avenue or do we support self-directed process-based work teams?

   At this time, some of you are seething and saying, Why not support both? Can't we get the best out of our heroes and yet get the best out of our teams? I guess someone could build a case where the heroes would *never* come in conflict with a team activity and the team activity would never be compromised by the hero effort. Yet, we're talking about cultural style and a culture must promote the good of the whole. How could the good of the whole reward those few without compromising the team? I guess we could lift up the marvelous national welfare system as an example. In the welfare model, those incented to work and choose to work end up supporting those who don't work and who choose not to work. Relating the welfare model to the hero, the hero could do all the work and the welfare recipients (teams) could rest upon the laurels of the hero. Now, is that productive? Are we getting the highest return on our investment using this model? Hardly! Yet every time we reward the hero and the efforts of the rest of the group are compromised, we broadcast our true nature to the entire organization.

   Now, the rest of you skeptics are asking, however, how about the commissioned sales heroes? Aren't they successful? First, the top commissioned sales performers are usually the ones who exploit every advantage they can get from their support team and they affirm the importance of team exploitation. Second, they have a "performance-based compensation" incentive, which we will be looking at this more closely later.

8.20.1 Attributes of Effective Action Plans

There are various critical success elements essential to making a quality transition from the As Is to the To Be. These elements include the following:

- Ownership
- Flexibility
- Simplicity
- Service
- Responsiveness
- Cost

Let's look at these a bit closer.

8.20.1.1 Ownership

The highest quality transition efforts come from those committed to the To Be. These visionary architects of *a better way* are able to contribute significant leadership effort. Marshaling a vested interest in the change creates a spirit that is undaunted by resistance to change. If working-level employees embrace the To Be vision, they are able to see the benefits to themselves, the benefits to their peers, and the benefits to the company, and are encouraged to make the change. The result is an *attitude* and *motivation*, which is difficult to constrain.

8.20.1.2 Flexibility

Transitioning, with the magnitude of change most organizations must make, will be *exciting*. Because the vision consists of unexplored territory for most companies, the rocks in the middle of the road may be many and large. Consequently, latitudes should be broad and flexibility maximized. Let's make sure that we understand what flexibility means. Just like **accuracy** represents freedom from error, **flexibility,** likewise, represents **freedom from constraint**!!

What an opportunity ... Management is frequently so "conditioned" to control that a knee-jerk reaction to flexibility is okay. We'll control the level of freedom and impose "our" influence upon the process!! Freedom from constraints means just that ... imposing management's preconditioned biased and deeply rooted magnetism to the As Is (probably much of their design) is a kiss of death toward accomplishing the To Be vision.

If only we could step out of our skins, for a moment, and view the To Be without the fears and shackles of our experience, what a refreshing journey that would be. This, stepping out, then, is that freedom from constraint that flexibility infuses. It would be like starting the company "brand new" with a clean slate, and amassing our energy and motivations, which are entirely focused upon viewing every individual resource as an integral ingredient in the productivity solution. Can we handle that!

We will discuss flexibility more later, but at this point we need to freely admit that flexibility can function as the blood transfusion to add new vitality to a tired machine.

8.20.1.3 Simplicity

What more can be said than elegance is highlighted by simplicity. If we pursue the KISS (keep it simple, stupid) principle we can certainly save time, money, and effort. Brevity, conciseness, thrifty, and nimble conveys the spirit of simplicity. It's opposite is complexity. Most of us and our companies have mastered the art of making the simplest things complex. Peeling back the layers of complexity and stripping the rust (warm and comfortable) away are more than a challenge to most organizations. We need a dose of refreshed mentality. We need to reverse engineer most of those layers of *can't do's* and invoke fluidity in the *can do*, and drive this mentality throughout the bowels of the organization. Just as Igor Stravinsky related "freedom comes from (our) restrictions," mastering organizational virility comes from pushing the envelope of simplicity throughout the talent base, IT tools, and processes of the company.

8.20.1.4 Service

The new To Be culture is the epitome of optimized service. Every employee must be focused upon the nested internal and external customer needs. Therefore, employing a "service" attitude in the transition process manifests the commitment to service. A definition of "transition service" may be appropriate. We may begin the process by asking those who will be impacted by change such questions as *What would make you more productive?* or *What approach can be used to minimize distractions in your sphere of influence?* or *What information will assist you in effecting change more smoothly?* A service orientation subordinates the traditional parochial perspective (our interests) in favor of exploiting the customer perspective. At this point, those of you traditionalists and mentally retired bureaucrats feel your blood pressure rising! You can't possibly suggest that working-level personnel are worthy of being provided service!! That's heresy to command and control principles. Recall that command and control specify that those in authority *dictate* and everyone else must kowtow. Otherwise anarchy will prevail and "those of us who have earned our rights to rule" will be compromised!! We're at a serious juncture now. We must decide whether tradition is to reign or are we really committed to propose revolutionary changes to the traditional hierarchical organizational culture? If the past is to dictate how we will run the business of the future, don't waste your time reading any further. Additional thoughts along these lines will cause you to become irate, at best, and some of you diehard bureaucrats may form hives or experience other uncomfortable physical reactions.

Let's reiterate, I am really proposing that the working-level perspective, not only be considered, but allow it to become a driving priority in the transition process. I'm advocating that the working-level personnel be allowed a voice and represent a significant impact to the transition decision process. I'm further advocating that the working-level personnel be afforded true leadership respect and that their impacts be given equal, if not greater weight, than those who have become elevated in the management (bureaucratic) ranks of the organization.

Service goes beyond involving the right resources. It also encompasses the quality of the transition. The cross section of companies I've been most exposed to over the years could benefit by a quick implementation. At this juncture, we must be a little cautious. *Implementing radical change* quickly requires the change process to be brought into by every level of the business. Even if we are only piloting a process, the change has such a broad impact that we are looking for a common vision, agrees upon goals, explicits performance measures, and requires constant monitoring. Our executive champions, transition leaders, and consultive champions must be acutely focused upon quality results.

The service process also requires managing expectations and perceptions properly. The contributing factors involve the level of reengineering, the level of flexibility, the budgetary constraints, and the magnitude of effort. Compromising any of the critical success elements limits the potential results and jeopardizes the overall success. The Executive Change Counsel must be especially sensitive to signs, motives, and roadblocks as the process proceeds. If any cracks seem to surface, the issue must be hit head on and resolved immediately, and the transition process reinspired. Any lingering delays, procrastination, or distractions can fester into a canker that can significantly affect the *hard fought turf* gained. It's like football, you can gain great ground, but if a fumble occurs, you can even lose possession of the ball. Swiftness can result in recovery of the ball, and a touchdown can result in loss of ball and can be a turning point in the game with WIN or LOSS potential.

8.20.1.5 Responsiveness

I split responsiveness out from service because I wanted to stress the "agility" perspective. Earlier, agility was developed as an important key to the To Be design. The whole concept of fast and adaptive response is central to reaping significant productivity gains.

Demonstrating the ability to perform an activity faster with the same resource level of inputs without compromising quality is true responsiveness. Our external customers are expecting quicker response. Our internal capabilities are paced by the weakest link in responsiveness. Our systems are expected to be more responsive.

Consequently, responsiveness should be a mind-set in our transition profile. The global competitiveness drumbeat stresses "Better, Faster, Cheaper." We can be sure of one thing: the future dictates that companies become *more responsive*. The survivors of the next decade will become more acutely responsive. We can either conform to a continuously improved responsiveness standard or lose ground to our competitors. The competitor factions are becoming aligned within responsive markets. You're either a predator or a prey. Each company must choose its plight. Posturing today will influence position tomorrow.

8.20.1.6 Cost

The transition process has cost, investment, and benefit impacts. I've deliberately separated cost from investment. Costs are out of pocket expenditures to affect the transition process, whereas investment has longer term cost and is more opportunity oriented than necessarily out of pocket oriented. Benefits can be a result of cost expenditures, investment in opportunities, or windfalls from changing practices.

Let's take a case at point. We can choose one of three strategies to pursue if we were to proceed with radical reengineering.

Strategy 1: *Downsizing*—This involves identifying areas of waste, redundant personnel, and immediate cost avoidance.

Strategy 2: *Same sizing and redeploy*—This involves the same waste identification, redundancies, and cost avoidance but personnel normally lost (downsized) in executing this strategy are redeployed into other areas of the company.

Strategy 3: *Growth and eating competitor's lunch*—This involves the same waste identification, redundancy, and cost avoidance, but the savings are converted into an aggressive competitive posture and an organization mind-set to eliminate competition.

Let's look at these strategies a little closer.

Downsizing (strategy 1)—It is by far the most popularly pursued course. Companies remove waste, use information tools to replace manual activities, and pursue aggressive cost avoidance. However, if these companies don't affect a change in culture, the cost savings will most likely be short-term oriented and waste will creep back in slowly over time. If not handled properly, this is just another cost reduction program, which seems to reduce cost but adds very little long-term return (cost avoidance analogy).

Same sizing and redeploy—Companies make "investment" by retraining the redundant personnel and placing them into open personnel requisition slots or assigning the personnel to other locations. These companies are most likely more competitive as a result of their strategy deployment and they will slowly gain market share (investment analogy).

Growth and eating competitor's lunch—Companies are out to conquer their competitors and, as much as possible, eliminate competition. These companies are the ones that pursue radical change and are expecting robust exploitation of their talent pool. These are the companies that are empowering their organizations and are challenging the hierarchical organization. They are fully benefits postured and regularly exploit windows of opportunity (benefits analogy).

The strategy selected is a matter of style and those maximizing competitive posture will be the future captains of industry.

This section focused upon successfully transitioning from the As Is to the To Be model. Specific points to be considered include the following:

- *Process mapping*—Defining the *entities*, *attributes*, and *relationships* and developing a logical flow in the new process design.
- *14-Point checklist*—To help ensure that nonvalue-added steps are removed from the new process design, this checklist challenges the project team to ensure that all aspects of the new process design are crisp and streamlined.
- *Prioritizing*—Discussions included priority along *tactical* as well as *strategic* planes and outlined the method to generate an effective pilot approach.
- *Empowerment*—An important aspect of exploiting the critical mass resource base centers upon flowdown accountability and included such concepts as follows:
  - Exploiting teamwork
  - Establishing agreed-upon goals, objectives, and deliverables
  - Dissolving hierarchical structures in favor of process (horizontal)-oriented structures, and then converting the business goals and objectives into measurements understood by "working-level" personnel
  - Developing continuous improvement approach to meaningful working-level goals and objectives
  - Diffusing the hero impact and leveraging team productivity throughput

The attributes of an effective action plan include the following:

- Ownership
- Flexibility
- Simplicity
- Service
- Responsiveness
- Cost

The transition will require a *quality* approach focused upon *maximizing productivity potential from every worker*, instilling the *vision* and *values* that enable perpetuation of success and *retention* of the *best people*. The new role of executive leadership will mandate a method to transfer responsibility from management into working-level hands. The transition itself would exercise the greatest latitude of team involvement and decentralized decision authority. The challenge is great, but the potential is gigantic.

8.21 Emerging Natural Work Teams

It's difficult to read a productivity improvement article today, which does not emphasize the importance of "team effort." Information truly is the lifeblood of business process effectiveness. However, most organizations wallow in the mire of data and few rise above the "data grind."

A critical assessment of many companies that have "automated" their business process shows that they have merely taken their manual system and put it onto electronic screens. Data is electronically processed, yet very few use computers as decision support tools. Why is this?

As a bit of a systems recap, one of the inherent flaws of system implementation projects was that the "automated system" architecture, transaction processing approach, and output media designs have emulated the manual system. Essentially, the systems conversion approach was to "take what we do today and do it electronically." With affordable desktop tools such as Windows, Excel, PowerPoint, and a host of others, the common systems user abandoned their "wait for data processing" response delays and opened their own "hotdog stands" and started mining data, which could be converted to decision support and productivity enhancement capability. However, these functions were not integrated with host applications, resulting in "various sources" generating core information. Management teams started recognizing that the lack of integration of various source data created numerous communication problems and host systems started talking (downloaded data) to spreadsheet, word processors, graphics, and a host of other software packages.

The trend continues to be decentralized data sourcing and that is the design architecture of the "client server" platform.

I believe that transferring data accountability "closer to the source" of activity is essential for data ownership. Data must then be converted into information productivity. Consequently, the conversion of data to information will play an important role if information management is to live up to its potential.

However, how do we *best* approach the conversion of data, information tools, and so on into productivity enhancement? First, we could certainly encourage each user to autonomously search the information "universe" and become independently skilled at technical information skills. Second, we could allow our information technicians or IT team members (IT) to define the tools in a vacuum on behalf of our users and then turn the tools over to them. Third, we could team up information experts (IT) with their nested internal customers (users) and form natural work teams. Let's look at each alternative a bit closer.

Alternative 1: Encourage each user to become a technical information expert and each autonomously becomes a pioneer.

On the surface, this certainly sounds appealing. In fact, in many companies, because of inadequate support from the IT department, many users have started this journey. Some users were intellectually intrigued about pursuing a technical adventure. Others were cautious but peer pressure forced them into reaching out. Some had to do it to survive the rigors and pressures of the job.

However, we must ask the question: Dispatching a vast number of pioneers who will each blaze their own trail, is this the best use of our resources? And, although they are attempting to become technologically current on use of information tools, can they adequately discharge their line responsibilities with the same zeal, energy, and commitment needed to meet the business goals?

Again, our challenge is to find the *best* approach, and I believe that launching a bunch of autonomous pioneers is not the best utility of resource. Let's look at the second alternative.

Alternative 2: Allow information technical resources (IT) develop tools for users in a vacuum.

This alternative has been practiced extensively over the years with dreary results. There have been many problems:

The IT resource doesn't understand the intimate needs of the users (they lack user skill).

Users have difficulty conveying their needs sufficiently for IT resources to convert into effective decision support tools.

IT resources tend to design solutions that are technically elegant but less than user friendly.

IT resources seldom possess the vision needed to place the information tools on the cutting edge of competitive capability.

This alternative certainly does not appear to be the best approach, so let's look at the third alternative.

Alternative 3: Form a natural work team by blending IT resources with their subject matter expert internal customers.

The concept of natural work teams is a powerful blend of expertise. The users can continue to perform their technical skills, yet they are teamed up with IT expert. The IT resource can observe the day-to-day practices of their internal customer (they live and breathe in the trenches with their internal customers). This alternative looks like the best solution. Let's look at it in more detail.

8.22 Natural Work Teams

A natural work team is a blend of the functional expert working closely with their IT nested partner. The synergy that surfaces from this alliance, if structured properly, is powerful. The rules are simple: The IT resource is chartered with full-time service inside the ranks of their internal customer base. Normally, the IT resource is assigned to a business process and typically crosses traditional organizational boundaries. Their report card comes from their internal customer deliverables.

Let's look at a practical blend of sustaining IT functionality with customer-focused activity. First, we must always keep in mind that whatever the resource expended, it must be focused upon external customer needs. This includes these natural work teams. The nested internal customer and IT resource must be sensitive to external customer value-added processes. Second, I will presume that reengineering activities have already been accomplished as previously portrayed. Third, we must develop the rules by which priority can be balanced between natural work team partners.

The process of decentralizing IT resources will involve a variety of internal customers sharing the same IT resource. Therefore, a priority assessment strategy and an arbitration process must be defined in advance. Traditionally, priority was given based upon hierarchical rank or who yelled the loudest. Under the natural work team concept, the user community should agree upon a priority strategy whenever possible. However, if a conflict rises, the functional system champion becomes the adjudicator.

Let's recap the role of the IT resource. They are to work side by side with their internal customers and help identify ways to use information so that users are more productive. Recall that productivity is getting increasingly more throughput with the same resource base. As a refresher, ideally, we would like to redeploy our increased productivity into gaining market share from our competitors. Consequently, revenue per employee continuously increases over time.

Let's look at this side-by-side partnership with users. How does the relationship function so that the IT resource does not distract from the functional role of the user, yet it is able to identify productivity enhancement opportunities? There are four distinct functions used by the IT resources:

- Consultive review of process
- Consultation with the natural work team downstream nested internal customer
- Daily consultation with internal customer users
- Debrief with functional system champions

Let's look at these processes more clearly.

8.22.1 Consultive Review of Process (Independent Observation)

A key role of the IT team member is to transform from a technical subject matter expert into a generalist role. A good technical resource is able to observe the data highway and recommend changes, which will

1. Reduce the time it takes to be converted from data to information.
2. Consolidate data conversion modes and eliminate redundancy.
3. Eliminate nonvalue-added data occurrences.
4. Identify methods to make information processing more user-friendly.
5. Deploy technology tools that enhance the information highway.

Observation and development of productivity enhancement opportunities is essential. However, the user may not necessarily be receptive to the recommendations that arise. Consequently, the recommendations are proposed before a weekly process collaboration forum consisting of user, user's internal customer representation, functional systems champion, and with ad hoc visits from the Executive Change Review Counsel member.

The objective of this process collaboration approach is to avoid bureaucratic inhibitors while focusing upon opportunity exploitation. Making a presentation before a collaboration team develops a healthy peer review of the proposal. I likened this process to the Operation Performance Review process, whereby each management team member must present results to the team as a whole. As long as operating results are favorable, the presenter is comfortable. However, if the trend is unfavorable, the peer pressure usually weeds out nonperforming elements. Our objective here is not necessarily to weed our nonperformance, but rather to leverage the intellectual energy of the natural work teams and process collaborators to affect the necessary change to produce ongoing productivity improvements. Observations have proved that when individuals in the trenches start improving, productivity improvement is a by-product. Our natural work team objective is to continue to increase individual and business process productivity over time.

8.22.2 Consultation with Natural Work Team Downstream Nested Internal Customer

One of the best resources to obtain productivity enhancement opportunities is to ask internal customers. Therefore, one of the roles of the IT team member is to identify process customers and glean their opinions, observations, and recommendations. Being free from direct business process involvement, the IT resource can function as an independent observer of the process activities and identify information generation opportunities. Together with the IT team member's independence and the downstream customer inputs, the opportunity exploitative farming can be extensive.

In this role, the IT resource becomes a catalyst and facilitator. They are able to take the necessary time to develop the design intent, not just the words conveyed, by the downstream internal customer. As a supplier of service, it is frequently difficult for a user to observe events through the eyes of the internal customer. However, with third-party participation, true communication can begin to surface removing the barriers to change that previously inhibited productivity enhancements. The IT resource functions as catalyst and facilitator to bring meaningful communication and ultimately process-oriented performance measurements to the table. I will be discussing developing joint process goals and objectives in more detail later.

8.22.3 Daily Consultation with Internal Customer Users

As an independent observer, a facilitator, and a catalyst, the IT resource must be provided with a forum to present opportunities to their internal customer. Frequently, this becomes a "sales meeting" where the role of the IT resource is to sell his or her "customer" (users they support) on the productivity improvement opportunities. During these daily debrief time slots, the IT team member spends time briefing his or her internal customer users. This is the time when the internal customer users can provide critical comments from "their" perspective.

This is the time when cost of change can be explored, alternatives pursued, and action plans developed. These daily forums are preparation for the weekly process collaboration forum. At a large computer chip company, this style of process collaboration is a significant part of their company culture. Inherent to this company, process collaboration consumes over 80% of the day-to-day management time commitment. Because of the multiplant nature of this company, they deploy teleconferencing and videoconferencing techniques to achieve the cross-functional participation needed for quality decisions. However, preparing for these public forums require the team members to do their homework. Relating back to our discussion, "doing their homework" will require a substantial investment in internal salesmanship between the IT team member and their internal customer user team member.

Without a daily debrief, the preparedness of the internal customer user team member can jeopardize the weekly forums potential for productivity enhancement.

8.22.4 Debrief with Functional Systems Champion

On numerous occasions, I have conveyed the importance of the functional systems champion to process-based management. The functional systems champion is an integral team member here as well. He or she has mastered the business process and serves as a strategic independent observer. To call upon a cliché, "You can't see the forest for the trees," is appropriate at this point. Internal customer users may be so entrenched in their day-to-day activities they may not be able to see the value of the proposal presented by the IT team member. Therefore, relying upon sage advice from another functional expert can become extremely beneficial.

The IT team member will meet regularly with the functional systems champion and recap findings and relate proposals. If appropriate, the functional systems champion will attend the daily debrief sessions between the IT team member and his or her internal customer user. His or her primary role is to prevent barriers from surfacing and help ensure that the time investment is productive. He or she provides a forum for both the IT resource and user team members to discuss potential problems and exploit alternatives.

This section introduces the concept of decentralizing 80% of the IT resource base and redeploying these resources into the user trenches. The result of this redeployment process is the surfacing of natural work teams. We looked at the three alternative approaches to information productivity enhancement and determined that natural work teams were the best alternative.

In our discussions of the natural work team, we identified various roles of the IT team member, including the following:

- Side-by-side worker
- Catalyst
- Facilitator
- Internal concept salesman

As an IT expert, we discussed the role of observing the data highway and

1. Reducing time to convert data to information.
2. Eliminating data redundancy.
3. Eliminating nonvalue-added data.
4. Making information user-friendly.
5. Using technology tools to increase productivity.

As we looked at the specific function that the IT resource would perform, we found that there are four distinct functions, which are as follows:

- Independent observer
- Consultation with downstream internal customers
- Daily recap with internal customer users
- Debrief with functional systems champion

Breaking down the hierarchical structure and redeploying resources along process lines require a sensitivity to ways to improve individual productivity. Increasing every individual's use of information tools is essential for company-wide productivity to be improved.

8.23 Process-Based Performance Measurements

In subsection 8.22, we discussed the importance of deploying IT personnel into the trenches to work side by side with their internal customer. A key function of the team is to develop joint goals and objectives, which are focused upon managing business processes. Referring back, we discussed the deficiencies associated with performance management in a hierarchical organizational structure. A central premise is that hierarchical performance measures tend to be parochial serving the interests of the political hierarchy rather than being "customer focused."

In this section, we want to focus upon two parallel themes: (1) performing services that optimize the external customer needs and (2) performing services that optimize the internal customer needs. Let's look at these parallel themes in more detail. To delve deeply into this subject, we need to ask a few pointed questions.

1. How much time and intellectual energy are focused solely upon customer value-added activities?
2. Who is responsible to represent the customer needs versus perceived needs and internal political agendas?
3. How often is the customer consulted, and seriously challenged, on the value-added activities performed by your organization?

I would like to relate, as an example, a pet peeve, I, as a customer, am forced to tolerate. While being a consultant, I had occasion to travel frequently, flying between 50,000–300,000 miles per year. I was somewhat consistent in this pattern for over 12 years. Now I doubt that this is a record, but it's not peanuts either. Due to my destinations and travel schedules, I used a single airline 80+% of the time. Now, wouldn't you think that I would be a viable target for airline's business. Would you believe that I accumulated over 1,000,000 miles with this one airline and have yet to be called and seriously asked my opinion on how the airline could better service my needs. Not only that, with the former intense travel schedule, you would think that there would be an airline marketing opportunity, and not one of its management personnel have called me to "contract" a travel agenda with me. This example is not isolated to the airline industry. In fact, there were a few of my clients who could say they are any better and my client base represented a large variety of industries.

Inasmuch as I've given an airline industry example, let's go a little deeper and relate how I might have responded and helped the airline become more customer service oriented.

1. As a frequent traveler, why should I have to be restricted to the inadequate menu items on dinner flights? I see no reason why I couldn't call up the airline the night before, be given electronic menu selection over the telephone and customize a specific menu to my needs by selecting options on the telephone keypad. In the computer age, this certainly seems like a minor investment to provide an "excellent service item" and it certainly would differentiate the airline.
2. As a frequent traveler, I could be issued an electronic read card, which automatically announces my arrival at the airport, thereby eliminating the need to wait in lines. The card would register my arrival and subsequently confirm my boarding the airplane, thus eliminating the need to generate a boarding pass, avoid my need to stand in lines, and reduce the need for a service agent to perform a "non-value-added" task. This streamlining would not eliminate the need for the service agent (e.g., not all flyers are frequent flyers) but would significantly reduce redundant processes. Why do the airlines/travel agents issue a boarding pass with the ticket and then reissue (or stamp) the boarding pass at the counter or gate? Seems redundant, doesn't it?
3. Travelers could be issued bar-coded "baggage boarding passes" and along with their tickets eliminate the need to stand in line to check baggage and obtain seat assignments. This would radically streamline the process.
4. As a frequent flyer, I have filled out an extensive "personnel preference profile," which, to my recollection, has never been referred to service or preferences followed.

I could continue to provide an exhaustive list of other "customer-oriented" enhancements. However, my objective is to relate how simple it is to exploit significant customer needs and desires. In the case of the airline, they only needed to ask. I'm confident that most companies' customers would be equally responsive if they were only given an opportunity.

Now, let me give you a few cautions in the case of airlines.

1. I have been given "bingo cards" to fill out. However, the surveys were very limited in scope and were geared at getting responses the airline wanted to hear (not really interested in the customer's interests, needs, and wants). Bingo card surveys may help with gathering statistics, but they seldom provide the product/service differentiation necessary. A better approach is to sit eyeball to eyeball with customers, focusing upon becoming true business partners. Then deep piercing questions need to be presented, which frequently expose "lack-of-service" vulnerabilities. This task is not a job for the suave sales representative, who hates rejection, who is then forced to communicate with the bombastic purchasing agent, and who may be either a brow beater or a pansy. This is a management job.

   At this point, naysayers are beating your chest and saying things such as "My customers are distributors and can't possible help me" or "I don't have any customers who represent any significant market share" or "My customers won't spend quality time." We can all find justifications and excuses that will prevent us from exposing our vulnerabilities and prevent us from getting closer to our customers. Those predatory companies that become global survivors find ways to penetrate getting closer to the customer. Is your company a winner in the global market or have you conceded defeat by an unwillingness to try? As the old adage goes, "You can't win if you fail to play the game."

2. Developing a relationship with your customers, who can withstand the test of time, requires focus and commitment. Customers are interested in quality and reliable service at a low price. The winners of this "quality/price" strategy will be the captains of industry as we dawn the next decade. Let's take, for example, Mr. Patrick Ng, CEO, at Grand Mart Warehouse Club. He had a vision of providing quality products at the lowest price. He observed the business operating culture of two highly successful discount retailers and found that they provided high-quality product, controlled their overhead costs, turned their inventory 20+ times per year, and started eliminating their competition. Mr. Ng saw an opportunity to do the same. He did some fundamental things when he started the business. He understood that cash flow could end his dream and, consequently, managed it carefully. He avoided long-term debt obligations. He dedicated two solid weeks designing his information system to
   a. Rely upon the latest technology.
   b. Ensure that it was customer and employee responsive.
   c. Ensure that the information tools were decision support oriented and designed to optimize employee effort.
   d. Provide a significant competitive advantage.

   After Mr. Ng invested two weeks in developing a synchronized business and information flow, it took programmers four months to fully customize the system to meet the business needs. His key supplier was significantly larger than Grand Mart, yet he recognized the weakness in their information systems (constipation due to internal politics, magnitude of effort, and lack of "fast cycle" responsiveness). Mr. Ng developed a subsystem (solely at his cost) that the supplier could use, which would reduce processing time. The benefit was the ability for his company to obtain "faster response" from this key supplier, which would then convert into a competitive edge for Grand Mart. Mr. Ng admitted that business timing was right and many variables came together well, but the results were phenomenal. In the first one and a half years, he went from $0 to $100 million in revenues and became $200 million within the next year. He survived the cash flow challenge, had only short-term debt, had funded growth from operations, and began eliminating competition. Now I like his spirit and the results speak for itself.
3. Getting closer to your customers is a concept that must be breathed into the fabric of every employee. Superficial activities will generate in "lip service" and superficial results. *Customer-attuned focus mandates continuous monitoring.* Beginning with the CEO, through the most meager pay grade individual, the sum of corporate effort must be geared toward "customer value-added process." Any activity that is not focused upon customer value-added process must be challenged. Hopefully, at this point, a clearer definition of "customer focus" should be apparent, providing the highest quality product/service, when the customer wants it, at the lowest price and listening to the wants and needs of the customers. Now that we can visualize customer focus, how do we execute our day-to-day activities to ensure that our intellectual energy, goals, and objectives deliver optimized "customer-focused" results?

8.24 Process-Based Performance Management

At this point, we need to redirect our attention to "internal customer" activities. Every internal activity must be "customer focused" as well. If we can fully exploit our internal customer focus and all of our internal customer effort exploits external customer focus, then we're postured to deliver "full strength" customer value-added process. Now this is a monumental challenge in most organizations because so much energy is spent on discharging the desires and feeding the vertical organizational structure.

Let's be so bold as to dream of an organization that is ***process based*** rather than hierarchical. In this organization, our internal customers would define, for our internal suppliers of services, what performance expectations are needed. To truly be effective, we must begin at seriously understanding our external customer and step backward through our organization.

The following is a recap of premises established earlier:

- Our internal customers, not hierarchical supervisors, should provide the significant input for our report card for performance evaluation reviews.
- Any arbitration goes to the functional systems champion from the business process first and then, if unresolved quickly, to an Executive Review Counsel member who is distinctly separate from our hierarchical structure influence.
- One primary role of executive staff is to not only make these business processes function successfully but continuously improve over time.
- To leverage the potential results, compensation that is performance based, rather than hourly based, salary based, or personnel department based, is best.
- Job descriptions are replaced with a "mission-critical triage" mind-set. Do what's necessary to meet quality customer needs within the least amount of time and at the least cost.

In addition to the five points listed above, our measurements must be conveyed in terms understood by every employee. Relating back to our shop floor machine operator, expecting that individual to understand how they are contributing to RONA is unrealistic. I submit that asking a "financially refined" top executive to specify how a shop floor operator contributes to profitability may even be a challenge. Therefore, how do we get every individual, within the organization, to clearly understand how they are contributing to profitability? We can approach this in various ways.

1. We could break down financially oriented objectives into work standards at every job level. This would be equivalent to engineered time and motion studies for factory employees, but applied to every job function (including management positions). Each unit of work would have an "engineered value." Then production credit could be given to every unit of work produced and variances to standards measured. However, this model quickly breaks down when we get a grasp on the magnitude of this effort; for example, every document produced would have an engineered standard. The cost of collecting performance data would soon exceed the value the data could provide.
2. We could allow our business process owners (internal customers) to specify the "acceptable performance measurements" for their service providers. Our primary business measurements would focus upon meeting "internal customer needs." If all our internal customer needs were then focused upon meeting our external customer needs, then performing well in every business process would result in optimized financial results.

Remember that our customer is expecting quality product, service when needed, at a low cost. This is certainly intriguing and naysayers out there are defeated by muttering such phrases as "That's impossible" or "Those working-level people couldn't possibly develop meaningful measurements" or "We'll lose all control if we give that type of 'power' to the 'great masses.' However, those hearty global oriented overachievers are accepting the challenge and starting to visualize how this can happen. The future captains of industry say, "Yes, there are a plethora of obstacles to overcome, but it's doable ... and we're going to start implementing today!"

3. The most radical of all is a performance-based compensation approach. A popular and recognizable form of performance-based compensation is a 100% commission-based sales representative. The commissioned individual does not get a paycheck unless he or she deliver results. Translating this concept to the broad-based performance compensated work force ... paychecks would come only as a result of profitability being generated. Now, everyone in the organization is incented to be
   a. Throughput efficient (converting booked sales orders/effort into collected revenue dollars).
   b. Acutely sensitive to continuous cost reduction opportunities.
   c. Sensitive to how well the entire team performs (dead wood is eliminated quickly; if everyone's compensation is based upon "everyone" carrying his or her own weight, then those who don't perform are removed by peer pressure).
   d. Acutely focused upon ensuring that this external customer is fully satisfied. If the customer is not satisfied, the revenue is not generated. Without revenue, profits are difficult to attain.

Isn't that exciting!!! Every employee has a "vested interest" in generating profits. Profits come from a balanced synchronization of customer satisfaction, cost management, and control, and throughput excellence. In effect, every employee is like an owner. The difference between an employee and a stockholder (if a stock-held organization) is that the employee is a performance stakeholder and an investing stockholder is an equity stakeholder.

Let's regress a bit:

In order to effectively manage a performance-based compensation approach, it is mandatory that nested internal customer and supplier of services agree upon measurements.

The external customer must never be compromised by greed of profit sharing.

Prudent expenditure in such things as marketing, product development, tooling and equipment, and information tools must not be compromised; otherwise, the result would be "mortgaging the future."

Some employees may not want to participate ... What do you "fairly" do with them?

The transition from the existing compensation program to the new plan must be developed comprehensively and fairly.

An investment in IT tools may be necessary to maximize productivity from every employee. Remember that our goal should be to get every employee the necessary information and decision support capability to radically improve his or her "individual" productivity.

There must be a way to fairly define levels of responsibility and how compensation "shares" are divided. Overperformance must also be rewarded in a fair manner.

Business vision, values, and ethics must be maintained at the most robust levels. Pride should be a natural by-product of successful productivity results. Yet, pride should not prevent the asking of critical questions, which helps ensure that the organization remains a finely honed competitive entity.

Continuous product enhancement is essential to capture or retain market share. The goal should be focused upon the elimination of competition by providing the highest quality and lowest price product/service deployed when the customer wants it.

The better operating entities are "balanced" businesses. Balanced translates into a variety of dynamics, not the least of which include the following:

- Synchronized flow of effort—Upstream effort cascades into continuous flow downstream. Obstacles are immediately resolved and the business engine resource is highly tuned.
- Workers are mission-critical sensitive. Consequently, the workforce is extensively cross-trained. When capacity bottlenecks arise, they can be immediately resourced from cross-trained employees.
- "Fast cycle" responsiveness is the watchword of every employee. The body of the workforce is "helping" oriented. When a peer needs assistance, someone gives them the aid needed immediately.
- Policies, procedures, and cultural nuances are formulated in a way that precludes less than acceptable quality performance, for example, deploying SPC, an early warning technique that signals quality degradation. Although SPC is admirable, a better technique is to predicatively design the process whereby producing other than quality products/services is impossible (*poka–yoke*).
- IT tools must transition from their traditional historical orientation (typical accounting practices) toward preventing the deviation from expected results. Tools such as real-time decision support, extensive simulation, cost of change modeling, and projected future variances must support minute-by-minute activities. Management must be able to monitor every key business activity and see real-time performance results.

Now that we've established our mind-set baseline, let's regroup and look more closely at performance-based compensation.

8.25 Performance-Based Compensation

The ability to properly engineer performance-based compensation expects total peer participation, cooperation, and coordination. If our business process owners (internal customers) have the latitude to define what "acceptable performance measurements" consist of from their service providers, and if every business process is focused upon meeting the external customer needs, then we have the embryo from which a workable performance-based compensation process may be developed.

Providing the internal customers the latitude (1) to define expected results from their service provider(s) and (2) then to empower them to generate subsequent "report cards" requires careful design. First, the typical employee probably doesn't know how to develop performance measurements. He or she will require tutoring and mentoring to become proficient at the process. Second, the typical employee was, most likely, not consulted or involved in the current process design and does not share "ownership" in the process activities. Remember the "ham and eggs" for breakfast story. The hen *participated*, but the pig was *committed*. Commitment and ownership share an entrepreneurial energy essential for sustained results. The objective is to convert our employee base from "robotic task doers" to *highly refined overachievers*. Every empowered entrepreneur must visualize themselves as if he or she was CEO over his or her specific business process responsibilities. They must be allowed to be immersed into the day-to-day reward/risk dynamics of the business. This journey can be scary for both management partner and employee partner. Our naysayer's are overjoyed at this point. Those individuals who are threatened internal politicians, bureaucrats, and power seekers will surface a plethora of reasons why this process should never be pursued. The challenge is that are we going to be global winners and figure out how this can be done, or are we going to admit defeat and act like a retreating puppy, whimpering with its tail between its legs as it backs away from the danger? Third, our management partners must become *process success catalysts*, immediately removing the barriers along the journey, functioning as cheerleaders and advocates, and blazing the path toward success. Our management partners must demonstrate acute leadership acumen by creatively making this transition a reality. Fourth, the hierarchical organization structure must fade away as the process-based structure evolves. Vertically oriented goals, objectives, and measurements must transition into process goals, process objectives, and process measurements. During this transition, we must never lose sight of our "external" customer needs. In fact, the acid test demonstrating success of this metamorphosis is that customer service improvements become logarithmic in nature. They explode into "quantum" jumps. Now that's exciting!!! Fifth, our employee skill levels will require significant improvement. Ongoing training for every empowered employee will become a driving priority and a critical business success factor. Exploiting the full complement of our employee talent base requires an assertive strategy and aggressive execution. Our objective is to transform every employee into *radical productivity producers*. We want to realize an "order of magnitude" of demonstrated empowerment from our entire employee base. We want to exploit our talent base so radically that we put our competition out of business overnight! We become the highest quality producer of product/service at the lowest cost and continuously empower both ingredients every day!

This section focused upon two parallel themes:

1. Performing services that optimize the external customer needs by providing the highest quality product/service at the lowest cost.
2. Performing services that optimize the internal customer needs by providing the highest quality service at the lowest cost.

The essence of the section is as follows:

An example was given which demonstrated a technique to get closer to the external customer by critically probing into the customer needs. We recognize that this step makes us vulnerable, but is essential if we are serious about a true business partnership with our customers.

We discussed developing a customer relationship that withstands the "test of time" and we looked at Mr. Patrick Ng's success story where he built a business from scratch to over $100 million through providing quality products at the lowest price. The result of Mr. Ng's strategy is that he was eliminating competition.

We recognize that staying close to our customers requires that every employee become external customer service oriented and that this attitude be breathed into the fabric of every employee.

We looked at three possible ways to discern how employees contribute to profitability:

1. Engineered standards for every employee
2. Process owners (internal customers) who specify service providers' measurements
3. Performance-based compensation

The most radical view is *performance-based compensation*. This view encourages every employee to have a vested interest in company-wide performance.

We looked at nine foundational requirements needed for performance-based compensation to work:

1. Being in agreement regarding measurements.
2. Never compromise external customer needs.
3. Never mortgage the future.
4. Consider all employee viewpoints.
5. Develop a comprehensive transition plan.
6. Establish responsibility-level "shares" of compensation.
7. Business values must not be compromised.
8. Ensure products/services exceed market needs.
9. Keep a "balanced" business perspective.

Enabling performance-based compensation requires the following:

- Employee mentoring.
- Employee *process ownership*.
- Executives must become a catalyst in the change process.
- The organizational structure will migrate from hierarchical to *process oriented* with results validated by logarithmic improvements in customer service level.
- Training will become a critical business success factor.

The synthesis of these concepts results in exploiting the talent base of every employee. Referenced earlier was Valve Corporation, which exemplifies this principle (see *Valve Handbook**).

To quote Jay Elliot again about Steve Jobs philosophy on aligning performance to ownership,

> Employee Owners—Rewarding the Product Stakeholders
>
> You can't talk about profit; you have to talk about emotional experiences.
>
> **Steve Jobs**
>
> On our long walks together around the Apple buildings, Steve hardly ever talked about himself (except, on rare occasions, about his relationship problems), and he talked about other people almost exclusively in terms of their work and ideas, or about how to deal with some particular person who was in some way not measuring up. Mostly, the conversation was focused on products and on Apple.
>
> During one of these walks in the early days, Steve and I explored the possibility of turning Apple into an employee-owned company. I thought, *"Wow, will the board go crazy over that idea!" But I had to agree that this would make Steve's emphasis on the product even more successful.* However, based on the feedback we got, it seemed almost impossible to make this happen, particularly with large blocks of Apple stock held by a few people. But it certainly made sense for the type of company we wanted to build.

---

* Robert Blake and Mouton, Jane, *The Managerial Grid*, Gulf Publishing, Houston, TX, 1966.

Steve always admired United Parcel Service (UPS), the global delivery company with the familiar brown trucks. (In Venice, deliveries are made by UPS boats painted in the same brown color.) The attraction for Steve was that UPS is one of the most notable employee-owned companies. Originally started by two teenagers as a message delivery service in 1907—before most people had telephones—it has since 1945 been mostly owned by its employees.

Early ownership was mostly management and supervisory personnel, but UPS eventually included the entire workforce that now numbers over 300,000. The company expanded its shares to all full-time employees through share purchase programs and the ability to convert retirement funds to stock. The purpose of these programs was to have its employees take direct responsibility for the customers' satisfaction. UPS found that the program allowed them to give employees greater decision-making latitude, which in turn allowed for reducing the number of supervisors. It also allowed building more time into the day for customer needs. Again, a product-driven approach to ownership.

I happened to be in Rye, New York once on a business trip in 1999 when UPS went public. The pub I was at was a local center for employee celebrations of the public offering. I thought, Damn, if we had done this at Apple, the employee owners would have voted to make Steve the head of the company, and he would never have left.

Stakeholders

When people talk about the stakeholders in a company, they are referring to the stockholders and, sometimes, the holders of options. At UPS, Apple, and many other companies, stakeholders also include those employees who own shares of stock or options in the company. That's standard, of course, for the top level; having a large part of the workers as stakeholders is much less common.

Unlike almost any other executive, Steve Jobs was never focused on profits, share price, or whether the stock market price of company shares was going up or down. His focus was instead on which product ideas to pursue and then making those few products as near perfect as humanly possible in every way.

In a product-driven company like Apple, the product stakeholders are incredibly important to its success. But the attitude at Apple is significantly different: Focus on making the products successful, and financial success will follow.

Making sure that Apple employees benefited financially from their role in helping to create great products was a very important issue to Steve. We spent a lot of time figuring out how to implement programs that made the product success everyone's success. In Steve's words, "With great products will come great rewards to all of us."

All employees who went to work at Apple got stock options on their first day of employment. Also all employees were eligible for profit sharing and bonuses—so all Apple employees were stakeholders. From the receptionist to the senior engineers to the senior vice presidents, we all had a stake in the company. By this model, companies provide value for those involved in adding value to the company.

What's important about a product stakeholder reward system is that it is truly tied to the accomplishments of individuals and the product. This is not about years of service to the company or your birthday; it's about meaningful accomplishments that push the overall strategy of the company ahead.

As an example of taking product performance into account, Steve got a bonus program implemented in 2002, despite the company missing its stated revenue and profit targets for the bonus program. Apple awarded employees in its incentive program a special recognition bonus amounting to between 3 and 5 percent of their base salary. Top management including Steve were excluded from the program.

All the rewards and bonuses were based on product achievements, not on financials.*

8.26 Committing to the Journey

We have weaved a thread of change that defines the vision of the stellar global captains in industry. Conducting business using a hierarchical organization structure is the management style of those companies that follow rather than lead significant productivity enhancements. Dramatic performance comes from those companies that make dramatic changes and focus of their efforts on process-based productivity alignment.

On our journey, we discussed PPM and accountability, looking at the process on an end-to-end basic rather than a snapshot. We discussed the attributes of a highly productive company and the need to manage change at a rapid pace. Finally, we discussed PPM and a new way to operate leveraging the full complement of resources within the organization.

The question now becomes, who are those companies that are willing to become the captains of their industry? Defining the future rather than merely operating as the future unfolds is a daunting challenge to any organization. So goes the challenge ... up rises the captains of industry!

* Robert Blake and Mouton, Jane, *The Managerial Grid*, Gulf Publishing, Houston, TX, 1966.

Chapter 9

Snags, Traps, and Black Holes

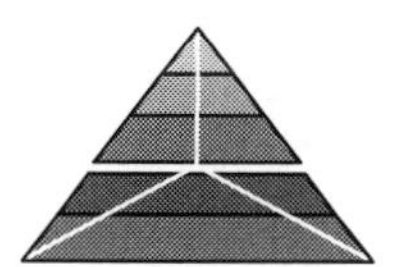

Experience has shown that many enterprise resource planning (ERP) projects just are not successful. This chapter addresses how to convert potential failure attributes into critical success factors (CSFs). It explores such topics as follows:

- GO/NO GO voting decision—Looking at the technical review and recommendations, functional review and recommendations, open issues, cutover plan, transition to production strategy, and other criteria for successful cutover
- How to tell when the project is going off the rails
- How to decide and prioritize what aspects of the system need tuning

We have spent considerable effort in laying the groundwork and guidelines for a successful ERP implementation project. During the journey, we have offered cameos and various snippets that described deviations from the prescribed process and their ramifications. At this point, I reiterate that an ERP implementation project is typically different from other company projects in the following ways:

- ERP implementation projects are seldom undertaken. Attempting to model what was done right and lessons learned from the last ERP implementation is valuable, but not necessarily helpful inasmuch as the rules change so substantially with every subsequent effort.
- This is likely the largest project, with the broadest employee involvement, of any project in the company's history.
- Because of the magnitude of the project, and frequently the long time frame, there are a plethora of stakeholders impacting the project.
- Resources engaged in the success formula tend to be very broad and deep, and frequently require one or more third-party consultants.

- The project usually has a variety of variabilities in its pursuit. Variables such as simultaneous software, hardware, database, business process, policy and procedure impacts, broad education and training impacts, and multiplane critical resource impact.
- There may or may not be a capital component, but if it does include capital, this adds complexity.
- ERP projects tend to attract a large "political" component, which frequently shapes and influences deliverable outcomes.

There are a host of other factors as well. The key is to recognize that these factors exist. I have found that documenting such a list and merging the list as part of the risk management process, with mitigation strategies, help as the project progresses. In addition, ensure that an arbitration and adjudication process is clearly detailed, early in the project, which is also a key to success. One additional safeguard is to assign the task of surveillance and resolution to a key executive, maybe even a board-level member. A prevailing theme, so far, is the importance of "engineering" solutions; this area is not an exception. Factors must be managed properly; otherwise, it will likely fall into the black hole distracter arena.

As advertised in this chapter title, let's look at snags, traps, and black holes (ROT), which may have a negative impact on the project vitality and ultimate success.

To begin with, if you choose **not** to adhere to the precepts already prescribed in Chapters 1 through 8, you will surely be a victim of snags, traps, and black holes (ROT). Particular vulnerabilities include, but are not limited to, the following:

- Software
  - Strategic concerns
  - Modifications
  - Interfaces and integrations
  - Customizations
- Hardware
  - Business interruption
  - Backup and recovery
  - Storage
  - History retention
- Database
  - Database sizing and space allocation
  - Performance benchmarking and tuning
- Business process
  - The big package deal
  - Lack of ownership
  - Cross-functional matrix management and stalls
- Managing third-party relations and internal/external **statement of work** (SoW)
- Portfolio management
  - Project management
  - ERP performance management
  - Timely decision management
- Miscellaneous

Let's take a stroll-down vulnerability lane and see why there are snags, traps, and black holes (ROT).

9.1 Software

An integral component of an ERP implementation is the software chosen to provide the application functionality essential to bring successful business results. We previously discussed the importance of performing a requirements generation process. Distilling sound requirements is a foundational precept discussed, in detail, in Chapter 2. Not only is it important to disclose pertinent requirements to be organically competitive today, but it must facilitate the future competitive posturing for at least the ERP effectiveness life span. This effectiveness life cycle is variable based on industry and a variety of other factors. A good rule of thumb is 5–10 years, based on who is giving the opinion and frequently based on the using companies' affordability parameters. As discussed in Chapter 8, an astute company wanting to maximize its ERP investment will be pushing the envelope on such capability as cost of change, projecting future variances, management dashboard, and other pertinent functionality.

9.1.1 Strategic Concerns

In addition to meeting the minimum needed requirements, software must represent the strategic interests of the using company's operating culture. The software client company needs to form a strategic alliance with the software supplier. Just as trading partners are critical to the success of a business, so too is the importance of ERP software to the future viability of the firm. The degree of integration, especially of multiple trading partner tiers that are involved, becomes a strategic factor of the software used and the end of software life decisions. It may be a very expensive proposition to change ERP solutions, if significant integration of software and business processes has occurred between trading partners. Therefore, to avoid an affordability and functionality trap, software strategic interests need to be thoroughly vetted out when selecting the right software partner.

9.1.2 Modifications

All companies operate their business differently. Consequently, there is a propensity to modify the software to match the business to a tee. This modification decision needs to give serious critical thought. Let's define a modification … It is changing the ERP author's source code. There are alternate solutions to modifications that do not have a major impact on the software, namely, customizations and report writing, which is merely configuring the software, but does not have a major impact to the software. Back to modification … On the one hand, it seems logical to modify the software so that users are comfortable with the software, just like they are used to (legacy). On the other hand, modifications come with a price … whenever there is a software update to be installed; it requires that the source code be modified again. This may be a significant ongoing cost. Even more important, modifying software source code has legitimate software failure potential. This may be disastrous and a potential black hole.

9.1.3 Interfaces and Integrations

Allowing third-party software to seamlessly flow (integration) with the ERP software suite, on the surface, seems to be a reasonable expectation. However, in reality, there are a plethora of variables that may influence the success of the combined software. A few of the variables include software architecture difference, data mapping differences, software upgrade timing differences, real-time versus batch processing differences, and a host of other potential roadblocks. Getting two or more disparate software solutions to work properly is challenging and requires a team's best resources to design, build, test, and maintain the interoperability properly. There are application program interfaces (API) that theoretically improve design and deployment timing. These are "off-the-shelf" solutions between two or more committed software trading partners who tend to need the connectivity on a frequent basis. However, depending on a variety of factors, the API may be a workable solution, or not. Because of the complexity associated with getting the software to reliably work together, interfaces and integrations are highly vulnerable to technical issues that may impact the ERP implementation schedule and cost results.

9.1.4 Customizations

The ERP software authors recognize the need for their ERP user community to configure its product for a wide span of users, so they provide the ability to customize their software, based on parameters, rules, and predetermined logic. As discussed in subsection 9.1.2, a customization is not considered a modification. Typically, this is a great alternative to modifications, but there is still risk associated with deployment. Although customization changes are considered "minor," it is still a change. Depending on the ERP author, this process may be well documented and easily supported, or it may be a highly risk-oriented task. The software authors typically expect the user resources performing the customizations to be highly skilled, seasoned on performing customizations on their software, and are adept at troubleshooting. Understanding the customization depicted attributes and resource assignments will likely point to the level of risk that this effort is to "cost and schedule" fidelity. It could turn out to be a trap, snag, black hole, or, refreshingly, successful.

As discussed earlier, conducting a robust requirements definition, then selecting the appropriate ERP software supplier, is one of the key CSFs in the ERP implementation journey. The correct software becomes even more pronounced when one or more tiers of trading partners are integrated.

To recap the software ERP implementation vulnerabilities, software is a very important ingredient to the ERP implementation journey. Choosing the right software to perform the right capabilities within the allotted cost and schedule is extremely important. With regard to the chosen software's impact on project derailment, it happens to be a major cog in fulfillment of the ERP implementation failure scorecard.

> I recall a project I was called in to assist in turning around a stalled and highly cost overrun project. The leadership and project stakeholders had been very aggressive in their interfacing expectations of a variety of third-party software with their core ERP solution. It was obvious that continuing down their current design path would become exhaustively costly and an increasingly risky path. The project stakeholders decided to take a new look at the interfaces and elected to scale the scope back to include the two most impactful interfaces in the current phase and delay the vast majority of interfaces to a later phase of the project. Deploying this new strategy allowed the project to recover and get back on track to remaining project deliverables. However, because of the scope change, cost variances for this aspect of the project were unrecoverable.

9.2 Hardware

Ensuring that there is sufficient reliable hardware horsepower to operate the new ERP environment should not be underestimated. As part of the project plan, stress, or regression testing, the load of the system should be a prominent deliverable. Any system upgrade should result in equal to or better response time performance and this should be an element in the system acceptance criteria.

9.2.1 Business Interruption

A preemptive posture describing the allowable downtime, as part of the service-level agreement (SLA), is a wise strategy. Due to the pervasiveness of ERP, any system downtime becomes a significant business productivity factor. To that end, a business interruption timetable needs to be defined and agreed to by the user community. The key questions are as follows: what the business impact is if the ERP system were down and what investment is needed to achieve the specified service level? There is a variety of capabilities available to help ensure optimized service level. Such capabilities as failover, fault-tolerant redundancy, high reliability optimizers, and mirrored disaster recovery environments are available at a price. The CSF is defining the acceptable environment and then executing a fulfillment strategy.

9.2.2 Backup and Recovery

An integral aspect of a reliable business interruption strategy is backup and recovery. This precept addresses the question: if there is a disaster that occurs, what is the exposure to data loss and how long will it take to recover? Having a preemptive agreement, again, is a wise strategy.

9.2.3 Storage

ERP tentacles tend to be very pervasive, frequently resulting in an exceptional amount of data and related storage capacity requirement. Depending upon the business interruption timetable, SLA commitment, and a variety of other factors, there are storage options that may be pursued. The purpose for bringing the topic up is to make your project stakeholders aware of the need for collaboration and agreement on a go-forward strategy.

9.2.4 History Retention

As stated earlier, the ERP data span is typically very broad and there needs to be a serious discussion on history retention. The question here invokes a critical thought to how much historical data will be real-time accessible and how much will be archived. The real issue, that typically surfaces, is how long archived data will take to retrieve and in what readable form will it be presented in. Another key issue, if you are transitioning from a legacy system to a new ERP system, is how the legacy data will be retrieved. Basically, there are two primary approaches dealing with legacy history: (1) copy the legacy data into the new ERP system data repository and (2) convert the history data into a generic retrieval mechanism so that once the legacy system is retired, the data are still available. Both of these approaches have their challenges. The key issue is to address it in advance of needing retrieval.

To recap the key hardware ERP implementation vulnerabilities, having adequate hardware capacity is central to providing the user community with acceptable service performance. Elements such as backup and recovery or data retention become a snag if not addressed as part of the transition to production activities.

9.3 Database

ERP software continues to increase in complexity, especially as it relates to tables, data dictionary, data security, and performance monitoring. With ERP software authors rearchitecting their systems, this becomes especially challenging for database administration. A typical ERP upgrade has both an application system database dimension and the database management system (e.g., Oracle) dimension to navigate through. The harmonious balancing of software, hardware, and database elements is at the heart of system performance.

9.3.1 Database Sizing and Space Allocation

Depending upon the number of environments and the volume of transaction activity, space requirements are rather dynamic and require monitoring to help ensure adequacy. Log files tend to grow with activity and a data management strategy is essential.

9.3.2 Performance Benchmarking and Tuning

The consistent and optimized performance of the database supports the SLA. Tuning helps ensure optimal dataflow. The purpose is to help ensure that the user community experiences maximum performance potential whenever possible.

Recapping the database vulnerabilities, monitoring and managing database size and performance is essential to both ERP implementation and ongoing maintenance.

9.4 Business Process

We have previously discussed the importance of the business process perspective to the ERP implementation journey. The business process element of the ERP implementation is frequently a neglected opportunity area. We also discussed the business process performance potential to an ERP implementation. The CSF aspect of business process is to help ensure that throughput is optimized on a nested end-to-end process basis. Triaging business process health and vitality may open up a can of worms for the overall ERP implementation. If a small number of business processes need reengineering, then it is likely that including it in the ERP project schedule will not overly extend the ERP implementation timetable. However, if extensive reengineering is required, it might be necessary to create parallel projects, allowing each project to be managed independently with a separate timetable. Regardless, it needs to be recognized that the same resources are needed to support each implementation effort.

9.4.1 The Big Package Deal

If the company's business processes require significant reengineering, this element may take several years of effort in itself. Therefore, careful planning and resource management is essential to keep fidelity of the ERP project schedule and complexity manageable. However, as the ERP return on investment (ROI) is concerned, if the business processes require significant reengineering and it is not performed, then the ERP ROI will be delayed accordingly. This decision is critical to the leadership expectations. It has become rather common to hear of an ERP implementation failure as it relates to ROI. Again, merely replacing legacy software with contemporary software does not yield much, if any, ROI. This area has the potential for a full derailment of a project.

9.4.2 Lack of Ownership

We discussed the importance of the roles and responsibilities (RACI) earlier in Section 1.1. Without clearly defined roles and responsibilities being defined, the ERP implementation is vulnerable to confusion, deliverable stalling, and likely derailment. There is both project RACI and business process RACI functioning concurrently during the ERP implementation. By far, the greatest impact area is the business process area, as new functionality is introduced to the business environment. Stakeholders need to step up and take ownership by ensuring that there is a good balance between RACI and business process changes, and education and training deployment is engineered into the ERP implementation portfolio.

Chapter 8 discussed the value of process-based performance management. If a process-based approach is taken, then the ownership concerns become moot. If the organizational performance measures remain in the legacy hierarchical structure, then ownership remains fuzzy, thus impacting ERP implementation effectiveness. Regardless, it is the stakeholder community that must adjudicate, arbitrate, and engineer clear business process ownership guidelines if ERP is going to have any positive impact on the business. Process ownership is an ERP CSF and, along with management commitment, is a frequent reason for project derailment and failure.

9.4.3 Cross-Functional, Matrix Management, and Stalls

Due to the pervasiveness of the ERP business impact, how the project principles perform and the business processes perform have distinct susceptibility to the ERP value proposition. Business processes tend to be cross-functional in nature, meaning that no individual or group is an island unto itself, requiring collaboration, cooperation, and coordination to operate on a fine-tuned basis if optimal performance is to be achieved. Therefore, the ERP culture expects high-output effort across the functional lines. When you add to this precept a matrix management structure, which impacts functional deliverables, the end result may be a loss of clarity and infusion of murkiness in the process authority and deliverable. Netting this out, the typical nested end-to-end process may have a multitude of dynamics and variability, impacting the ability to realize their deliverables successfully. Be aware that snags, process stalls, and deliverable delays may surface resulting from the lack of business process accountability and clarity.

To recap the key business process ERP implementation vulnerabilities, elements such as ownership, size of business process SoW, and organizational structure may result in derailment or project failure if not addressed properly.

9.5 Managing Third-Party Relations and SoW

Depending on the size of company, the size of the project, the size of the ERP implementation budget, the degree of trading partner involvement, and the amount of integration that is to occur, a third party may be contracted to successfully complete deliverables. When outsourcing tasks, it is essential that the third party fully understand what is expected of them ... the deliverable, the quality and acceptance criteria, the timetable, and their contribution to the success of the ERP implementation effort. The appropriate instrument to use to formally communicate these expectations is a SoW that is agreed to by both sides. Like any resource deployed on the ERP implementation project, this SoW needs to be monitored and managed by the project core team. An agreed-upon communication strategy needs to be created and the SoW statused at least weekly through the commitment period and successful completion of their SoW. Just because the third party is not part of your company does not lessen the value they contribute to the ERP implementation effort. Keep third-party team members as actively involved as the least of internal resources are involved. If per chance, the third party is shared with other clients and schedule impacts become an issue, this needs to be hit head-on and an aggressive mitigation strategy enforced. As part of SoW negotiations, schedule commitment is a cornerstone element, and if the third party is not able to achieve their schedule, then management needs to step in and take necessary corrective action. No resource is excluded from achieving their schedule commitment. Remediation may require that the third party become micro-managed (e.g., four-hour schedule updates); otherwise, a new third party, with adequate resource span, needs to take over. It does bring up an interesting point. As the SoW is negotiated, what are the third-party consequences if they are unable to perform according to the schedule? The third-party rules of engagement need to be very clear and monitored appropriately.

9.6 Portfolio Management

Due to the length and depth of the ERP implementation, facets may be delegated out to form subprojects or miniprojects. Any task, deliverable, resource, or other value streaming element attendant to the ERP implementation is important. Just like a third-party SoW may be outsourced; portions of the ERP implementation schedule may be delegated to a responsible entity if the size of the ERP implementation effort becomes too big. When this occurs, the ERP implementation project manager may have other project managers reporting to them. Regardless of the project structure, the ERP implementation project manager is still responsible for the project. To that end, if there are subprojects working, there needs to be close coordination between each entity to help ensure project harmony, schedule fidelity, and task completion as expected.

9.6.1 Project Management

The delegation of authority for these subprojects needs to be comprehensive, which means that the subproject project manager needs autonomy to properly manage their SoW, yet they must be under the authority of the overall ERP implementation project manager. The autonomy is to facilitate project agility, quick response, flexibility, and other essentials to project success. Each ERP implementation project will have its own project culture; however, the project focus and expectations must be central to the operations of the project.

In my past, I found that I delegated SoW fully to the project core members. They are truly the subject matter experts and should be empowered to perform as autonomously as possible without compromising project fidelity. Therefore, when one of their SoW elements comes to forefront during a briefing, they are encouraged to lead the briefing. Sharing stage with these high-output team members not only builds confidence in their accomplishments, but gives the lead project manager backups that may be relied upon. I believe that it is good stewardship to spread the wealth, as it were.

9.6.2 ERP Performance Management

An independent observer of the ERP implementation may ponder the question: How would ERP performance be properly managed? That's a loaded question. If you consider ERP performance, there is a vast arena to consider, including, but not limited to, the following:

- *Software module*—for example, inventory management
- *User*—for example, response time as stated in the SLA
- *Data access*—for example, security profile
- *Web or portal*—for example, remote user
- *Reporting*—for example, report run time
- *Project schedule*—for example, task completions timetable
- *Network reliability*—for example, hardware equipment, software services, human factors, accessibility, availability
- *Remote device*—for example, iPhone and iPad
- *Trading partner integration*—for example, real-time drill-down into supplier/customer production and release schedule
- *Graphics*—for example, export to pivot tools and charts
- *24/7 operating support*—for example, international site support, maintenance window, and so on
- *Legacy history record viewing on demand*
- *Real-time* material requirements planning (*MRP*) *decision support based upon exception triggering*

It is obvious that ERP performance has a wide berth of dynamic and variable elements. The performance management **best practice** is to clearly define ERP performance elements at the forefront of the project, engineer the scorecards in advance, define baseline measures from which to calculate performance variances, document performance objectives, and so on. As a reader would analyze the ERP failure causes, it is clear that performance expected results versus performance actual results may have a serious gap. Whether we are analyzing technical causes, user-oriented causes, or environmental causes, it really doesn't matter. It is an area where snags seem to be a recurring event and impacts all partners of the ERP implementation journey.

9.6.3 Timely Decision Management

Posturing an ERP implementation effort, for success, has a host of factors that need consideration. Let's begin a partial list of qualifiers, which are as follows:

- Do we have the "right and sufficient" talent on the project team?
- Are the team members competent to discharge their responsibilities?
- Do we have adequate support resources and capacity to achieve the ERP deliverables?

- Do we have adequate budget, cash flow, and line of credit to support the effort?
- Is our technology stack able to support the effort in the long run?
- How much of a distraction will the ERP project be on achieving the annual operating goals and objectives? What is the impact if we miss these goals and objectives?

If an astute critical thinker were to list all the dynamic factors and qualifiers, the result may be overwhelming to the extent that you might even question why a reasonable intelligent leadership team would want to risk their business on even going down this ERP path ... not a bad question! There certainly needs to be a compelling reason to "launch" an ERP implementation journey. But even more important is the fortitude to sustain the journey and deliver business ROI.

We have navigated through the qualifiers and now we need to address the importance of timely decisions to the outcomes of the ERP implementation. We discussed the need for a change management process, which will assist the project core to stave off changes in scope and arbitrary whims. Based on the company culture, management style, and cost impacts, the leadership must be prepared to be nimble in their decision-making practices. I doubt that the team members assembling the project plan, project schedule, and other time-sensitive supporting documents placed a lead-time offset for the decision process. Consequently, decisions should be executed within hours and even better within minutes to support a harmonious workflow of project task and deliverable completions. Taking a month to make a decision will likely negatively impact the project schedule. To that end, the project core needs the ability to call together a decision support forum, at their discretion, to make timely decisions. Attempting to work within the decision boards and other constraints of the typical day-to-day operating culture will likely gain a negative cost and schedule impact. If a decision needs escalation, this process should be supportive of a "within minutes" escalation and resolve process for the ERP implementation activities ... *an author's commentary ... why isn't this lean practice norm for day-to-day operating decisions as well?* The laborious constraints of a typical day-to-day bureaucratic process cannot be tolerated in an ERP implementation life cycle. We have discussed the need for agility, nimbleness, deftness, and the like to operate the ERP implementation; any compromise from operating within these lean practices will likely result in snags, derailment, or actual failure of the project. Companies that actually conduct an in-depth ERP postmortem, at the conclusion of the project, frequently point to this cultural disease as a primary contributor to missing the mark in their ERP project journey. For some of you, I need to say quit whining and just do it! Just as design is central to the ERP implementation, lean execution is an equally important contributor.

9.7 Data Accuracy

As discussed in Chapter 4, the importance of data accuracy is paramount to ERP implementation integrity. The computer does not care whether information is accurate or not; however, management cannot live with inaccuracy. Therefore, it is recognized that information accuracy is not a system problem, but rather a management problem. The standard then is 100% of the records, 100% accuracy, and 100% of the time (see Figure 4.4). The impact of poor data then has substantial consequences and permeates to the core of the ERP integrity. As mentioned frequently, ERP has extensive tentacles across the business. Consequently, with the broad impact, any data accuracy issues impact the business greatly. With the tools available today, to help improve data accuracy (bar codes, Radio Frequency Identifications [RFIDs]), data filtering, triggers, etc.), data accuracy should be nearing the 100% level.

A few years back I was quality reviewer on a data warehouse project that was being deployed, and although the software worked according to specification, the inputs and outputs of the tool were gibberish. As the technical team dissected the data flows, they found that there were data conflicts that were irreconcilable. Not only were there duplicate and disparate source data feeds, but they uncovered data transformation algorithms that actually corrupted good data. In peeling back the onion, data accuracy was at the heart of the issue. With all the data conflicts, it was estimated that the data feeding the data warehouse was at a 43% accuracy level, when all factors were melded in. Corrective action required a 100% review of all the source data records, a decision on which source data flow was the master and a cleanup of all records. Data transformation algorithms were suspended until the data warehouse contained pure source data. It was also discovered that there were some input processes that introduced record error; these were reengineered. When data cleanup was completed (after seven months), the data accuracy was pronounced at a 97% level. Still short of the goal of 100% accuracy, but accurate enough to begin feeding the data warehouse application.

9.8 Resource Commitment Breaches

Regardless the engineered thoroughness of the planning process, to obviate schedule delays, there will be episodes where resource commitments are compromised. As previously discussed, as the resource area commits to their SoW, the resource provider needs to create a detailed workplan (roadmap) on the approach they will use to help ensure integrity of schedule fulfillment (see Chapters 4 and 8). If monitored properly, there is likely an early warning trigger that the resource area is falling behind. As part of an escalation event, the project leadership needs to work with the affected resource area and mitigate the resource shortcomings, by using overtime, contracting with a third party resource, modifying the expected deliverable, or offsetting in another manner. The use of a RED project health status may be very effective in instituting a quick recovery corrective action. Regardless of the method used, it needs to be addressed early and decisively to avoid potential project derailment. This is especially salient in small companies that are typically resource constrained at the onset of their ERP pursuits. Scaling back the expected deliverable might be the best option for smaller companies, inasmuch as budget may not be available for overtime work or third-party outsourcing.

9.9 ERP for the First Time

There are still a plethora of companies that operate on legacy systems (spreadsheets and flat files) that are not complex as an ERP model. The transformation from a simple (although it may be an extensive environment of application one-offs and spreadsheets) to a fully integrated ERP environment is not an easy task. A few elements that need be considered are as follows:

- ERP is a philosophy for operating a business model. If your company does not want to adapt to this philosophy, save yourself the headache and don't pursue ERP.
- ERP impacts all aspects of the business and clarity of objective needs to be committed to at the onset. Don't try to implement all the modules at the onset, rather taking a small number of core competencies and deploy them up-front, and schedule expanded capability at a later date.

- Limit the number of disparate system integration and interface connection points. Keep it to the bare essentials at the onset and expand capability over time.
- ERP spans hardware, software, human capital, and the business process environment as integral baseline functionality. It also assumes management commitment, considerable financial vesting, and a change to how you operate the business from today. If any of these baseline elements is not part of the plan, you are at substantial risk for failure.
- ERP implementation and operating the day-to-day business are parallel and equally important efforts. Approaching ERP in a haphazard fashion dooms the project for failure, or at least suboptimized results. This is especially susceptible for small companies that believe they can be successful at implementing ERP during their lunch hours.
- ERP brings a totally new mind-set to approaching business processes. Whereas a non-ERP operating environment relies upon a significant manual and/or brute force operation, ERP expects users to accept a totally different approach. This is frequently conveyed as resistance to change; however, it is even more insidious inasmuch as the user needs to exhibit faith that the inherent ERP business processes will work without the starts and stops of manual processing. In addition, many of the manual (spreadsheet) processes were elegantly simple; the details of the equivalent ERP process may be very complex.

I could continue to detail the vast differences between the ERP and non-ERP but would like to convey a few practical examples instead.

Non-ERP	*ERP*
Financials posted separately—Material issue value posted only when financials are updated	Financials posted dynamically—This means that as you do an inventory issue to a work order (WO), the value is posted immediately to the WO material issues
Sales order (SO) posted separately—Material issue value posted only when financials and SO are updated	SO posted dynamically—This means that if you issue material to a SO, it is immediately posted
Cost of sales updated periodically (usually monthly)	Cost of sales posted immediately
WO material issue posting occurs periodically—Issue reconciliation can take place for multiple assembly levels simultaneously and moved into finished goods inventory (FGI) at any time	WO reconciliation requires that every material issue associated with each and every relevant WO has to be closed out before assembly is received into inventory to move to the next higher assembly WO. If many levels of the bill of material, then all levels need reconciliation until you can receive the finished product into FGI
Purchase order (PO) material issue value posted only when financials are updated	POs posted dynamically—This means that as you do an inventory issue to a PO, the value is posted immediately to the PO material issues
Poor man's MRP—Usually works one level at a time and the run-out schedule for POs and WOs do not net and time-phase dynamically requiring manual tracking	MRP performs PO, WO creation on the fly, netting of on-hand and on-order on the fly, and lead-time offsetting on the fly

There is a significant difference between timing and integration between the two approaches. There are various benefits for the non-ERP in processing timing (typically less rigors, but less control) versus ERP, whereas an integrated ERP gives immediate postings to different modules, but typically requires rigorous reconciliation along the way. There are pros and cons for each tool, the key takeaway; they are overwhelmingly different and the ERP learning curve may be substantial when you dig into the details and logic of the tool.

9.10 Miscellaneous

We have been reviewing opportunities for traps, snags, black holes as well as various CSFs, when discharged, which will obviate the snares. There is a plethora of opportunities to derail an ERP implementation. Let's drill-down a bit to identify some of these conditions, which are as follows:

- **Modify the software source code** instead of adapting to the new inherent business philosophy proffered
  - Hanging on to the bad habits from legacy systems, not only suboptimizes the ERP benefits but perpetuates bad practices.
  - Requires reinstallation with every upgrade, increasing the total cost of ownership of the ERP software.
  - Is typically a distracter to the value stream.

  There are few instances when "we've always done it that way" improves the new ERP model. These improvements typically fall into the following categories:
  - Customer requirement
  - Contractual requirement

  Except for the above criteria, these modifications tend to become a trap that will perpetuate ERP discord.
- **Short cutting the user education and training**—Granted, educating and training end users tends to be a costly line item. However, the quaint phrase "you pay me now, or you pay me later" really does apply.

  I'm reminded of the proverb "For want of a nail,"* which goes as follows:

  For want of a nail the shoe was lost.
  For want of a shoe the horse was lost.
  For want of a horse the rider was lost.
  For want of a rider the message was lost.
  For want of a message the battle was lost.
  For want of a battle the kingdom was lost.
  And all for the want of a horseshoe nail.

  Being shortsighted in your ERP investment becomes a bit foolish!
- **User community unable or unwilling to create desk instructions and/or procedures**—A desk instruction is a "how to" recipe that guides the user on using the system to perform a specific task. The lack of user documentation tends to surface gremlins when the principal

* Wikipedia, the free encyclopedia.

user becomes sick, goes on vacation, or leaves the company. This is a trap that will impact you later.

- **No time to document the AS IS business processes**—We discussed this earlier; the AS IS functions as the operational baseline (what we are doing in current system). Adapting to a new design without understanding what we're doing today is shortsighted.
- **Performing inadequate or taking shortcuts in testing activities**—This is frequently on the list of reasons that the ERP project failed.
- **Ill-equipped or unqualified project manager**—The ERP project manager should be a seasoned top leader within the company; with great communication skills, the ability to troubleshoot is a great team player, willing to take on any foe, and tenacious in pursuit with a passion for excellence.
- **The purchased software does not work for critical business functionality**—This is likely the failed result of the software selection due to diligence and reference checking process. Not only can this be costly, but may have a significant impact on business operating results.
- **Take too long to implement**—Biting off more than you can chew may derail what had been expected to be a successful implementation.
- **Too much connectivity**—Rather than being realistic about the number of integration or interface connections to get operational at the onset, it overloads the project resulting in a failure to launch event.
- **Selected the wrong consultant to assist in the implementation SoW**—Every resource deployed on the ERP implementation is critical to ERP results.
- **Did not adequately detail the conversion weekend cutover tasks.**
- At the beginning of the project, management not only funded the project but also functioned as a key resource. However, **as time lapsed, not only did the funding begin to waffle, but their hands-on support became nonexistent**.
- The already aggressive project plan was imposed with **scope creep**—What started off as good project change control became watered down and ineffective as time passed.
- At the onset of the project, our nested trading partners were active with our project core team. As time passed, and we ran into technical difficulties, as well as general business pressures, the **trading partners commitment softened and ultimately withered away**.
- **Project costs grew at an order of magnitude** and wobbled out of control.
- **There was a change in key project personnel after the first six months**. The project could not recover from the loss.
- **The project was too complex, the project team consisted of junior-level talent, and we short-changed the level of software company support**. This project resulted in a failure.
- Rather than time-phasing the module GO LIVE over various timetables, **the entire system went live all at once**. Risk was not properly assessed nor managed. The company recovered from a full failure, but it took three years to recover.
- **The project was fully an IT effort**—With little or no user involvement, integration testing was sparse and stakeholders waffled in their project allegiance. This failure caused a rollback to the legacy system and the company had to have a restart the next year.
- **The project Go Live was scheduled for year end**—This was the most user resource contentious time, resulting in only part of the system to go live at the onset. Recovery lasted almost an entire year.
- Although the company engaged the ERP software firm to assist in the implementation, the **knowledge transfer process was sorely inadequate** resulting in a GO LIVE with an ill-prepared user community and a fiasco. Stabilization took months as a result.

- **Project goals and objectives were fuzzy**—See Appendix F.1 by McKinsey & Company for an exhibit of the four dimensions of the value assurance methodology concerning goals.
- **Technical complexity and variability was too great to overcome** given the allotted time and budget.
- The project reeked with **extreme levels of toxic contentiousness**. Not only did leadership not adjudicate, but they tended to be the most active participants.
- **The selected software required thousands of control file settings to be configured** and the user community was confused as to the optimal interactivity on their configuration. Although the system functioned adequately, the resultant settings required months of fine-tuning, after GO LIVE, before the "look and feel" of the system supported harmony within the user community processes.
- **The IT infrastructure was inadequate** to handle the new ERP environment resulting in a poor user experience and lethargic business process performance.
- **The security model was configured so tightly that it took the user community weeks to be fully operational** ... the tag given ... THE **CAN'T DO** SYSTEM.
- The internal software **transaction matrix (interoperability of module interfaces) was not configured properly and the message broker did not alert users "that transactions only 'partially' completed processing."** The result was lack of fidelity between the operational data integrity and the financial bookings. It took months to troubleshoot the issue and required a significant financial restatement of earnings.

Unfortunately, this list could go on and on. Needless to say, there are numerous traps, snags, and black holes that may bring ERP implementation discontent. Having said that, **it is possible to have a good ERP implementation experience**, given proper planning, adequate budgeting, management commitment, rigorous education and training, abundance of system testing, and a bundle of common sense such as adhering to the precepts presented in this book.

9.11 CSFs while Approaching GO LIVE

We have already discussed various CSFs, as part of the subject matter content, presented earlier. However, I would like to focus on a few very specific CSFs that may provide guidelines for a company to avoid going off the rails, especially as GO LIVE (GOLD—Go-Live Date) nears. This is the spot in the project life cycle where test plans have been approved and being deployed (data validation, application setup, unit test, system integration testing, user acceptance testing, process testing, and regression testing); interfaces and integrations are being validated/certified; the user community is almost complete on their knowledge transfer and training activities; policies and procedures have been approved; information workmanship standards have been approved; stakeholders have begun signing off on deliverable completions; and senior management is starting to count their chickens in the pen (ROI results are being validated). There have been scope changes addressed and a few minor ones were actually approved. There were various technical challenges that required remediation. There were budget pressures that surfaced. Some of our trading partner integrations required hourly monitoring through completion, and a variety of other variabilities that needed triage and corrective action. Although there were stormy events, the company has weathered the swirl and proven that the project leadership had the fortitude and passion to surface as winners. However, we have now come to a very important short-interval period juncture, preparing for GO LIVE, which will become the icing on the cake, when completed successfully. Let's take a look.

- *Technical review and recommendations*—This may or may not be a toll gate, regardless it is a very important event. This is the time when the technical team goes through their internal formal checklists and ensures that their interfaces/integrations are documented properly (design, test, functional specification, peer review, etc.), their customizations are tested and documented, their configurations are tested and documented, technology stack is fully operational, batch and online processes are operational and documented, database is fully operational and documented, and any other technical element is fully operational and documented (e.g., legacy system retirement plan). Open issues are fully remediated, or a waiver is in place to address after cutover.
- *Functional review and recommendations*—This may or may not be a toll gate, regardless it is also a very important event. This is the time when the stakeholders and functional team go through their internal formal checklists and ensures that their policies and procedures, desk instructions, business processes, work flows, and approvals are fully operational and documented. The To-Be system actual result testing is fully aligned to the expected results. The user community knowledge transfer is completed, users are certified, if applicable, and the information workmanship standards are fully operational and certified, if applicable. Open user issues are fully remediated, or a waiver is in place to address after cutover. In other words, the user community is fully trained and postured for a "100% quality" GO LIVE.
- *Transition to production strategy*—Preparing for support of the new ERP system as the production instance requires thorough critical thought. A checklist that includes key ingredients after GO LIVE is as follows:
  - Help desk or triage resource for the start-up of the new software
  - Security profile—The user access rights
  - Software configuration management—Managing change
  - Software quality assurance—Keeping your best foot forward
  - Test plan completion verification
  - Requirements traceability validation
  - Communication plan validation
  - Data conversion strategy, if needed
  - Start-up activities profile
  - Post implementation support model
  - GO LIVE contingency plan, if rollback is needed (how to)
  - Cutover plan validation

  Adhering to a transition to production strategy is a project GO LIVE **best practice**.
- *Issues log*—There is a need to record software bugs, hardware mishaps, procedural shortfalls, and other GO LIVE preparatory anomalies. The beginning of the log is early in the project; however, these issues need to be resolved by the time of cutover (GOLD). There are situations where a few minor issues may be carried over until after GO LIVE. The purpose of the log is to track their status until resolved.

  Managing the issues log to completion is a project GO LIVE **best practice**.
- *GO LIVE cutover plan*—The cutover plan is a detailed (minute-by-minute) series of steps that are to be accomplished, with start times and end times, during GO LIVE weekend and the immediate time period leading up to GO LIVE. As preparatory for GO LIVE, a cutover team roster needs to be created which lists every GO LIVE resource participating

in GO LIVE contact information. This would include all technical resources, system acceptance testers (users), and third-party participants (trading partners, consultants). There may be a cutover plan review checklist that may include such things as follows:

- Have we reviewed the issues tracker for cutover "placeholders?"
- Are there any cutover setups needed?
- Are there any module defaults needed?
- Are there any table setups needed?
- Are there any triggers, flags, exceptions, and e-mail notifications needed?
- Are the any supporting toolset setups needed?
- Anything external to the company needed (e.g., banks)?
- Are there any database setups needed?
- Are there any custom report setups needed?
  - Any other long-running reports need fixing?
- Have we adequately accounted for legacy history retention and archives?
- Is the needed documentation in place?
- Initiate the system change request.
- Send out e-mail alerts to the user community a couple of days before GO LIVE.
- Send out GO LIVE warning 15 minutes before legacy system power down.

This is by no means a definitive list, just a mental stimulator. Now that we have done the preparatory SoW, the following are a sample of cutover weekend elements that need to be defined:

- Printer setup
- Database management system setups
- Exports
- Imports
- Backups (this will occur various times through cutover)
- Disable users
- Create restore point
- Close out legacy software elements, as needed
- Create various checkpoints, in case rollback is needed
- Follow the conversion/installation script (may be one or more)
- Security updates
- Reconnect test users
- Conduct user testing and acceptance testing
- GO LIVE startup
- Rebuild tables

The above steps will vary depending upon the software deployed, the risk factors, and the complexity of interfaces and testing. It could be anywhere from 100 steps to over 1000 steps. Once the cutover begins, the project manager, or designate, needs to trigger status updates to interested team members throughout the cutover process, about every four to six hours. The typical cutover weekend process may take one to five days.

Adhering to a GO LIVE cutover plan is a project GO LIVE **best practice**.

■ *GO/NO GO voting*—The entire extended team (stakeholders, sponsors, technical team, key users, and trading partners) need to be represented in voting that the company is ready to proceed with GO LIVE. It is a formal briefing and typically includes the following:
  - Project overview, goals, and objective
  - Project schedule review—Validating deliverable attainments

- Technical readiness review—Production readiness from a technical perspective and recommendation on how to proceed (GO/NO GO)
- Functional readiness—Test methodology validation, training and certification validation, procedure and desk instruction validation, process validation, and recommendation on how to proceed (GO/NO GO)
- Open issue review and validation
- Formal vote by stakeholders (recorded and published)

Adhering to a GO/NO GO vote of stakeholders is a project GO LIVE **best practice.**

9.12 Stabilization

There is a time period (after GO LIVE) where the system is being validated to ensure integrity of the system. This time period is called stabilization. During stabilization, the user community validates that their system is working according to their expected results (transactions are processing correctly; response time is as expected, etc.). The stabilization period varies by company. Typically the shortest time is about a month, but it could run from one to six months or more. The stabilization period should span at least a month-end close period of time to help ensure that periodic processes occur as expected in addition to the day-to-day processes. During stabilization, any open issues on the issues log should be resolved. You know that you are done when the following checklist is completed:

- All TO BE processes are working as expected.
- Issue log items are completed.
- Users have signed off on their ability to do all their assigned work using the new toolset.
- All the technical tasks are functioning properly and the system performance meets SLA criteria (this includes hardware, software, network, database administrator [DBA], etc.).
- Cleanup of interim database instances is completed.
- Old servers are retired, if applicable.
- Customers (stakeholders) have signed off on the system functionality (requirements and traceability).
- Project controller has created a pro forma budget versus actual and variance analysis report and documented the ROI to date.

Depending on the scope of the project, one or more core project team members may remain engaged in the maintenance and stabilization effort (one to four weeks). However, most of the core project team is reassigned after a couple of weeks. If there is a phase 2 to the ERP implementation, they may be quickly reassigned and support stabilization on an *on-call* basis.

9.13 System Tuning

Regardless of the amount of effort expended designing, creating, and testing the new ERP production environment, the postproduction cutover environment needs monitoring, adjusting, and performance tune-up. This monitoring is not an event, but a process requiring various remediation injections. Technical tuning refers to the ERP software application tuning, database management system tuning, hardware tuning, and the like. In addition, there is typically the need to perform functional tuning, including tweaking desk instructions, procedures, end-to-end business processes, information workmanship standards, and business process performance metrics, and augment the training model and a host of other supporting elements. The objective of tuning is to continuously tweak the principle elements until there is a harmonious and optimized business management system. Each of these tunings will require prioritizing to help ensure that variability is controlled in an engineered manner; the most disharmonious elements get the highest priority.

9.14 ROI Tracking

As discussed in Section 3.1, an ERP implementation should have targeted ROI goals and objectives. An astute leadership team would have agreed upon where the ROI would be coming from at the earliest of ERP implementation project days. To help track the realization of the ROI, a financial analyst would likely be assigned to the project and progressively publishing ROI progress to the goals. Depending on the degree of business process disharmony and other factors, I have seen an ERP implementation project attain 70%–80% of its planned ROI months before the ERP GO LIVE.

Another aspect of ROI performance monitoring, it need not, and should not, necessarily be limited to financial ROI. Lean should be a business driver that permeates all aspects of the business. With the new ERP environment, lean effectiveness, along with a process performance focus, should be continually squeezing improved productivity from every minuscule aspect of the business. The result should yield improved agility, deftness, increased flexibility, quick response, and progressively improvement of throughput. Recall our discussions in Section 7.6 on throughput—the conversion of a booked order into collected revenue. Recall also that maximum throughput is the instantaneous conversion of a booked order into collected revenue. The business performance potential is the difference between our current realized throughput performance and instantaneous performance results.

Chapter 10

Conclusion

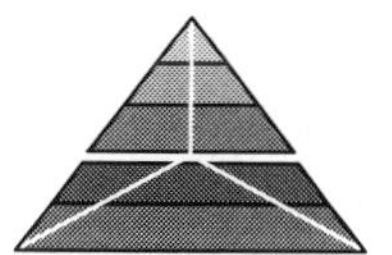

Keeping sanity yet achieving exponential results on time and on budget. That is the prevailing theme as we wrap-up. My hope is that this is a book of encouragement. We hear so much chatter about the enterprise resource planning (ERP) failures. I wanted to help change the approach ERP projects were taking so that every ERP project implementation would generate stellar return on investment (ROI).

Let's look at a few recurring themes:

- Anything that will impact results needs to have an engineered approach. Things don't just happen; they happen correctly by design.
- Much of which is engineered required testing, so test plans are essential as well. As you create test plans, define the expected results, then perform the test, and compare actual results to expected results. Understand WHY there are differences.
- Pay attention to detail, and then pay more attention to that same detail. Follow roadmaps. Design deliverables and the process used to ensure deliverable completion ON TIME and within BUDGET. Monitor deliverable completion using intense follow-up, down to the HOURLY basis, if needed.
- In a world that focuses on FASTER, BETTER, CHEAPER, time is usually a diffuser of optimal productivity. Maximize throughput by eliminating nonvalue-added processes and treat TIME as a most important strategic asset. Throughput is enriched by doing things that are visionary with agility, applying innovation, while being flexible, exercising ingenuity, in a nimble and deft manner.
- Leadership must lead—not merely stand on the sidelines and watch things happen.
- Use best practices and best processes to achieve exceptional results. Don't settle for commonplace.
- Process-based performance measurements allow a company to achieve order-of-magnitude better results.

No conclusion would be complete without looking at best practice highlights!

- Deliverable fulfillment: EVERY DELIVERABLE requires that the fulfillment resource send an artifact to the project core team.

Note: The multicolored triangle is the Best Process Logo and represents both best practice and best process precepts.

- Developing a risk management strategy and mitigation plan is a leadership **best practice.**
- Performing a comprehensive job of distilling good ERP project requirements (needs and expectations) is one of the hardest tasks within a project and functions as a documentation **best practice** in the ERP implementation process.
- An aspect of requirements generation is the productivity dashboard, which is a reporting **best practice.**
- Effective change management that incorporates cost of change element is a leadership **best practice.**
- The statement of work becomes a defining leadership **best practice**, which facilitates the achievement of stellar ERP ROI results.
- Establish a change environment that facilitates the generation of pertinent information in support of quality and timely decisions is a leadership **best practice.**
- The ability to tie the business drivers (unique aspects differentiating our company from our competitors) in such a way that the resource base (users), benefitting from ERP-delivered functionality, may excel in their individual job performance is a performance **best practice.**
- The ability to trace requirements flow from their source (originator), through the various project phases (design, prototyping, customization, testing, piloting, and delivery) is a requirements generation **best practice.**
- Use of triggers, drill-downs, and simulations/projections is a reporting **best practice.**
- The use of the scorecard, to rigorously track results, is a reporting **best practice.**
- Engineer an agreed-upon process that yields an order of magnitude of business improvement results to transform the business into a lean, mean, high-productivity business operation is a performance **best practice.**
- Populating a broad library of deliverable artifacts is a documentation **best practice.**
- Fostering lean change management practices is a leadership **best practice.**
- Documenting a robust requirements list (including business drivers and a traceability matrix) aligned to expected results tracking is a documentation **best practice.**
- Creating a robust library of documentation that supports the requirements and process deployment is a documentation **best practice.**
- Use of out-of-the-box visionary tools is a reporting **best practice.**
- Obtaining a commitment to expected results and validating their accomplishment is a performance **best practice.**
- Creating a comprehensive education, training, and implementation framework is an ERP implementation **best practice.**
- Setting goals and monitoring to help ensure their achievement is a project monitoring **best practice.**
- Mastery of the ERP software and related process solution impacts system performance results and is an operational **best practice.**
- Senior management must lead, and be examples for, the entire organization, which is a leadership **best practice.**

- The engineering of robust functional specification is a documentation **best practice.**
- Creating and managing to a project rules of engagement is a leadership **best practice.**
- Using visionary, innovative, and flexible tactics is a leadership **best practice.**
- Using exceptional throughput and nimble tactics is a leadership **best practice.**
- Adhering to the project initiation process is a project management **best practice.**
- Mastery of the ERP deliverable roadmap is a project management **best practice.**
- Exploiting project core team member talent roadmap is a project management **best practice.**
- Aggressive barrier removal is a leadership **best practice.**
- Adhering to the project review and project status framework is a leadership **best practice.**
- Managing project capacity and load balance is a project management **best practice.**
- Adhering to process performance management tactics is a leadership **best practice.**
- Engineering best processes is a leadership **best practice.**
- Maximizing nested internal customer/service provider performance elements along profitability lines is a leadership **best practice.**
- Leadership-led continuous improvement practice is a leadership **best practice.**
- Adhering to performance workmanship standards is a leadership **best practice.**
- Optimizing intellectual energy is a leadership **best practice.**
- Adhering to fast cycle change tactics is a leadership **best practice.**
- Adhering to optimized throughput tactics is a leadership **best practice.**
- Adhering to a transition to production strategy is a project management **best practice.**
- Managing the issues log to completion is a project management **best practice.**
- Adhering to a GO LIVE cut-over plan is a project management **best practice.**
- Adhering to a GO/NO GO vote of stakeholders is a leadership **best practice.**

We've had an exciting journey looking into the frameworks and precepts essential for a truly world-class ERP implementation. It became obvious from the beginning that an ERP implementation was not merely another company project to deploy. Within its tentacles, the ERP span of authority is broad, to say the least. However, if a company adheres to the precepts described, then success will be forthcoming.

Not only did we collaborate on foundational concepts, but my passion has been to scintillate your critical thinking to proceed on a path forward that will generate an order of magnitude of improved business results. I know, you believe that some of the precepts contained herein are a bit difficult to sell within your management leadership team. To you folks, I say, subscribe to those precepts that are sellable. Your company will not be the captains of industry in the next decade, but you will at least complete the ERP implementation successfully. To those of you who are bold and willing to take on leadership stodginess, you have an opportunity to break new ground, turn up the heat, and make things happen! If you are with a company that either failed in their ERP implementation quest or did not experience the ROI that your company was expecting, you may want to launch a process improvement initiative and begin a best practice ERP renewal effort. Therefore, where do we go from here?

For you who have not begun the ERP journey yet, pass this book around, or better yet, get more copies and give them to the movers and shakers within the organization, and read it thoroughly (maybe even a couple of times). Make an outline of the precepts that you are willing to fight for and start your journey toward a best practice ERP implementation.

For you who are still on the ERP implementation path, but are on the edge of derailment or just not happy with the direction you have pursued. Stop and reengineer your ERP implementation to incorporate as many of the best practices discussed in the book as you can and jump start the project again ... revitalized and focused upon doing it right!

For you who want to do it again, but, do it right, this time.

- Use this book as a guideline to sell the leadership team on a new ERP implementation effort.
- Kick off the ERP implementation II (ERP-I_2) with a collaboration workshop and gain leadership commitment based on the expected results that will be realized.
- Structure your ERP implementation II (ERP-I_2) adhering to the framework and precepts presented.
- Staff the ERP implementation II team with the best and brightest resources. We want the A+++++ team to lead the best practice ERP-I_2 turnaround effort.

I hope the rest of you enjoyed reading how to do it right. You seem to be diffused, out of energy, and don't mind wallowing in the mire of mediocrity and willing to accept failure as your destiny. Go ahead and run away like a whimpering puppy, with its tail between its legs. I truly hope there are only a small number of you folks in this defeatist category.

Appendix of Terms

The *terms* in the appendix will help ensure that definitions that might be leading or misunderstood are clearly defined.

Term Definition

ABC costing	Activity-based costing—a cost management tool that allows a company to identify its cost drivers, which may be used to reduce cost
API	Application programming interface—a set of procedures that allows disparate components to interact together in a standard manner
Artifact	Used to describe a deliverable, key form, document, or other project-sensitive element
CEO	Chief executive officer
CM	Configuration management—a term used to manage consistency using structured change control processes
CMMI	Capability maturity model integration—a process improvement training and appraisal program and service administered and marketed by Carnegie Mellon University (Wiki definition)
COS	Cost of sales
COTS	Commercial off-the-shelf software—ERP solutions that are purchased compared to developing your own software
CSF	Critical success factor—an element essential for the success of a project
CTO	Configure to order—a term used by a company to describe selling and assembling specific options requested by a customer
DBA	Database administration—responsibility includes installation, configuration, upgrade, monitoring, maintenance, and securing database

DILO	Day in the life of—a reflection upon the role, processes, and function of a team member and what they do in a day
GAAP	Generally accepted accounting principles
GOLD	Go-Live Date—the date that the system starts up
HR	Human resources
IT	Information technology—spanning hardware, software, and technical standards
IWS	Information workmanship standard—the minimum acceptable quality level for transactions, job functions, work processes, and ultimately the resulting information
KPI	Key performance indicator—a set of values from which to measure performance
Mat'l	Material
Ovhd	Overhead
PMBOK	Project Management Body of Knowledge—a guide to standard project management terminology and practices
PPV	Purchase price variance—the difference between the standard and the actual cost
RACI	Responsible, accountable, consulted, informed—a role and responsibility matrix used to define project roles
RMA	Return material authorization—number given to a customer authorizing them to return damaged goods
ROI	Return on investment—the composite of cost savings, performance improvements, and related productivity enhancement associated with making the investment on the ERP project
RONA	Return on net assets—a financial performance measure that considers asset usage
RTV	Return to vendor—product returns
SE	Software engineer—a developer or programmer
SLA	Service-level agreement—service performance level between two or more parties
SME	Subject matter expert—individuals who know most about specific processes
SOP	Standard operating practice
SR	Standard rate—in a standard costing system, it is the standard
SyE	Systems engineer—a technical resource focusing upon risk
TQM	Total quality management
Txn matrix	Transaction matrix—a roadmap that ties cross-functional transaction and reason code events to specific financial chart of account
WIP	Work in process
WO	Work order

WRAP-UP

This concludes Section III where we examined the importance of **Project Management** to the overall success of the enterprise resource planning (ERP) implementation effort. We described the attributes that are necessary to help ensure achieving timely deliverables. We explored the necessity of commitments, reporting the status, invoking Steering Committee guidance, and day-to-day issue/decision management, are the tried and true practices of good project management. We peeled back the onion to describe precepts such as behind-the-scenes salesmanship, removing risk barriers, and executive ownership process practices, monitoring rules of engagement, and other vital critical success factors (CSFs) needed to guide the ERP project implementation to stay the course.

We took a close look at **Process Performance Management** and the transition to use **best processes** as a competitive weapon. Attributes such as **flexible, agile, nimble, lean, quick response, fast cycle, adept,** and **deft** are watchwords for the captains of industry over the next decade. We provided the leadership team a formula that may be used to gain *order-of-magnitude* performance results, compared to mere incremental productivity enhancements. We delved into the need for an **end-to-end process** perspective, looked at **performance goals and objectives**, and emphasized the essential for **performance accountability** as well as the need to **manage performance expectations.** We drilled into the heart of the chapter—the precept of **process performance measurements**... looking at an **organizational perspective, parochial performance objectives, value streaming**, and capstoned the chapter with the **vision of the business process** in the future. Finally, we pursued **process-based performance measurements** and even merits of **performance-based compensation.**

Finally, we examined the **Snags, Traps, and Black Holes**. Here we surfaced contributing elements that lead to departures from planned activities, derailment gremlins, and bad practices. We examined software, hardware, database, third party, and a variety of other ingredients, where, improperly deployed, led to what we hear so frequently, "sour" ERP implementation results. We concluded this chapter with a positive perspective by looking at **CSFs while approaching GO LIVE** and **system tuning** and **return-on-investment tracking**.

As discussed in recap of Sections I and II, obtaining improved business results will align with doing a good job implementing the ERP solution. However, to obtain stellar results (even order-of-magnitude improvements) will only result from managing CSFs and adhering to best practices. These include the following:

- Using visionary, innovative, and flexible tactics
- Using exceptional throughput and nimble tactics
- Adhering to the project initiation process
- Mastery of the ERP deliverable roadmap
- Exploiting project core team member talent
- Aggressive barrier removal
- Adhering to the project review and project status framework
- Managing project capacity and load balance
- Adhering to process performance management tactics
- Engineering best processes
- Maximizing nested internal customer/service provider performance elements along profitability lines
- Leadership-led continuous improvement practice
- Adhering to performance workmanship standards
- Optimizing intellectual energy
- Adhering to fast cycle change tactics
- Adhering to optimized throughput tactics
- Adhering to a transition to production strategy
- Managing the issues log to completion
- Adhering to a GO LIVE cut-over plan
- Adhering to a GO/NO GO vote of stakeholders

The next chapter is the Conclusion ... where we keep our sanity yet achieve exponential results ... on time and on budget.

Appendix A (Chapter 1)

A.1 Communication Plan

Communication is essential for project success. There are varying degrees of communication requirements, depending upon the nature of the role within the project. Differing roles and a communication proposed strategy are discussed in this chapter.

A.1.1 Communication Strategy

Executive briefing—Steering Committee and executive leadership
Operational leadership—Key organizational stakeholders, systems champions, core project team members, and interdependence/interface owners
Key users and subject matter experts—Day-to-day prominence
Casual participants—Leaders and users who have minor roles in system usage and yet have some ownership in system success

A.1.2 Purpose

The purpose of the communication plan is to contribute to the successful implementation of the project with right communication delivered to the right audience at the right time.

- Spread knowledge and status about the key deliverable and upcoming process changes.
- Facilitate the "ownership" process by end users.
- Provide information and ideas for greater productivity in the future.

A.1.3 Objectives

- Enable leadership advocacy—Provide information to enable leaders to be advocates of the project.
- Build synergy within the core project team and leadership.
- Provide communication that helps foster the team's effectiveness.
- Prepare management and staff—Create awareness and understanding of project impacts and implications.

- Enroll stakeholders—Generate interest and buy-in for the project deliverables. Brief them on project developments so that they are involved, have an opportunity to give feedback, and are acknowledged for their contributions.
- Manage expectations—Reinforce the scope of the project realistically (under promise/over deliver) to manage perceptions/expectations and to ensure staff understand that there will be (temporary) takeaways prior to long-term gains being realized.

A.1.4 Format

Ad hoc—Impromptu meetings for fast cycle decisions and change control and attendees rallied as needed

Weekly/biweekly project briefing—Regularly scheduled meeting of core project members to discuss status, issues, and upcoming activities and WAS/IS Project Change Control

Company newsletter—Overview of project with a traffic meter (R, Y, G)

Focus group—A synergy event aimed at such deliverables as creating test plans, conference room piloting, issue resolution, and the like

A.1.5 Communication Principles

The following guidelines outline the preferred way that communication is developed and delivered. They are assumptions that govern how communication activities take place.

- Tailor communications to discreet audiences according to needs analysis.
- Design communication using fact-based information and deliver openly, regularly, and in a straightforward manner.
- Ensure that communications contain consistent core messages.
- Deliver face to face/one on one where messages contain job-sensitive information (i.e., an individual job will change).
- Continually reinforce the business reasons for change.
- Consistently ask for feedback and involvement, and acknowledge the same.
- Evaluate at predetermined points to ensure that message is understood.
- Pursue communication opportunities at involvement activities (focus groups, workshops, training sessions, etc.).
- Consult with corporate communications through various phases of the project to ensure communications-related decisions.
- Broadcast to a wide audience key deliverable and milestone achievements.

A.2 Sample Risk Management Log, Mitigation Strategy, and Contingency Plan

Sample risk management log

Risk#	Risk Description	Stage	Owner	Probability	Severity	Priority	Mitigation	Status
	User resource constrained	Testing/ desk instructions		Highly likely 75%	Significant	High	Vanilla implementation	
	Resources have no backup	All			Moderate	High		
	Inadequate resource to master PS tools	All				Medium	Triage toolset and Pareto priority	
	Inadequate PS technical documentation	All				Medium		
	Some resources have no upgrade experience	All		Likely 50%		Medium		
	Resources not dedicated to project	All			Minor	Medium		
	Requirements definition presumed	All			Moderate	Medium		
	Simultaneous dynamics (Oracle upgrade, PS 9.2 upgrade and migration to the blade)	All		Likely 50%	Moderate	Medium		
	Crystal reports	All		Low	Moderate	Low		

Sample risk management mitigation strategy

Risk#	Action/Event	Start Date	End Date	Success Criteria	Mitigated Risk-Level	Status

Sample risk management contingency plan

Risk#	Action	Triggers

A.3 Sample Risk Action Plan

Risk Action Plan

PART A. RISK IDENTIFICATION			
Risk Number:		Identification Date:	
Statement of Risk (below):			
Risk Context (below):			
▪			
▪			

PART B. RISK ANALYSIS					
Risk Impact Value:		Catastrophic (value = 4)	Critical (value = 3)	Marginal (value = 2)	Negligible (value = 1)
Risk Probability Value:		Near Certain (value = 4)	Probable (value = 3)	Possible (value = 2)	Improbable (value = 1)
Risk Exposure Value: (Risk Impact × Risk Probability)					
Risk Impact Time Frame /Affected Phase:		Near Term	Mid Term	Long Term	

Risk Classification (below)*					
CLASS:		ELEMENT:		ATTRIBUTE:	

PART C. RISK MITIGATION PLAN			
Step #	Responsibility	Due Date	Activity
1			
***			

Contingency Plan: (Contingency plans should identify a "trigger" that would cause the contingency to be executed and high-level cost estimates)
▪
▪

* There are various Risk Elements, popular elements include process, integration, culture, and infrastructure. http://www.corporatecomplianceinsights.com/key-elements-of-the-risk-management-process/.

Appendix B (Chapter 4)

B.1 Sample Job Function Information Workmanship Standard

Document by Job Function Information Workmanship Standard (IWS) matrix is as follows:

Documents	Departments Affected
Authorization for Disposal	(PC, MP, IC, PUR)
Cycle Count Sheet	(IC)
Debit/Credit Memo	(PUR)
Engineering Change Order	(PC, MP, PUR)
Inter-Stockroom Transfer	(SR)
Inventory Adjustment	(IC)
Issue	(PC, IC, PUR)
Item Master Update—Lead Time	(PUR, MP, PC)
Order Policy	(PUR, MP, PC)
Pick List	(PC, PUR)
Manufacturing Order	(PC)
Material Order Release	(PUR)
Material Requisition	(PC, IC, MP)
Material Transfer	(SR, PC, IC, RC)
MRB Disposition—Rework in House	(PC)
Outside Processing Order	(PUR, PC)
Purchase Order	(PUR)
Receipts-in-Process Locator	(RC)
Receiving Memo	(RC)
Re-Inspection Document	(RC, PC)
Reject Ticket	(PC)
RTV Shipping	(RC)
Returned Goods	(RMA) (PC)
Route Sheet (Order-Dependent Routing)	(PC)
Rework in House Order	(PC)
Scrap Ticket	(RC, PC, IC)
Shipper	(PC)
Stock Location Maintenance	(SR)

Job Function
"Receiving as a segment of the process"

B.1.1 Receiving Associate

B.1.1.1 Background

The receiving department is a crucial player in assets management. Most companies recognize inventory at the point of receipt. Therefore, properly identifying the asset and its quantity is essential to the book inventory valuation. Accurately counting the balance at the point of origin (receiving) improves the probability that the count will be accurate downstream. To improve count quality, the blind tally approach will be used to obtain an independent quantity check. The packing slip quantity will be reconciled to the blank tally.

B.1.2 Information Workmanship Standard

The receiving associate is responsible to accurately process the following documents:

- Receiving memo (receipt at dock)
- Material transfer (dock-to-stock)
- Return to vendor (RTV) shipping document (RTV credit)
- Receipts in process
- Receipts in process locator

B.1.2.1 Receiving Memo

B.1.2.1.1 Standard

1. All receiving memos will be processed within 15 minutes of physical receipt of goods measured by the date/time stamp of the packing slip.

> **Monitor technique:**
>
> - Random review by the receiving department supervisor.

2. The data entered will reflect 100% accuracy level on the following data elements:
   a. Item number
   b. Receipt quantity
   c. Unit of measure
   d. Purchase order number
   e. Vendor
3. All other data elements will be within 97% accuracy.
   Accuracy baseline
   a. Packing slip
   b. Blind tally
   c. Purchase order

d. Item master file
e. Accounts payable (AP) data (invoice)

Audit technique

a. On-line edits
b. Receiving inspection review
c. Receipts exception report
d. AP three-way match
e. Cycle inventory reconciliation
f. Departmental sample audit
g. Internal audit spot check

Measurement criteria

a. Truck arrival time versus receiving memo transaction processed

Best practice or benchmark

a. Utilize scanning rather than key entry
b. Minimize data entry by using electronic vendor advance shipping notice as receiver
c. Backflush vendor pay point at time of shipping finished goods to eliminate "all" receiving and AP invoice paperwork

B.1.2.2 Material Transfer (Dock-to-Stock)

B.1.2.1.2 Standard

1. Dock-to-stock material transfers will be processed within 15 minutes of "release" disposition by receiving inspection measured by the date/time stamp affixed by receiving inspection.

Monitor technique:

- Random review by receiving department supervisor
- Planning department peer review
- Quality department peer review

2. The move quantity entered will reflect 99% accuracy level.

Accuracy baseline

a. Receiving inspection released document
b. Dock-to-stock exception report

Audit technique

a. On-line edits
b. Transaction register
c. Receipt quantity variance report
d. Cycle inventory reconciliation
e. Departmental sample audit
f. Internal audit spot check

Measurement criteria

a. Receiving inspection release versus material transfer transaction processed

Best practice or benchmark

a. Utilize scanning rather than key entry

B.1.2.3 RTV Shipping Document (RTV Credit)

Receiving will initiate a RTV shipping document upon receipt of a debit/credit memo from purchasing.

1. All RTV shipping documents will be processed within 30 minutes upon receipt of debit/credit memo from purchasing measured by the date/time stamp entered on the debit/credit memo.

Monitor technique:

- Random review by receiving department supervisor

2. The quantity shipped data entered will reflect 99% accuracy level.
   Accuracy baseline
   a. Debit/credit memo
   b. Purchase order
   c. Vendor invoice credit document
   Audit technique
   a. On-line edits
   b. Transaction register
   c. Receipts-in-process exception report
   d. RTV audit report
   e. Cycle inventory reconciliation
   f. Departmental sample audit
   g. Internal audit spot check

B.1.2.4 Receipts-in-Process Locator

All receiving lots will have location control.

1. The receiving location will be transacted within 15 minutes of physical storing measured by the date/time stamp on the receipts-in-process locator document.

Monitor technique:

- Random review by receiving department supervisor

2. The item/quantity/location data entered will reflect 99% accuracy level.
   Accuracy baseline
   a. Sample audit
   b. Cycle inventory audit

Audit techniques

a. On-line edits
b. Transaction register
c. Dock-to-stock location report
d. Departmental sample audit

B.1.3 General Accuracy Guideline

The receiving associate will maintain a receipts-in-process minimum accuracy level on a receipt-by-receipt (item level) basis of 99% on all production inventories. All other receipts will reflect 97% minimum accuracy level.

The primary validation activity will result from receiving inspection counts, stockroom counts, material planner review, and cycle inventory audits.

B.1.4 Departmental Certification

Operational certification, on *all* of the above documents, will be *mandatory* at a *minimum acceptable quality level of 95%* prior to being allowed to process transactions within the production database. A certification exam (operational, verbal, and written) will be given to all new or transfer employees. Passing certification at the 95% level is required within the 30-day probation period. The inability to pass certification is grounds for immediate dismissal or reassignment.

1. **Operational certification guidelines**
   Within the certification database (use of a conference room pilot may be acceptable), the individual must display a **mastery** of the following software transactions:
   a. Receipt screen (normal and revised)
   b. Material transfer
   c. RTV shipping document
   d. Receipts-in-process locator
      i. The individual will be required to process 5–10 representative transactions within each of the above screens at a 98% accuracy level within 30 minutes.
      ii. Two typical operation problems will be introduced and transactionally solved by the user.
      iii. The individual will be treated for log on/log off, navigation, help screens, and other interactive skills.
2. **Verbal certification guidelines**
   The individual will be verbally quizzed on the associated procedural issues affecting the above-noted screens. The individual will exhibit a competence in understanding the logic and interdepartmental cause and effects of processing each transaction. The acceptable level must be within 95% confidence level. The individual will exhibit a positive attitude toward information accuracy.
3. **Written certification guidelines**
   The individual will be required to pass a written exam spanning a 30-minute time period. The exam will be passed at a minimum 95% correct level.

Note: See subsection B.3 for sample written certification exams.

B.1.5 Recertification

Each receiving associate will be recertified annually. Recertification will consist of passing the operational and verbal sections. The proctor may choose to execute the written exam if there is any doubt as to the proficiency level of the individual.

B.2 Sample Department IWS

B.2.1 Material Management Department

B.2.1.1 Background

The material management department is an integral resource in the material planning, shop floor scheduling, and asset management functions. The material management department has responsibilities, which include the following:

- Inventory control (IC)
- Material handling (MH)
- Material planning (MP)
- Production control (PC)
- Purchasing (PUR)
- Receiving (RC)
- Shipping (SH)
- Stockroom (SR)

Each of the functions plays a key role in enterprise resource planning (ERP) and execution. Each function has a variety of documents that require a high degree of accuracy if quality information is to be attained.

B.2.2 Information Workmanship Standard

Material management is an integral catalyst to implement *supply chain management* principles. The material content of many manufacturing companies represents a significant investment in the overall cost of manufacturing. Therefore, material management must drive continuous improvement and increase *throughput* in order to stay competitive in a global economy. *Throughput shall be defined as the conversion of raw materials into collected revenue.*

In its role as catalyst to achieving quantum leaps in throughput improvement, material management must drive the vision of managing the supply chain through all tiers of vendor relationships and extend out to the ultimate consumer. This multi-tier influence can have a dramatic impact upon the long-term viability of the company as well as their suppliers and customers. An agile and flexible supply chain can significantly impact the overall economic posture of the participating members of the partnership.

The material management department in aggregate is responsible to accurately control the following documents:

Documents	Departments Affected
Authorization for Disposal	(PC, MP, IC, PUR)
Cycle Count Sheet	(IC)
Debit/Credit Memo	(PUR)
EngineeringChange Order	(PC, MP, PUR)
Inter-Stockroom Transfer	(SR)
Inventory Adjustment	(IC)
Issue	(PC, IC, PUR)
Item Master Update—Lead Time	(PUR, MP, PC)
Order Policy	(PUR, MP, PC)
Pick List	(PC, PUR)
Manufacturing Order	(PC)
Material Order Release	(PUR)
Material Requisition	(PC, IC, MP)
Material Transfer	(SR, PC, IC, RC)
MRB Disposition—Rework in House	(PC)
Outside Processing Order	(PUR, PC)
Purchase Order	(PUR)
Receipts-in-Process Locator	(RC)
Receiving Memo	(RC)
Re-Inspection Document	(RC, PC)
Reject Ticket	(PC)
RTV Shipping	(RC)
Returned Goods	(RMA) (PC)
Route Sheet (Order-Dependent Routing)	(PC)
Rework in House Order	(PC)
Scrap Ticket	(RC, PC, IC)
Shipper	(PC)
Stock Location Maintenance	(SR)

The following documents are the primary responsibility of accounting:

- Authorization for disposal
- Debit/credit memo

Engineering has primary responsibility for the engineering change order. The material management department provides integral input to these documents and will be responsible to ensure a 99% accuracy level. The primary area of responsibility will be the monitor points.

All other documents are the material management department's primary responsibility for either initiation or primary care-taking functions. Detailed in each job function's IWS is the specific accuracy baseline and audit technique for each document.

However, globally for the department, the accuracy-level objective for the composite of all documents is 99%. IC, accounting, and quality control will perform the genuine baseline for accuracy audit. The source for obtaining accurate performance will be the manufacturing planning and control system exception reports, reconciliation, and departmental and/or internal spot audit.

Performance to measurements will be graphically tracked. The performance package is posted outside the Vice President (VP), Materials Department office as well as within individual units' areas. The performance package is also an integral part of the VP, Materials Department monthly progress reports to the President.

In addition to documents, the material management department is responsible for maintaining proper levels of exceptions. The following exception levels will be maintained and monitored by the manufacturing planning and control system:

Exception	Current Goal Level	Future Goal Level
Supply Demand Exceptions (aggregate)	<2%	<1%
Order Status Exception (aggregate)	<2%	<1%
ECO Effectivity Implementation	>98%	>100%
Closed Order Demand Recon. Exception (units)	<½%	<¼%
Closed Order Demand Recon. Exception	<$200	<$50
Receipts Exceptions	<2%	<1%
Receipts-in-Process Reconciliation Exceptions	<2%	<1%
WIP Order Reconciliation Exceptions	<2%	<1%
Bulk Issue Reconciliation	<2%	<1%
Shortages:		
• Purchase	<5%	<2%
• Manufacturing	<2%	<1%
• Rework	<5 %	<2%
On-Hand Negative	<20	<2
Projected Negative Inside Lead time	<50	<5
Demand Past Due	<25	<10
Firm Past Due	<100	<20
Cut Past Due	<50	<20
Release Past Due	<50	<20

In addition, exception conditions are an integral part of combined system data integrity. These exceptions outlined below will be tracked and reported in the same manner as documents.

Exception	Current Goal Level	Future Goal Level
Pick Past Due	<10	<2
Completion Past Due	<20	<10
PO Place Past Due	<20	<20
Dock Past Due	<25	<20
Customer Order Past Due (line items)	<5	<1
Floor Stock Negative	<20	<20
Items Missing Lead time	<10	<1
Items Missing Order Policy	<10	<2
Items Missing Cost—Manufacturing	<2	0
Items Missing Cost—Purchased	<5	0
Purch. Items Pend. MRB Disposition (line items)	<10	<5
Purch. Items Pend. MRB Disposition	<$5000	<$1000
Mfg. Items Pend. MRB Disposition (line items)	<2	<2
Mfg. Items Pend. MRB Disposition	<$5000	<$2000
On-Hand Accuracy (units)	>98%	100%
Inspection Accuracy (units)	>98%	>99%
Floor Stock Accuracy (units)	>95%	>98%
MRB Accuracy (units)	>99%	>99.5%
Exception	Current Goal Level	Future Goal
WIP Accuracy (units)	>98%	>99%
Material at Vendor Accuracy (units)	>98%	>99%
Engineering Changes Which Increase Prod. Cost	<$20K (Y-T-D)	<$5K

B.2.3 Certification

Certification (operational, verbal, and written) is mandatory as a condition for employment within the material management department. Individuals unable to maintain a "current" certification level on aspects affecting their unit will be candidates for immediate transfer or dismissal. In addition to software certification, the planners are strongly encouraged with incentives to achieve APICS certification, and the buyers are strongly encouraged with incentives to achieve Institute for Supply Management (ISM) Certified Professional in Supply Management (CPSM) certification.

B.3 Sample IWS Written Exam

Date: **Name:**

Material system practicum certification test

As exceptions to the material plan occur, orders must be changed to reflect new quantity requirements and realistic due dates at all levels of the bill of material. In order to demonstrate your understanding of the effect of changes, the orders displayed must be put back in balance. The certification units covered include the following:

MCS-1: How the system monitors the material plan and reports exceptions
MCS-2: How changes can restore balance to the material requirements plan

The subunits covered include the following:

MCS 10/11

A. Pick date changes and downward rescheduling
B. Order status changes
C. Projected inventory negative outside lead time
D. Projected inventory negative inside lead time
E. Past due exception messages
F. Order changes due to effectivity dates

The bill of material for the unit is as follows:

X
Y Z D
K L

Item	LT	Demand Order	Supply Order	Quantity	Due Date
X		C100	M100	75	4/30/XX
Y		M100	M200	150	4/16/XX
Z		M100	P100	75	4/16/XX
D		M100	P200	75	4/16/XX
K		M200	P300	300	4/6/XX
L		M200	P400	450	4/6/XX

LT = Lead time

Today is 4/1/XX
Fill answers in the blanks: ***(Some answers provided in italicized parentheses)***

A. Pick date changes and downward rescheduling subunit

1. Set hot key for supply/demand inquiry and hot key for manufacturing order inquiry.
2. View manufacturing orders M100 and chart the dates; also chart order M200.

M100		*M200*	
Due		Due	
Start		Start	
Pick		Pick	
Cut		Cut	
Firm		Firm	

3. Change the pick date of M100 to 4/10/XX.
   Why? (*This allows an extra four days for the pick/build cycle.*)
   What does the order status screen for order M100 advise for the new due date of component Y? (*They have changed.*)
   Should components Z and D also have new due dates? (*Yes*)
4. Change the due date for item Y on order M200 as required.
   Now status order M200 to see if components K and L need changes in due dates.
   Order status inquiry shows K due date =? (*4/6/XX*)
   Supply demand inquiry shows supply due date =? (*4/6/XX*)
5. Etc.

B. Status changes subunit

1. View manufacturing order M100.
   The order status code is now (*planned*)?
   Which means (*under MRP control*)?
2. Can you change the order status code from "planned" to "firm?" (*Yes*)
   What does order status "firm" mean? (*under planner control*)?
   What does the "force cut" indicator show? (*Off*)
3. Etc.

C. Etc.

Date: **Name:**
MCS/MRP CERTIFICATION EXAMINATION

1. The function of material control is to have the right (*material*) in house at the right (*time*) in the right (*quantity*).
2. List three of the files that make up the material control/MRP systems:
   a. (Item Master, Bill of Material, Inventory, Purchasing)
   b. Etc.
   c. Etc.

3. In the following list, match the documentation on the left with the phrase on the right that best describes the contents. Match the two by putting the correct letter in the blank.

Program specification	(f)		a.	*Defines the database, control file, and common procedures*
General specification	(a)		b.	Describes general nonspecific module edits; front-end editing
Edit	(b)		c.	Describes to data processing how to use the on-line system
Operator's manual	(c)		d.	Defines the data elements
Data element description	(d)		e.	Presents the information by procedures
User manual	(e)		f.	Contains detailed logic and operating programs

4. Problem 1: Complete the following supply demand problem.

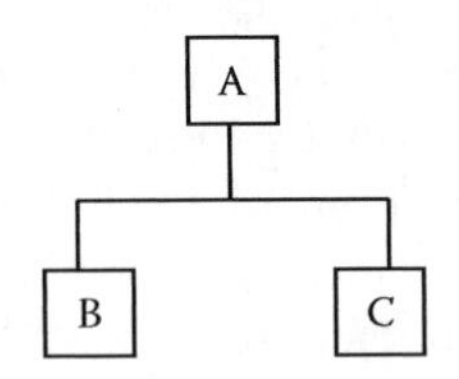

On-hand balance of A = 20
On-hand balance of B = 50
On-hand balance of C = 250
Order policy = Discrete

	Item	*Supply*	*Demand*
Customer order # C100	A		100
Manufacturing order # M100	A	80	
Dependent demand # M100	B		80
Dependent demand # M100	C		240

5. Etc.

Manufacturing orders

Land Times (1)	Planner Time	Release Time	Pick Time	Build Time
Exceptions (2) Past Due	Firm Past	Cut DuePast	Pick Due	Completion Past Due
Order Action (3) Firm Date Order Status (4) Planned	Cut Date Firm	Pick Date Cut	Start Date Pick	Completion Date

- What *exception* occurs if you are late cutting an order? (*Cut past due*)
- What is order *status* while production is building product? (*Pick*)
- What *order action* is prevalent immediately after cut status? (*Start*)
- What is the order life cycle? (*Planned* ⥣ *Firm* ⥣ *Cut* ⥣ *Pick*)

Purchase orders

Lead Times (1)	Planner Time	Place Time	Vendor Time	Insp Time
Expectations (2) Past Due	Firm Past Due	Place Past Due	Receipt Past Due	Stocking
Order Action (3) Firm Date Order Status (4) Planned	Place Date Firm	Vendor Date Place	Insp Date Inspect	To Stock Date

- Which lead-time elements do manufacturing orders and purchase orders have in common?
- Describe how order status relates to order action date. Give an example.

This section was purposely shortened due to so many variables, options, and understanding levels. However, it is clear that written tests are one effective tool to accomplish credentials that support a high-impact ERP implementation project.

Appendix C (Chapter 5)

C.1 Sample Education and Training Matrices

C.1.1 Organizational Function versus Business Process Matrix

This matrix grossly depicts which functional areas will have either business process or software module involvement. For example,

→ Business process →

	BoM	INV	SOE	PUR	MRP	MFG	SFC	WIP	CMS	G/L	A/P	A/R	PAY
Engineering	x		x	x			x	x	x				
Material control	x	x	x	x	x	x	x	x	x				
Prod control	x	x	x		x	x	x	x	x				
Order admin	x	x	x						x				
Marketing	x	x	x		x				x				
Human resource													x
Accounting	x	x		x				x	x	x	x	x	x

Legend:

BoM = Bill of Material
INV = Inventory
SOE = Sales Order Entry
PUR = Purchasing
MRP = Material Requirements Planning
MFG = Manufacturing
SFC = Shop Floor Control
WIP = Work in Process
CMS = Cost Management
G/L = General Ledger
A/P = Accounts Payable
A/R = Accounts Receivable
PAY = Payroll

Note: We have elected to display three different matrix examples. Other matrixes may include personnel-to-session, mix-to-session, feature-to-module, and so on. A separate series of matrixes are needed for *Education* and another for *Training*.

C.1.2 Individual by Transaction/Feature Matrix

Another matrix consists of detailing each business process/module into functionality and listing personnel attendance. For example,

Inventory module features →

	Receipt	Issue	Move	Adjustment	Pick	Txn matrix	Cyc inv	Cost	Location	Reject	Database
Gerry W.	x	x			x	x		x	x	x	x
Robert L.				x			x		x		x
Michael M.	x	x	x	x	x	x	x	x	x	x	x
Joan P.	x	x			x	x			x		x
Bobbie S.		x			x		x		x		x
Boris K.	x		x							x	x
Donny O.			x		x	x				x	x

C.1.3 Transaction/Feature by Session Matrix

Another matrix involves functions with exercise sessions. The objective of this matrix is to tie a specific exercise session to different features. For example,

Session number →

	1	2	3	4	5	6	7	8	9	10	11	12
Txn matrix	x	x	x		x					x		x
Receipt		x	x					x			x	
Issue			x	x				x				x
Move				x					x			
Adjustment					x				x			
Pick					x	x			x			
Cycle inv									x			x
Cost	x						x			x		
Location	x									x		
Reject						x		x			x	
Database										x		x

Legend:
Txn = Transaction INV = Inventory

C.2 Example of a Conference Room Pilot

C.2.1 Overview

The cost buildup system (CBS) structures and develops the standard or actual costs for a company. We have selected standard cost method for our example and the cost types include the following:

Frozen Standard cost
Current Manufacturing cost
Current Engineering cost

Next Period budget cost
Last Year's Frozen Standard cost

The Frozen Standard cost will be used as follows:

Inventory valuation
Target: Means to measure purchase cost variance
Target: Means to measure manufacturing usage cost variance
Target: Means to measure engineering cost variance

The cost elements consist of the following:

Direct material
Direct production labor
Direct test labor
Material overhead
Labor overhead
Outside processing
Scrap
Freight

The cross-functional team participants are as follows:

Manufacturing engineering
Database administration
Materials manager
Cost accounting

C.2.2 Session Deliverables

Session 1: Validate Current Manufacturing cost and roll over into Frozen Standard. The objective of this session is to simulate the rollover activities when Current Manufacturing costs roll over into Frozen Standard.

The cross-functional team will enact the necessary controls and procedures as if this were the "real" event.

The edit and validation of all costs and performance of the proper sequence of transactions are central to the success of the session.

Session 2: Understanding the effects of Engineering Changes on costs, when Current Manufacturing costs roll over into the Frozen Standard.

The objective of this session is to simulate the effects that the Engineering Changes will functionally have on various cost types.

The exercise will involve creating a new part on the item master file, changing the Bill of Material structure with a future effectivity date, and comparing the cost differences. The team will then make the change effective (change effectivity date) and note the results of the cost transfer from Current Engineering to Current Manufacturing.

Session 3: Effects of a pseudo on cost buildups.
The objective of this session is to simulate effects of converting an assembly to a "pseudo" (or blowthrough) category.
The cost buildup impact is to be noted.
Session 4: Effects of changing the make/buy code for an item. The objective of this session is to simulate a change in the make/buy status of an item and observe the effects on cost.
The transition should invoke various cost analysis exception messages. The team should understand the action necessary to resolve exception conditions.

C.2.3 CBS Conference Room Pilot Guidelines

1. The conference room pilot (CRP) will simulate as much as possible normal operations and employ normal practices. However, due to the time constraints, processes that normally require overnight processing (e.g., rollover) may be run within a given cycle. These normal overnight runs will be flagged as "cycle" on the pilot activity.
2. Prior to conducting the session, each team member will review the exercise and individually write down what their "expected results" will be. If this is a group exercise, discuss the entire group's expected results before beginning the exercise.
3. After conducting the exercise, compare "actual results" to the individual's expected results and write down the reason for the difference. This diary should be kept for future reference. *The act of comparing expected to actual results is an extremely important step. This is where the mind-set of the team members rests and spending quality time reconciling the differences is where true process change logic will be uncovered.*
4. As a result of each session, the team will list activities that will require policy change, procedural augmenting or change, and any noteworthy security or process issues.
5. At the end of the CRP, the team will develop a functional specification that addresses the following:
   a. Make recommended parameter (control file) settings.
   b. Recommend software processing sequence (job streams).
   c. Identify where policy, procedure, and desk instructions are needed.
   d. Identify process workflow, functional responsibilities, and inputs and outputs.
   e. Recommend a security (update vs. query) table by functional area.
   f. Define the criteria essential to classify a user as a "certified expert" for training and process deployment certification.
   g. Modify the CRP exercises for training sessions.
   h. Define criteria for final sign-off of business process and walkthrough parameters.

C.2.4 CBS CRP Scenarios

1. Validate Current Manufacturing costs for products in test database.
   a. Run cost generation job for Current Manufacturing.
   b. If costs agree, roll Current Manufacturing into Next Period type.
   c. Validate Next Period and roll into Frozen Standard cost type.
   d. Compare Frozen Standard to Current Manufacturing and note any differences.
   e. Override Value Added, Total, and set Cost Control = Buy on an item and attempt to change the standard and note any difficulties.

2. Generate an Engineering Change on a parent with Effectivity 1 month into the future. Replace one component with a newly added item that has no cost.
   a. Compare the Frozen Standard to Current Engineering cost. Cycle 1
   b. Add cost to the newly added part and run another cost comparison, comparing Current Engineering to Frozen Standard.
   c. Change effectivity date to yesterday's date. Cycle 2
   d. Compare Frozen Standard to Current Manufacturing. Cycle 3
   e. Delete (attempt) the component that was phased out from the item master file.
   f. Generate a Product Cost Sheet for product as of "point in establishing new standards."

Note: Keep a diary of the cost changes and highlight differences between "expected results" and "actual results."

3. Create a "pseudo" bill relationship.
   a. On pseudo item, create a value-added cost.
   b. Change Cost Control of pseudo parent to *KIT* item.
   c. Review cost of Frozen Standard and Current Manufacturing.
   d. Change Cost Control code of pseudo parent to MAKE and review cost effect.
   e. Change Cost Control of parent to *KIT.* Cycle 1
   f. Change Cost Control of pseudo item to *KIT* and review pseudo cost for Frozen Standard and Current Manufacturing. Cycle 2
   g. Change Cost Control of pseudo to *MAKE* and review Frozen Standard and Current Manufacturing costs.
   h. Change cost rollup code of pseudo item = "NO."
   i. Change value-added cost of a component of the pseudo. Cycle 3
   j. Run Cost Generation job for pseudo parent item. Cycle 4
   k. Review effect of cost rollup = "NO" for pseudo parent on Current Manufacturing cost.
   l. Change cost rollup for pseudo = "YES," Cost Control code = "Non-Costed." Cycle 5
   m. Rerun Cost Generation job and review effect of Current Manufacturing cost.
   n. Change Cost Control code = "Kit." Cycle 6
   o. Rerun Cost Generation and review effect on Current Manufacturing.

Note: Keep a diary of the cost changes and highlight differences between "expected results" and "actual results."

4. Change a parent's Make/Buy code from *MAKE* to *BUY.*
   a. Leave Cost Control code = "Make" for parent.
   b. Review cost effect of Make/Buy decision.
   c. Change one of parent's component cost. Cycle 1
   d. Change Cost Control code to *BUY* and review cost effect.
   e. Change Cost Control code to *KIT* and review cost effect. Cycle 2
   f. Change parent's Make/Buy code back to *MAKE.*
   g. Change a component's cost and review cost affect. Cycle 1
   h. Change Cost Control code of parent to *BUY* and review cost effect. Cycle 3
   i. Change Cost Control code of parent to *KIT* and review cost effect.
   j. Change a component's Cost Control to *NON-COSTED* and review Current Manufacturing cost effect. Cycle 4

Note: Keep a diary of the cost changes and highlight differences from "expected results" from "actual results."

5. Modify *Run Hours* on an Assembly Operation.
   a. Create Cost Type "Simulated."
   b. Rollover Cost Type Current Manufacturing into "Simulated" and then perform a Cost Generation.
   c. Modify *RUN HOURS* on an Operation and review cost effect.
   d. Delete an operation and review cost effect.
   e. Add a new operation and review cost effect.
   f. Run Cost Generation.
   g. Compare cost effect of routing changes on Current Manufacturing and Current Engineering cost types.
   h. On Cost Type "Simulated," set Cost Type usage to *CLEAR* and review cost effect of "Simulated" costs.
   i. Delete (attempt) Cost Type Simulated.

Note: Keep a diary of the cost changes and highlight differences between "expected results" and "actual results."

6. Simulate a year-end process.
   a. Rollover costs from Frozen Standard to Prior Year.
   b. Rollover costs from Current Manufacturing to Next Period.
   c. Rollover cost from Next Period to Frozen Standard.
   d. Set Cost Override code to *FROZEN* once you're happy with the result.

Note: Keep a diary of the cost changes and highlight differences between "expected results" and "actual results."

7. Create a Work Center Overhead Rate (WORC) Table (5000) with value of 10%. Create Work Center Overhead (WCOH) Table (90) and tie it to Work Center 18 and Table 5000.
   a. Create an existing Product Cost Sheet.

   Cycle 1

   b. Tie the newly created WCOH 90 table to the item "a" above and generate a new Product Cost Sheet.
   c. Note difference between the two Product Cost Sheets.

   Cycle 2

   d. Retie the original Item record and delete WCOH 90.

Note: Keep a diary of the cost changes and highlight differences between "expected results" and "actual results."

8. Open a Manufacturing Order.
   a. Print out a Product Cost Sheet.
   b. Modify Order Dependent bill (substitute a component).

   Cycle 1

   c. Process a receipt for a partial completion of the Order.

   Cycle 2

   d. Close the Order *Short* and review the *closed order costs*.
   e. Reopen the Order and modify the Frozen Standard costs (override).
   f. Process a receipt for the remainder of the Order.

   Cycle 3

   g. Review Closed Order Cost and compare to "d" above.
   h. Reopen the Order and change the Order Quantity to +10.
   i. Change the Issue Control code on one component to *BACKFLUSH*.
   j. Process a receipt for 10 units.
   k. Review Closed Order Cost and compare to "d" and "g" above.

9. Create a Material Overhead Rate Calculation (MORC) Table (5000) using a FIXED $ AMOUNT and tie it to a given part. Cycle 1
    a. Note the cost effect of processing the transaction. Cycle 2
    b. Remove Table 5000 from the part and note the cost effect of processing the transaction. Cycle 3
    c. Retie MORC Table 5000 to item along with creating Table 5010 with an override cost amount and note the cost effect of processing the transaction. Cycle 4
    d. Remove Tables 5000 and 5010 from the item.
    e. Delete Tables 5000 and 5010.

Note: Keep a diary of the cost changes and highlight differences between "expected results" and "actual results."

10. Develop a Product Line Overhead Type *A* to represent Product Line 960L.
    a. Select various parts to be included in the product line and encode these parts with Product Overhead Type = *A*.
    b. Create WORC Table 6000 using a text message to note Product Line.
    c. Set up a Product Line relationship by setting the WCOH Table 01 to Product Overhead Type A, code 6000. Cycle 1
    d. Review associated Product Cost Sheets and Costed Bills of Materials.
    e. Change one item and remove it from the Product Line. Cycle 2
    f. Rerun Product Cost Sheets and Costed Bills of Material and compare to (d) above, and note changes.
    g. Change the Cost Item code to 01 and delete WCOH Table 10. Cycle 3

Note: Keep a diary of the cost changes and highlight differences between "expected results" and "actual results."

Appendix D (Chapter 6)

D.1 Education Matrix Forms

D.1.2 Organizational Function versus Business Process Matrix

This matrix grossly depicts which functional areas will have either business process or software module involvement. For example,

→ Business process →

	BoM	INV	SOE	PUR	MRP	MFG	SFC	WIP	CMS	G/L	A/P	A/R	PAY
Engineering	x		x	x			x	x	x				
Material control	x	x	x	x	x	x	x	x	x				
Prod control	x	x	x		x	x	x	x	x				
Order admin	x	x	x						x				
Marketing	x	x	x		x				x				
Human resource													x
Accounting	x	x		x				x	x	x	x	x	x

Legend:

BoM = Bill of Material
INV = Inventory
SOE = Sales Order Entry
PUR = Purchasing
MRP = Material Requirements Planning
MFG = Manufacturing
SFC = Shop Floor Control
WIP = Work in Process
CMS = Cost Management
G/L = General Ledger
A/P = Accounts Payable
A/R = Accounts Receivable
PAY = Payroll

Note: We have elected to display three different matrix examples. Other matrixes may include personnel-to-session, mix-to-session, feature-to-module, and so on. A separate series of matrixes are needed for *education* and another for *training.*

D.2 Procedures Training and Education Planning Matrices

Individual by transaction/feature matrix

Another matrix consists of detailing each business process/module into functionality and listing personnel attendance. For example:

→ Inventory module features →

	Receipt	Issue	Move	Adjustment	Pick	Txn matrix	Cycle inv	Cost	Location	Reject	Database
Gerry W.	x	x			x	x		x	x	x	x
Robert L.				x			x		x		x
Michael M.	x	x	x	x	x	x	x	x	x	x	x
Joan P.	x	x			x	x			x		x
Bobbie S.		x			x		x		x		x
Boris K.	x		x							x	x
Donny O.			x		x	x				x	x

Transaction/feature by session matrix

Another matrix involves functions with exercise sessions. The objective of this matrix is to tie a specific exercise session to different features. For example:

→ Session number →

	1	2	3	4	5	6	7	8	9	10	11	12
Txn matrix	x	x	x		x					x		x
Receipt		x	x					x			x	
Issue			x	x				x				x
Move				x					x			
Adjustment					x				x			
Pick					x	x			x			
Cycle inv									x			x
Cost	x						x			x		
Location	x									x		
Reject						x		x			x	
Database										x		x

D.3 Developing Objectives and System Measures

As part of a total plan to successfully install and then manage a state-of-the-art enterprise resource planning (ERP) system, a balanced integrated set of objectives and system measures for every application area should be defined and monitored.

For each of the criteria in the following representative lists of application areas, a number of considerations should be addressed:

- Method of monitoring, including reporting method and frequency
- Who is responsible, either as a functional unit or as an individual, for the performance?
- Who should be responsible for monitoring?

- What are the specific goals (numbers, etc.), including an allowable error range?
- How do objectives and measures fit into a logical hierarchy of objectives, and to what level in that hierarchy should the reports be made?
- Is there a proper balance between management objectives and system measures?
- Should there be a separate report for different units of measure (i.e., dollars vs. quantity) or is it possible to combine them on one report?

A few representative objectives and measures to be considered during the ERP implementation are listed. An implementation plan best practice would be to expand these objectives and measures on a module-by-module basis for *every* ERP module to be implemented.

I. *Application areas: Master scheduling*
Production planning
Applicable module: Master production scheduling
  - Inventory levels, compared to sales volume by product groups
  - Production plan versus sales forecast by product groups
  - Ratio of total late shipments and on-time deliveries
  - Level and type of exception messages coming out of material requirements planning
  - Level of safety stock
  - Analysis of late orders
  - Analysis of lost orders
  - Lead-time compression

II. *Application area: Bill of material (BoM)*
Applicable modules: BoM System
Material requirements planning
  - Turnaround time to create BoM records
  - Levels of BoMs and throughput impedance
  - Speed of processing engineering changes
  - Accuracy of computer file data
  - Standardization efforts in engineering (i.e., fewer duplicated parts)
  - Level of maintenance/update efforts for all areas, both before and after implementation

III. *Application area: Routing*
Applicable module: Process and routing system
  - Time to create routings and amount of non value-added process
  - Accuracy of computer file data versus the way the part is actually made

IV. *Application area: Inventory*
Applicable module: Material control system
  - Accuracy of cycle count
  - Transaction count accuracy
  - Number of unplanned shortages
  - Inventory levels and reduction methodology
  - Productivity of personnel in stockroom
  - Amount (dollars, manpower) on annual physical inventory (should be a one-time-only benefit)
  - Financial summary of cycle counting
  - Document processing

V. *Application area: Material requirements planning*
Applicable module: Material requirements planning
- Material availability, including shortages
- Levels of inventory
- Productivity of all personnel involved in system, direct and indirect
- Exceptions generated (by type)
- Number of shop and purchase orders released
- Lead time, fences, and cumulative lead time

VI. *Application areas: Capacity requirements planning*
Shop floor control
Applicable module: Capacity requirements planning
Shop floor control
Product costing system
- Levels of work in process (WIP) and compression methodology
- Lead times, planned versus actual
- Analysis of queue and continuous improvement
- Amount of level of exception conditions, such as splits and priority overrides.
- Productivity of labor, both direct and indirect
- Analysis of work centers, utilization's planned, and actual input and output
- Make/buy ratios
- Overtime
- Analysis of order shortages
- Analysis of standard hours or cost to actual hours or cost
- Number of transaction errors
- Flow and timing of paperwork in system, for example, shop order release
- Schedule performance

VII. *Application area: Purchasing*
Applicable modules: Purchasing control system
Product costing system
Material requirements planning
- Delivery date performance
- Analysis of quality
- Number of exception messages in material requirements planning
- Variance of cost, including indirect elements such as shipping charges
- Open commitments and how they match master schedule

VIII. *Application area: Marketing*
Application modules: Sales order entry
- Customer service delivery performance
- Quality and returns analysis
- Number of order exception messages
- Variance of profit margins by account
- Open order time phase
- Performance to forecast

IX. *Application area: Accounting*
Application modules: General ledger
- Accounts payable

- Accounts receivable
- Cost management

Document:

- Cutoff standards
- Days sales outstanding
- Cash flow projections
- Margin by product line
- Budget versus actual expenditures
- Projecting future variances

D.4 Cutover Checklist for a Formal ERP System

The following list of questions should be answered before you make the decision to begin the cutover to a formal manufacturing system.

	Yes or No
Users	
Are conversion users available?	[ ]
Are validation users available?	[ ]
Are start-up users available?	[ ]
Are users available for daily operation of the system?	[ ]
Training	
Are conversion users trained?	[ ]
Are validation users trained?	[ ]
Are start-up users trained?	[ ]
Are daily users trained?	[ ]
Are other support people trained for daily system operation?	[ ]
Documentation	
Are conversion procedures published?	[ ]
Were conversion procedures reviewed and approved by key users?	[ ]
Are instructions published for conversion data?	[ ]
Are instructions published for daily operating data?	[ ]
Are start-up procedures published?	[ ]
Were start-up procedures reviewed and approved by key users?	[ ]
Are cutover procedures published?	[ ]
Are daily operating procedures published?	[ ]
Are technical procedures published?	[ ]
Were daily system policies and procedures reviewed with users?	[ ]
Is a user reference manual published and thoroughly indexed?	[ ]
Software	
Were conversion programs system tested for all known conditions?	[ ]
Were conversion programs tested with a representative sample of data?	[ ]
Were conversion procedures used for a representative sample of data?	[ ]
Were new system programs system tested for all known conditions?	[ ]
Were new system programs tested with a representative sample of data?	[ ]
Are conversion job streams tested?	[ ]
Are new system job streams tested?	[ ]

Hardware

Is computer conversion time available? [ ]

Is data entry conversion time available? [ ]

Is new system run time available? [ ]

Is new system data entry time available? [ ]

Cutoff

Are you preparing transactions in the new format? [ ]

Are you cleaning old data out of the "pipeline?" [ ]

If you have answered "No" to any of the above questions, you are not ready for cutover. Before making a decision to proceed in spite of one or more negative responses, you should evaluate the potential impact on the success of the cutover and daily system operation.

D.5 Start-Up Final Checklist

1. Test/training pilot completed and objectives met? Yes___No___
2. Conference room pilot completed and objectives met? Yes___No___
3. Conversion programs completed and tested using full production database? Yes___No___
4. All interface programs completed and signed off by the user? Yes___No___
5. Education and training completed for the key users at a minimum? Yes___No___
6. All forms and procedures completed and signed off? Yes___No___
7. System control parameter settings for production mode operation? Yes___No___
8. Production mode job control language established and tested? Yes___No___
9. Management information systems (MIS) operating schedule established? Yes___No___
10. Performance measurements identified and monitoring procedure in place? Yes___No___
11. Start-up plan documented, in detail, with names and dates (times)? Yes___No___
12. Fallback/recovery plan established and published? Yes___No___

D.6 ERP Implementation Guide

Although included in Chapter 6, this guide has a very broad span and may be used throughout the content discussed in the book.

D.6.1 Introduction

One of the more difficult tasks involved in implementing ERP philosophies is defining what tasks need to be accomplished for each module. As an aid to assist companies, we have prepared a series of questions and points to consider for various modules. These checklists were designed to function as a representative starting point for ERP modules. An implementation plan best practice would be to expand these checklists a module by module for every ERP module to be implemented.

In order to maximize the benefits of the checklists, each checklist item requiring action should be transferred to a separate page and detailed according to the requirements and desires of your company. In addition, the checklists may function as a work plan. Each checklist provides for assignment of responsibility and due dates.

We recommend that your project manager and pertinent subject matter expert sit down and review the use of the checklists and actually walk through the detailing and assignment of task due dates and responsibilities. If used properly, the checklists will guide the project down the path of success.

D.6.2 Overview of a Successful ERP Project

1. *Commitment*

   Prior to any other activity, a decision and commitment must be made which expresses the following:

   a. What is to be accomplished?
   b. What resources will be provided?
   c. Definition of major project event completions
   d. What means will be used to monitor progress?
   e. When the estimated payoff will grow to fruition?

   To attain a commitment, middle management most likely will submit a proposal to top management. Top management will modify the proposal as necessary, notify all members of the management team of the existence of the project, and then be prepared to give the project 100% backing.

   It is then the middle management's responsibility to "make it happen" and the top management's responsibility to observe progress (via steering committee activity) and intercede as necessary to ensure a successful completion.

   An ERP project should be measured like any other project: Performance evaluation should be discharged according to the degree of success or failure of the project (project profit and loss).

2. *Start-up activities*

   Once the commitment has been made and everyone understands the conditions of the commitment, a project team should be formed. The project should be headed by an influential individual from the user community, most desirably the manager from the supply chain group or master scheduling group. Depending on the timetable, areas of emphasis, and so on, the members of the project team will consist of the following:

   a. Materials expert
   b. Engineer*
   c. Marketing*
   d. Cost accountant*
   e. Procedures writer
   f. Outside consultant

   These members will typically be full-time members or intensely used only during the design and implementation involving their area of expertise. The intensity depends upon project completion goals.

Note: Concurrent with the formation of the project team, specifications should be written on what modules will be implemented first, tentative implementation schedules, and assignment of module implementation responsibility. Once the team has been assimilated, detail specifications should be defined. Detail specifications result from the following:

a. Interviews from top management
b. Defining functional interactivity
c. Flowcharting existing business system

---

* These areas are frequently optional on ERP projects ... whether to be included is at leadership discretion (based upon the amount of impact the project will have to these functions).

Note: Do not expect to have the first phase of the system operational for the first three to six months of the project. This period will be either increased or decreased according to the availability of accurate data, procedures to maintain this data, resources dedicated, and so on.

3. *Implementation*
   Consider each module as it is implemented according to schedule.
   a. Paperwork cutoff should have been detailed prior to implementation.
   b. Training should have already occurred on the following:
      i. Input documents
      ii. Reports detailed
      iii. Procedures
      iv. Processes
   c. Procedures should be written for any documents and functions in the module.
   d. The module software procedures and interfaces should have been tested.
   e. The audit control and document pickup procedure should have been established.
   If the above steps were completed prior to implementation, the tasks necessary during implementation would be the following:
   a. Initialize cutoff procedure
   b. Load data
   c. Verify data
   d. Establish input/output frequency and distribution
   e. Check timeliness of document flow
   Needless to say, if the pre-implementation steps are not accomplished prior to implementation, they need to be accomplished either during or after implementation. Waiting until implementation will tend to cause the following:
   a. Crisis
      i. Already limited resources must be diluted even more resulting in rushed decisions, inadequate research, and less than dedicated participation.
   b. Confusion
      i. If education has not occurred, users will either not know what to do, and do nothing, or do things incorrectly. This condition has a tendency to mushroom into data integrity problems quite rapidly.
   c. Failure
      i. If the pre-implementation steps are not adequately accomplished, the result will be the following:
         A. *Partial failure*—Not receiving planned benefits and, unless recognized quickly, the project will lean toward total failure
         B. *Total failure*—Abandonment of the project, resulting in personnel turnover, discouragement, and no benefits to the business bottom line
4. *Post-implementation review*
   Equally important to the commitment, preparation, and implementation is signifying that the module has integrity and that implementation is a success. This process should occur approximately six weeks from the date of the conversion cutoff. During the review, payback goals should be established by the users and a method of measuring progress should be defined. For example, after converting the inventory module, if one payback (return-on-investment) goal was to decrease the total inventory by 20% within the next four months,

this should be written down and a means of measuring results report defined. One method might be to follow the example of successful companies. Results-oriented companies attest to the fact that they have a higher success rate if goals are linked with performance analysis (pay raises) and Management by Objective (MBO)/Management by Results (MBR) objectives (see Chapter 3 for more detail).

D.6.3 General Considerations for Systems Design

In the design of any system (subsystem, module, or report), there are several basic questions that must be addressed and answered before the details can be specified. (Once they are answered, often the solution becomes obvious.) Some of these questions are as follows:

- What is the purpose of the system?
  - What decisions will be made from it?
  - What visibility will be gained?
  - What questions will be answered from it?
- What information is required to meet the purpose?
- Where will the data be coming from?
  - Database
  - Operational transaction
  - Input with report request
  - Other source(s)
- When must this information be made available?
  - How timely must the data be?
  - How fast a "turnaround time" between the inquiry and the answer?
  - What "surprises" have happened in the past?
  - What could we have known to avoid these surprises?
- What is the value of having this timely information?
- What are the costs (and complications) of the alternatives in achieving the purpose?

Experience has shown that all of Murphy's laws are applicable to ERP systems, and in general the KISS (keep it simple stupid) principle is the best starting point.

D.6.4 Maximize the Benefits from Using the Checklist

1. Use the appropriate checklist as the module is planned for implementation:
   a. Project team
   b. All department coordinators
   c. Steering Committee
   d. All departmental support staff from the module being planned

Actually, anyone within the company that can provide assistance for the particular module should review the checklist and supply comments, additions, and so on for review. For example, if the work order module were being considered, the following people would review the checklist:

Department	*Personnel Reviewing Checklist*
Production control	Materials manager, production control manager, all schedulers, all expediters, and associate responsible for issuing work order numbers and lot numbers
Production	Production manager, all production superintendents, all production supervisors, all foremen, and lead persons; a select group of production workers, timekeepers, departmental associates, and material coordinators
Engineering	Manufacturing engineer responsible for router; product/design engineer responsible for the bills of material
Stores	All persons handling shop traveler packets
Quality Control	All persons signing the work orders for quality approval

2. Each department coordinator will gather the comments, suggestions, added items, and so on from his group and summarize onto blank checklist
3. At project team meeting
   a. Summarize and discuss
      i. Items requiring action on checklist
      ii. Add-ons
   b. Assign due dates and responsibilities
   c. Transpose items requiring action (checked items) to a blank form
   d. Make a detailed plan for each item requiring action and assign due dates and responsibility
4. Obtain Steering Committee inputs and approval
5. Complete assigned tasks, formalize procedures, obtain approvals, and so on
6. Six weeks subsequent to implementing action item procedures, conduct a post implementation review and
   a. Add/modify as required
   b. Brief Steering Committee
7. Six months subsequent to implementing module, test for
   a. Increased efficiency
   b. Decreased costs
   c. Improved bottom line

D.6.5 Module Checklist

D.6.5.1 Item and Product Structure

Module Item and Product Structure

Responsibility ______________________
Compl Due Date ______________________
Approved By ______________________
Rev# __________ Date ______________

Implementation Checklist	Action Req'd	Start Date	Actual Start	Compl Date	Actual Compl
1. Are all item numbers loaded? Has this been verified item-by-item? 2. Have all data elements been filled in for all items? a. Who is responsible for this to happen? b. Has a due date been established? c. Is each responsible department aware of the source to obtain the necessary data? d. Is Top Management aware of any delays? i. Who is responsible to notify Top Management of progress and delays? ii. Is Top management being given a progress report at least biweekly? 3. Does the product structure reflect the way the item is being manufactured? (Manufacturing Bill of Material) 4. Is the product structure complete and accurate? a. Has each line item on the manual bill been compared to each line on the computer configuration of the bill? 5. Has quantity/assembly field been validated? a. Is the scrap factor accurate for each assembly/component relationship? 6. Have input forms been established?					

1

Module Item and Product Structure

Responsibility ______________________
Compl Due Date ______________________
Approved By ______________________
Rev# __________ Date ______________

(con't) Implementation Checklist	Action Req'd	Start Date	Actual Start	Compl Date	Actual Compl
7. Are engineering changes being processed expeditiously? a. Are forms available? 8. Are Audit Trails monitored and errors corrected? a. How are delinquent error corrections being expedited? 9. Are the product structures in sufficient shape to kit by? a. If not, how will they be made ready? b. How will they be verified? 10. Are parts which should be attached to a product structure, but not, identified? 11. Are assemblies, which should be attached to another assembly, but not, identified? 12. Will Item List/Product Structure be validated as 100% accurate? a. How? b. Who will be responsible? 13. Have structure records been assessed to remove nonvalue-added inclusions?					

2

D.6.5.2 Inventory

Module Inventory

Responsibility ___________
Compl Due Date ___________
Approved By ___________
Rev# ___________ Date ___________

Implementation Checklist	Action Req'd	Start Date	Actual Start	Compl Date	Actual Compl
1. Are all item numbers on file? What report will be used to verify this? 2. How will accuracy be verified for: a. On-hand balances b. Transactions c. Scrap d. Requisitions e. Move to Stock f. Receipts 3. Prior to on-hand conversion: a. Have documents, which affect the on-hand balance, been identified? b. Do you have "before" and "after" inventory rubber stamps available for use at the time of conversion? c. Have all "before" conversion documents been properly processed? i. How are you ensuring that this is happening? ii. Will a person "beat the bushes" to ensure that all documents are either posted or changed to "after?" d. Will "a before" stamping occur at least two days prior to conversion? 1. How long will "after" stamping occur?					

3

Module Inventory

Responsibility ___________
Compl Due Date ___________
Approved By ___________
Rev# ___________ Date ___________

(con't) Implementation Checklist	Action Req'd	Start Date	Actual Start	Compl Date	Actual Compl
4. Is the various field data available for input? a. Lead-time b. Class c. Cost Fields d. Bin-number e. Description f. Lot-size g. Revision h. Safety Stock i. Shrinkage 5. Who is responsible for gathering and ensuring that the data is current? 6. How will work-in-process (WIP) be input? a. What is the cutoff procedure? b. How will it be validated? (Monthly WIP inventory, visual check, etc.) 7. How will inventory adjustments be handled? a. Which departments will be required to approve the adjustment? b. How will adjustments be recorded? c. Are adjustment forms prenumbered? d. Who is authorized to initiate an adjustment? e. What is the criteria for items which require adjustments be cycle counted? f. Will the cycle change?					

4

Module Inventory

Responsibility ____________________
Compl Due Date ____________________
Approved By ____________________
Rev# __________ Date __________

(con't) Implementation Checklist	Action Req'd	Start Date	Actual Start	Compl Date	Actual Compl
8. How will items in Receiving be adjusted? a. How will balance discrepancies be adjusted? b. Who is responsible for Receipt Quantity values? c. How will rejects (MRB) be handled? d. Who is responsible for rejects? e. Who is responsible for the reject/hold (MRB) area? f. How will material be moved to and from the reject–hold area securely?					
9. How will rework-in-house be controlled? a. What field/file will the value be tracked? b. What form will be used to move into and out of that area? c. How will scrap be accounted for? d. How will time be billed back to the Vendor? e. What overhead factor will be applied?					
10. How will Scrap/Damaged goods be handled? a. WIP b. Receiving c. Stores d. In-Transit					
11. Is the Stockroom locked? a. Are the keys to the stockroom restricted? b. When will top management test the security of the stockroom?					

5

Module Inventory

Responsibility ____________________
Compl Due Date ____________________
Approved By ____________________
Rev# __________ Date __________

(con't) Implementation Checklist	Action Req'd	Start Date	Actual Start	Compl Date	Actual Compl
12. Are Bin-numbers assigned to each item? a. Are overflow bins identified? b. Is a valid up-to-date location list available?					
13. Will each transaction be dated and time stamped? a. When will date and time stamp equipment be available? b. Where will the stamping machines be located? c. When will training on the use of this equipment take place? d. Who will be trained?					
14. Are the procedures written for and forms available for the following: a. Cycle Counting b. Kitting c. Material Requisition d. Stock Return e. Scrap Notices f. Receiving Memo g. Reject Notice h. Rework-in-House i. Use-as-Is j. Inventory Adjustments k. Stock Location System l. Returned Goods m. Re-inspection Requests n. Others (list)					

6

Module Inventory

Responsibility ______________________
Compl Due Date ______________________
Approved By ______________________
Rev# ____________ Date ____________

(con't) Implementation Checklist	Action Req'd	Start Date	Actual Start	Compl Date	Actual Compl
15. Are withdrawals and receipts weigh counted? a. Is weigh count equipment being checked for accuracy periodically? b. Are the weigh count personnel being checked for accuracy periodically?					
16. Have stockroom personnel been trained on how to ensure that the objective of Zero Defects is met? a. Are Zero Defects signs posted in the stockroom?					
17. How will record accuracy be validated? a. How often? b. Who will ratify it? c. How will discrepancies be handled? d. Who will be notified? e. What methods of corrective action will be enforced? f. Who is responsible?					
18. How are error messages accounted for? a. Is there a follow-up system to expedite delinquent error corrections?					
19. Will an Audit File be maintained?					

7

D.6.5.3 Forms

Module Forms

Responsibility ______________________
Compl Due Date ______________________
Approved By ______________________
Rev# ____________ Date ____________

Implementation Checklist	Action Req'd	Start Date	Actual Start	Compl Date	Actual Compl
1. Cycle Count Sheet					
2. Inter-Department Material Transfers (production material)					
3. Inter-Stockroom Transfer					
4. Inventory Adjustment					
5. Kit List					
6. Labor Actual Hours					
7. Lot Number Control Adjustments					
8. Material Order Release (Blanket PO Release)					
9. Material Order Release Change					
10. Material Requisition					
11. Material Review Board Disposition a. Return to Vendor b. Rework-in-House c. Scrap d. Send to Another Vendor For Rework e. Use-as-Is					

8

Module Forms

Responsibility ____________________
Compl Due Date ____________________
Approved By ____________________
Rev# __________ Date __________

(con't) Implementation Checklist	Action Req'd	Start Date	Actual Start	Compl Date	Actual Compl
12. Move to Stock 13. Outside Processing Purchase Order a. Change b. Delete c. New 14. Purchase Order a. Change (CPO) b. Cancel PO c. Close PO d. Delete PO 15. Receiving a. Receipt At Dock b. Receipt Dock-to-Stock i. Partial Receipt Dock-to-Stock 16. Receipt of In-House Produced Goods a. Inspection Required (Fab, Molding, Extruded)					

9

Module Forms

Responsibility ____________________
Compl Due Date ____________________
Approved By ____________________
Rev# __________ Date __________

(con't) Implementation Checklist	Action Req'd	Start Date	Actual Start	Compl Date	Actual Compl
17. Release of One Shipment Having Multiple Vendors 18. Release to Stock of In-House Produced Goods (Fab, Molding, Extruded, etc.) 19. Receiving Memo Error Correction 20. Re-inspection Request 21. Rejections a. Receiving b. Credit c. Replacement 22. Notice of In-House Produced Goods a. Reject from Work Center b. Rejected Material Disposition c. Vendor Corrective Action 23. Request for Quotation 24. Returned Goods a. Defective: Repair b. Defective: Scrap c. Engineering Change					

10

Module Forms

Responsibility ____________
Compl Due Date ____________
Approved By ____________
Rev# ________ Date ________

(con't) Implementation Checklist	Action Req'd	Start Date	Actual Start	Compl Date	Actual Compl
25. Returned Goods					
a. Lot Recall					
b. Past Due: Shelf-Life					
c. Repair: Billable					
d. Repair: Warranty					
e. Repossession					
f. Return for Credit					
26. Route Sheet					
27. Scrap					
a. Production					
b. Receiving					
c. Stores					
28. Shipments					

11

Module Forms

Responsibility ____________
Compl Due Date ____________
Approved By ____________
Rev# ________ Date ________

(con't) Implementation Checklist	Action Req'd	Start Date	Actual Start	Compl Date	Actual Compl
29. Stock Locations					
a. Change					
b. Delete					
c. New					
30. Work Order					
a. Completion					
b. Issue					
c. Operation Completions (partial)					
d. Splits					

12

Module Forms

Responsibility ____________________
Compl Due Date ____________________
Approved By ____________________
Rev# __________ Date __________

(con't) Implementation Checklist	Action Req'd	Start Date	Actual Start	Compl Date	Actual Compl
31. Customer Data					
a. Change					
b. Delete					
c. New					
32. Database Fields Request					
a. Add					
b. Change					
c. Delete					
33. Error Correction Input Sheet (General Purpose)					
34. Master Schedule Data					
a. Change					
b. Delete					
c. New					

13

Module Forms

Responsibility ____________________
Compl Due Date ____________________
Approved By ____________________
Rev# __________ Date __________

(con't) Implementation Checklist	Action Req'd	Start Date	Actual Start	Compl Date	Actual Compl
35. Item Number Data					
a. Change					
b. Delete					
c. New					
36. Product Structure					
a. Change					
b. Copy					
c. Delete					
d. Engineering Change					
e. New					
f. Substitute					
37. Report Request (Form or Web)					
a. ABC Analysis					
b. Cost Report					
c. Costed Bill of Material					
d. Customer Due					

14

Module Forms

Responsibility ______________________
Compl Due Date ______________________
Approved By ______________________
Rev# ____________ Date ______________

(con't) Implementation Checklist	Action Req'd	Start Date	Actual Start	Compl Date	Actual Compl
37. Report Request (Form or Web)					
e. Cycle Count Report					
f. End-Use					
g. Explode Indented					
h. Inventory Valuation					
i. Kit List					
j. Open PO Report					
k. Open SO Report					
l. Open WO Report					
m. Item PO Report					
n. Item SO Report					
o. Item WO Report					
p. Item Status Report					
q. PO Due Reports (Exception)					
r. PO Overdue Reports (Exception)					
s. PO Status Report					
t. Router Book					

15

Module Forms

Responsibility ______________________
Compl Due Date ______________________
Approved By ______________________
Rev# ____________ Date ______________

(con't) Implementation Checklist	Action Req'd	Start Date	Actual Start	Compl Date	Actual Compl
37. Report Request (Form or Web)					
u. SO Commitment Report					
v. SO Due Report (Exception)					
w. SO Status Report					
x. Variance Reports					
y. Vendor Due Reports (Exception)					
z. Vendor Listing					
aa. Vendor Overdue Report (Exception)					
ab. Work Center Listings					
ac. Where Used Indented					
ad. Where Used Book					
ae. WO Completed Report					
af. WO Due Report (Exception)					
ag. WO Overdue Report (Exception)					
ah. WO Status Report					
ai. Work-in-Process Report					
aj. Yearly Inventory Report					

16

Module Forms

Responsibility ______________________
Compl Due Date ______________________
Approved By ______________________
Rev# ____________ Date ______________

(con't) Implementation Checklist	Action Req'd	Start Date	Actual Start	Compl Date	Actual Compl
38. Router					
a. Change					
b. Delete					
c. New					
39. Sales Order					
a. Change					
b. Delete					
c. New					
40. Unit-of-Measure					
a. Change					
b. Delete					
c. New					
41. Vendor					
a. Change					
b. Delete					
c. New					

17

Module Forms

Responsibility ______________________
Compl Due Date ______________________
Approved By ______________________
Rev# ____________ Date ______________

(con't) Implementation Checklist	Action Req'd	Start Date	Actual Start	Compl Date	Actual Compl
42. Work Order					
a. Change					
b. Delete					
c. New					
43. Work Order Item					
a. Change					
b. Delete					
c. New					

18

Module Forms

Responsibility ____________
Compl Due Date ____________
Approved By ____________
Rev# ________ Date ________

Implementation Checklist Form Should Be Tested For	Action Req'd	Start Date	Actual Start	Compl Date	Actual Compl
1. Is Form Required/Changing?					
2. Has document flow been flowcharted?					
3. Has a study been conducted on time between transaction occurrence and posting?					
4. When will Document Training occur?					
5. Who is responsible for training?					
6. Who will be trained? (include Supervisory personnel in each area)					
a. Stockroom Personnel					
b. Material Handlers					
c. Material Coordinators					
d. Stores Clerk (office)					
e. Cost Accounting					
f. Inventory Control					
g. Production Control					
h. Audit Control					
i. Office Services					
j. User Department (production, engineering, QC, etc.)					
k. Staging Area Personnel					
l. Time Keepers					
m. Production Supervision					
n. Receiving Clerk					
o. Marketing					
p. Salesmen					
q. Branch Offices					
r. Quality Control					

19

Module Forms

Responsibility ____________
Compl Due Date ____________
Approved By ____________
Rev# ________ Date ________

(con't) Implementation Checklist	Action Req'd	Start Date	Actual Start	Compl Date	Actual Compl
6. Who will be trained:					
s. Quality Assurance					
t. Engineering:					
i. product					
ii. manufacturing					
iii. design					
iv. industrial					
v. quality					
7. Will the document be assigned a control number?					
a. Are Documents preassigned to user departments each month?					
b. Are all lost and cancelled documents accounted for at the end of the month?					
c. Have audit and reconciliation procedures been formalized?					
d. Will the Control Number be pre-printed?					
8. Has Error Correction Procedure been formalized?					
a. Has form been designed?					
b. Who will correct errors?					
i. Data Entry Errors					
ii. Invalid Data Errors					
iii. Negative on-hand balances					

20

Module Forms

Responsibility ________________
Compl Due Date ________________
Approved By ________________
Rev#________Date________

(con't) Implementation Checklist Form Should Be Tested For	Action Req'd	Start Date	Actual Start	Compl Date	Actual Compl
9. What will be done to ensure all errors are corrected? a. Log b. Copy of error message in file c. Computer suspense file 10. Who will be responsible for error corrections? a. Audit Control b. Inventory Control c. User 11. Who will review document for legibility, pertinent data, proper authorization, etc., prior to going to data entry? 12. How will shortages be communicated? a. Who will be notified? b. When will they be notified? c. Who will be responsible for notifying interested parties? 13. Will document be date and time stamped? 14. What will the distribution be?					

21

Module Forms

Responsibility ________________
Compl Due Date ________________
Approved By ________________
Rev#________Date________

(con't) Implementation Checklist	Action Req'd	Start Date	Actual Start	Compl Date	Actual Compl
15. Will material be weigh counted? a. If there is a discrepancy between the paperwork and the count, what media will be used to notify . . . Initiator, purchasing, accounts payable, vendor, etc.? 16. Will Quality Control release be indicated on the document? 17. How will partial releases be handled? 18. Will document be photo reproduced? 19. Who is responsible for monitoring and expediting material flow? 20. Will the same form be used for: **a. Rejects** i. Receiving rejects A. Credit B. Replacement ii. Reject from work center iii. Notice of in-house rejection vi. Rejected material disposition v. Vendor Corrective Action **b. Returned Goods** i. Defective A. Repair B. Scrap ii. Return to stock as-is					

22

Module Forms

Responsibility ____________
Compl Due Date ____________
Approved By ____________
Rev# ________ Date ________

(con't) Implementation Checklist	Action Req'd	Start Date	Actual Start	Compl Date	Actual Compl
iii. Lot Recall iv. Past due shelf-life v. Repair A. Billable B. Warranty vi. Repossession vii. Return for Credit **c. Scrap** i. Scrap from Work Center ii. Scrap from Receiving iii. crap from Stores iv. Scrap from Material Review Board v. Scrap from Engineering **d. Purchasing** i. Maintenance, Repair, and Office (MRO) ii. Outside Processing iii. Blanket Order iv. Material Order Release v. Standard Order vi. Sister Division vii. International viii. Sister Plant ix. QC Release A. Vendor B. In-House produced goods (fab, molded, extruded) x. In-House produced goods					

23

Module Forms

Responsibility ____________
Compl Due Date ____________
Approved By ____________
Rev# ________ Date ________

(con't) Implementation Checklist	Action Req'd	Start Date	Actual Start	Compl Date	Actual Compl
21. Will the Stock Location assigned be verified by a second party; and a signature given? Spot-checked by Supervision as well? 22. What Performance tracking tool will be used to identify continuous improvement processes?					

24

D.6.5.4 Purchasing

Module Purchasing

Responsibility ______________________
Compl Due Date ______________________
Approved By ______________________
Rev# ____________ Date ______________

Implementation Checklist	Action Req'd	Start Date	Actual Start	Compl Date	Actual Compl
Vendor Data: 1. Is there Vendor Data on all current vendors? a. Is the data up to date and accurate? 2. Will the Vendor File contain: a. Purchase Vendor data only? b. Accounts Payable data only? c. Combination of the above? 3. Who will be responsible for verifying vendor data? a. Loading vendor data? b. How will accuracy be measured? 4. Will Buy Card History data be stored for analytical purposes? a. If yes, which data? 5. Is there a Vendor Code (number) assigned for every Open Purchase Order a. If not, who will ensure that it is accomplished prior to loading Purchase Orders? 6. Who will be responsible for assigning Vendor Codes in the future? a. How will this be done with adequate control, yet quick enough to respond to requirements for data entry? 7. Are the procedures in place for maintaining the Vendor File?					

25

Module Purchasing

Responsibility ______________________
Compl Due Date ______________________
Approved By ______________________
Rev# ____________ Date ______________

Implementation Checklist	Action Req'd	Start Date	Actual Start	Compl Date	Actual Compl
Purchase Order Data: 1. Has Purchase Order File been purged of cancelled, closed, etc., orders? a. If not, will it be purged prior to loading? b. Who is responsible for signing off the purged data? c. How will accuracy be measured? 2. Are the Purchase Orders in Purchasing, Receiving, and Accounts Payable totally in agreement? 3. Is a Vendor Code (number) on each Purchase Order? a. If not, when will it be assigned? b. Who is responsible? 4. How long does it take (elapsed time) between placing an order and posting it on the system? 5. Will nonstock, supplies, laundry, MRO, etc., POs be maintained on file? a. Who is responsible? 6. Will POs be closed prior to paying the invoice? a. What happens if there is a discrepancy between the invoice and PO?					

26

Module Purchasing

Responsibility ________________
Compl Due Date ________________
Approved By ________________
Rev# ________ Date ________

(con't) Implementation Checklist	Action Req'd	Start Date	Actual Start	Compl Date	Actual Compl
7. Who will edit the POs prior to data entry?					
8. How will Purchase Order changes take place?					
9. Who is ultimately responsible for closing and canceling POs? a. How will all other interested parties be notified of such occurrences?					
10. How often will file purging take place?					
11. How will discounts taken be recorded?					
12. How will purchase price variances be monitored? a. How will they be recorded (Buyer, monthly, etc.)?					
13. How will receiving be handled? a. Rejects					
14. What type of Receiving Audit Data is to be used? a. What information is to be retained? b. How long?					
15. Are procedures and forms in place for: a. Canceling POs b. Creating POs c. Changing POs d. Receiving (Dock)					

27

Module Purchasing

Responsibility ________________
Compl Due Date ________________
Approved By ________________
Rev# ________ Date ________

(con't) Implementation Checklist	Action Req'd	Start Date	Actual Start	Compl Date	Actual Compl
15. Are procedures and forms in place for: e. Vendor Reworks f. Scrapping g. Reclamation/Receiving (Inspection) h. Rejections i. MRB j. Vendor Returns Excess/Obsolete Material k. Deposition (Regrind, etc.)					
16. Will QC be notified for scheduling unusually large inspection requirements? a. How will they be notified?					
17. Have plant closing cutoff procedures been formalized?					
18. Have just-in-time philosophies been communicated to affected vendors? Have source inspection and certification programs been developed?					
19. Have continuous flow contract delivery schedules been negotiated?					
20. What is strategy for helping the vendors be successful J-I-T suppliers? What is communication mechanism? How will frustrations be managed?					

28

D.6.5.5 Sales Order Entry

Module Sales Order Entry

Responsibility ____________
Compl Due Date ____________
Approved By ____________
Rev# ________ Date ________

Implementation Checklist	Action Req'd	Start Date	Actual Start	Compl Date	Actual Compl
Customer Data 1. Is there customer data on all current customers? a. Is the data up-to-date and accurate? 2. Will the customer file contain: a. Sales customer data only? b. Accounts receivable data only? c. Combination of the above? 3. Who will be responsible for verifying customer data? a. Loading customer data? 4. Will sales order history data be stored for analytical purposes? a. If yes, which data? 5. Is there a customer code (number) assigned for every open sales order? a. If not, who will ensure that it is accomplished prior to loading sales orders? 6. Who will be responsible for assigning customer codes in the future? a. How will this be done with adequate control, yet quick enough to respond to requirements for immediate data entry?					

29

Module Sales Order Entry

Responsibility ____________
Compl Due Date ____________
Approved By ____________
Rev# ________ Date ________

(con't) Implementation Checklist	Action Req'd	Start Date	Actual Start	Compl Date	Actual Compl
7. Are the procedures in place to maintain the customer file? a. Is maintenance form(s) printed and ready for use? b. Who is responsible? **Sales Order Data** 1. Has sales order data been purged of cancelled, closed, etc., orders? a. If not, will it be purged prior to loading? b. Who is responsible for signing off the purged data? c. How will accuracy be measured? 2. Are the sales orders (SO) in customer service, shipping, and accounts receivable totally in agreement? a. How will this be verified? b. Who is responsible for getting the trio into agreement? 3. Who is ultimately responsible for closing and canceling sales orders? a. How will other interested parties be notified of such occurrences? 4. How often will file purging take place? 5. How will sales order changes take place? a. How will system be updated? (due dates, quantity, price, etc.)					

30

Module Sales Order Entry

Responsibility ____________
Compl Due Date ____________
Approved By ____________
Rev# ________ Date ________

(con't) Implementation Checklist	Action Req'd	Start Date	Actual Start	Compl Date	Actual Compl
6. How will discounts given be recorded? a. Monitored? 7. How will shipping be handled? a. Returns? 8. What type of shipping audit file is to be used? a. What information is to be retained? b. How long? 9. Are procedures and forms in place for: a. Creating SOs b. Changing SOs c. Shipping d. Returns e. Warranty Review Board f. Scrapping Reclamation/Excess/Obsolete g. Material Disposition 10. Is a customer code (number) on each SO? a. If not, when will it be assigned? b. Who is responsible? 11. Who will edit SOs prior to data entry? 12. How long does it take (elapsed time) between receiving an order and posting it on the system? a. How will this time be decreased? b. Who is responsible?					

31

Module Sales Order Entry

Responsibility ____________
Compl Due Date ____________
Approved By ____________
Rev# ________ Date ________

(con't) Implementation Checklist	Action Req'd	Start Date	Actual Start	Compl Date	Actual Compl
13. Are multiple distribution centers to be used? a. How will data be maintained? b. Who is responsible? 14. What method and data will be used in order to quote delivery dates? 15. How will real-time available-to-promise be calculated? 16. What is plan to reduce competitive lead time? 17. Who is responsible for measuring customer service and product quality improvements? 18. What steps are being taken to get closer to customers? 19. How will just-in-time philosophies affect customer commitments? 20. What steps are being performed to extend electronic data interchange (EDI) to key customers? 21. What is implementation methodology regarding statistical process controls? 22. What is plan to use manufacturing as a competitive strategy in pursuit of world class status?					

32

D.6.5.6 Master Scheduling

Module Master Production Schedule

Responsibility ______________________
Compl Due Date ______________________
Approved By ______________________
Rev# ____________ Date ______________

Implementation Checklist	Action Req'd	Start Date	Actual Start	Compl Date	Actual Compl
1. Will Master Scheduling be performed?					
2. Who is responsible for the Master Production Schedule?					
a. Who approves the Master Schedule?					
b. How will accuracy be measured?					
3. What are the primary sources from which to prepare the Master Schedule?					
a. Forecast					
i. Intrinsic Model					
ii. Extrinsic Model					
b. Marketing Manager					
c. Manufacturing Manager					
d. President					
e. MIN/MAX Report					
f. MRP (as a result of customer backlog)					
g. Other (specify)					
4. What purpose will the Master Schedule (or the MRP/CRP result of the Master Schedule) be used for:					
a. Purchasing Material					
b. Prioritizing Due Dates					
i. Purchase Parts					
ii. Assemblies					
c. Capacity Planning					
d. Costing Manufacturing Plan					
e. Cash Flow Analysis					
f. Customer Delivery Scheduling					
g. Coordinating Engineering Changes					

33

Module Master Production Schedule

Responsibility ______________________
Compl Due Date ______________________
Approved By ______________________
Rev# ____________ Date ______________

(con't) Implementation Checklist	Action Req'd	Start Date	Actual Start	Compl Date	Actual Compl
4. What purpose will the MPS be used for?					
a. Contract Quoting					
b. Communications Networking					
c. Budgeting					
d. Profit Planning					
e. Resource Planning					
f. Strategic Planning					
g. Distribution Center Loading					
h. Planning Transportation Loading and Expenses					
i. Top Management's Commitment to the Business Plan					
5. How will spares be Master Scheduled?					
a. Who is responsible?					
6. Will Master Schedule be frozen at any point?					
a. How will this be done?					
b. What periods will be frozen?					
c. Can this freeze be overridden?					
d. How?					
e. By whose approval?					
f. What is the cutoff?					
7. Will the Master Schedule include requirements for:					
a. International					
b. Sister Divisions					
c. Sister Plants					
8. How will these requirements be monitored?					
a. Who is responsible?					
b. What is cutoff?					

34

Module Master Production Schedule

Responsibility ____________
Compl Due Date ____________
Approved By ____________
Rev# ________ Date ________

(con't) Implementation Checklist	Action Req'd	Start Date	Actual Start	Compl Date	Actual Compl
9. Are forms and procedures in place for: a. Creating PS b. Changing PS c. Deleting PS					
10. Is there an authorization list for individuals who can request PS changes? a. Who maintains this list?					
11. Whose performance review is directly related to the integrity of the MPS?					
12. How will the MPS versus actual production performance be measured? a. How often? b. By whom?					
13. What is the horizon of the MPS?					
14. How will Master Schedule versus shipping be measured? a. How often? b. By whom?					
15. How will the Master Schedule performance be measured with respect to the Business Plan? a. Shipments b. Resource loading c. Dollars d. Profits					

35

Module Master Production Schedule

Responsibility ____________
Compl Due Date ____________
Approved By ____________
Rev# ________ Date ________

(con't) Implementation Checklist	Action Req'd	Start Date	Actual Start	Compl Date	Actual Compl
16. What mechanism is available to provide the Master Scheduler with timely feedback for rescheduling? a. Due to machine/tool downtime b. Due to efficiency problem c. Late delivery d. Insufficient Components (shortages) e. Cancelled order f. Other (specify)					
17. How will the rescheduled Master Schedule interface with: a. Finance b. Manufacturing c. Purchasing d. Marketing e. Business Plan f. Other (specify)					
18. What is plan to stabilize the MPS and conform to time fences?					
19. What is plan to compress lead time? a. Remove queue b. Streamline processes c. Reduce structure levels d. Reduce lot sizes e. Develop lead time review criteria f. Formulate continuous improvement methodology g. Develop "total quality management" approach and standards h. Formulate plan to remove waste					

36

D.6.5.7 Work Orders

Module Work Order

Responsibility ____________
Compl Due Date ____________
Approved By ____________
Rev# ________ Date ____________

Implementation Checklist	Action Req'd	Start Date	Actual Start	Compl Date	Actual Compl
1. Will Work Orders (WOs) be retained on the system a. Will WO file be purged prior to loading? 2. What is the source for the WO number? 3. Who will be responsible for assigning WO numbers? a. Will cancelled and closed WOs be reconciled each month? 4. How will accuracy be measured? a. Who will be responsible? 5. Will the shop floor be purged at the end of the month? (Will all WOs with zero work and behind schedule be reclaimed back into production control at the end of each month?) 6. Will splits be permitted? a. How will they be controlled? 7. Will WOs be assigned to: a. Fab b. Subassembly c. Assembly d. Final Assembly					

37

Module Work Order

Responsibility ____________
Compl Due Date ____________
Approved By ____________
Rev# ________ Date ____________

(con't) Implementation Checklist	Action Req'd	Start Date	Actual Start	Compl Date	Actual Compl
8. How will work order changes take place? a. How will system be updated? b. Will there be a work order audit file? 1. What information is to be retained? 2. How long? 9. Are forms and procedures in place for: a. Creating WOs b. Changing WOs c. Canceling WOs d. Deleting WOs e. Completing WOs f. Partially completing WOs g. Scrap against WOs h. Rework against WOs 10 How long does it take (elapsed time) between issuing a WO and posting it on the system? a. Can this time be decreased? b. Who is responsible for monitoring it? 11. How will completed quantity be verified? 12. Will completed quantity move to the stockroom? a. How will work-in-process be credited and stockroom be debited?					

38

Module Work Order

Responsibility ______________
Compl Due Date ______________
Approved By ______________
Rev# ________ Date ________

(con't) Implementation Checklist	Action Req'd	Start Date	Actual Start	Compl Date	Actual Compl
13. How will partially completed work orders be treated at month-end book closing?					
14. What level of detail will WO be tracked at? a. Work Order b. Work Center c. Router Operation					
15. How will work orders be prioritized?					
16. Will prioritization occur by: a. Work Order b. Work Center c. Machine d. Operation e. Skill					
17. Who is responsible for assigning priority? a. What media will be used to communicate priority to shop floor and dispatch list?					
18. What is plan to reduce level of WIP?					
19. What is plan to reduce rework?					
20. How will continuous improvement processes be implemented?					
21. How will queue be eliminated? What is plan?					

39

D.6.5.8 Work Center/Router

Module Work Center/Router

Responsibility ______________
Compl Due Date ______________
Approved By ______________
Rev# ________ Date ________

Implementation Checklist	Action Req'd	Start Date	Actual Start	Compl Date	Actual Compl
1. Will Work (cost) Centers be retained on the system? a. Who will control Work Center numbers?					
2. Will Operation Routers be retained on the system?					
3. How will Work Center Router accuracy be measured? a. Who is responsible?					
4. Will Routers be purged prior to loading? a. Who is responsible?					
5. Who will maintain the Work Center/Router data?					
6. Who will approve any changes to the Work Center Router Data (cost accounting, production, industrial engineering, etc.)?					
7. What will a Work Center (WC) represent for Capacity Planning?					
8. Will Vendor Capacity be monitored on the system?					
9. Are forms and procedures in place for: a. Creating WCs b. Deleting WCs c. Changing WCs d. Creating Operations e. Changing Operations f. Deleting Operations g. Interdepartmental material transfer					

40

Module Work Center/Router

Responsibility ____________
Compl Due Date ____________
Approved By ____________
Rev# ________ Date ________

(con't) Implementation Checklist	Action Req'd	Start Date	Actual Start	Compl Date	Actual Compl
10. If tracking is to be at the operation level, are forms and procedures in place for: a. Obtaining actual hours b. Completion of operation c. Completing a partial quantity through a specified operation d. Specifying an alternate operation/skill used to complete that operation					
11. How will completed operations be verified?					
12. What is plan to reduce number of work centers?					
13. What is plan to convert batch-oriented operations to process flow?					
14. What is plan to eliminate nonvalue-added operations?					
15. What is plan to challenge manufacturing roadblocks?					
16. What is plan to critically assess manufacturing competitiveness?					
17. What is guiding principles to ensure continuous improvement progresses according to plan?					
18. What is plan to ensure concurrent engineering efforts incorporate latest simplicity approaches?					
19. What is plan to significantly reduce setup?					

41

D.6.5.9 Material Requirements Planning

Module Material Requirements Planning

Responsibility ____________
Compl Due Date ____________
Approved By ____________
Rev# ________ Date ________

Implementation Checklist	Action Req'd	Start Date	Actual Start	Compl Date	Actual Compl
1. What is the planning horizon for the MRP?					
2. Will MRP be: a. Real-time b. Daily c. Weekly d. Weekly first eight periods the monthly next 10 periods with collapsed period encompassing the entire horizon? e. What horizon will be displayed real-time? f. Will a peg report be required? g. Other (specify)					
3. Will MRP be based upon: a. Master Production Schedule b. Customer Backlog c. Forecast d. Combination of the above (specify)					
4. Have quantity/assemblies (BoM) been verified?					
5. Will requirements consider: a. Economic Order Quantity b. Lot size 1. Fixed 2. Period Order Quantity 3. Part Period Balancing 4. Minimum Order Quantity 5. Lot for lot (Discrete)					

Module Material Requirements Planning

Responsibility ______________
Compl Due Date ______________
Approved By ______________
Rev# ______________ Date ______________

(con't) Implementation Checklist	Action Req'd	Start Date	Actual Start	Compl Date	Actual Compl
5. Will requirements consider: c. Scrap d. Shrinkage e. Safety Stock f. Lead Time Offset g. Engineering Change Effectivity h. Other (specify) 6. Will requirements be time-phased? a. Have lead times been verified? 7. Has Master Schedule been formalized? (See master schedule checklist) 8. Will on-order status be considered for purchase parts? a. Is purchasing data complete and valid? (See purchasing checklist) 9. Will on-order status be considered for manufactured parts? a. Is work order data complete and valid? (See work order checklist) 10. Will a Capacity Plan be run? a. Has Work Center/Router data been defined? (See work center/router checklist)					

43

Module Material Requirements Planning

Responsibility ______________
Compl Due Date ______________
Approved By ______________
Rev# ______________ Date ______________

(con't) Implementation Checklist	Action Req'd	Start Date	Actual Start	Compl Date	Actual Compl
11. Will MRP be used for other planning? a. Resource Planning b. Profit Planning c. Cash Flow Planning d. Labor Planning e. Business Planning f. Acquisition Planning g. Tool Planning h. Vendor Capacity Planning i. Market Planning i. Market Share ii. Geographical iii. R&D j. Other (specify) 12. Will one or more periods be used for: a. ABC analysis (scheduled/demand) b. Costing MRP output c. Kitting Allocation d. Projecting Future Variances e. Other (specify) 13. How often will MRP be generated: a. Real-time b. Daily c. Weekly d. Other (specify)					

44

Module Material Requirements Planning

Responsibility ______________________
Compl Due Date ______________________
Approved By ______________________
Rev# ____________ Date ______________

(con't) Implementation Checklist	Action Req'd	Start Date	Actual Start	Compl Date	Actual Compl
14. Will MRP be displayed in: a. Top management summary form b. Detail purchase/shop floor manner c. Sort order other than item number d. Selected manner: i. "A" items only ii. Action only iii. Purchased parts only iv. Manufactured parts only v. Other (specify) e. Exception only 15. What method will be used for buyers to inform planners when purchased parts are delayed? a. Is a procedure written? 16. What method will be used for shop supervision to inform planners when manufactured parts are delayed? a. Is a procedure written? 17. Once the buyers and shop supervisors inform the planners of delays, what procedures will the planners follow in order to reschedule? 18. What action will planners/buyers take on past due orders/ a. Leave as past due? i. Explain b. Reschedule? i. What criteria will be used for rescheduling?					

45

Module Material Requirements Planning

Responsibility ______________________
Compl Due Date ______________________
Approved By ______________________
Rev# ____________ Date ______________

(con't) Implementation Checklist	Action Req'd	Start Date	Actual Start	Compl Date	Actual Compl
18. What action will planners/buyers take on past due orders? c. Ignore? d. Other (specify) 19. Will all parts be controlled by MRP? If not, what is the basis for deciding which parts will be MRP'd? 20. If EOQ or Part Period Balance lot sizing techniques are used, have the setup and carrying costs been defined and set? 21. Who is responsible for ensuring that all transactions are processed according to the cut off procedures, prior to the running of the MRP? 22. How will MRP be benchmarked for effectiveness? a. Plan vs. Actual Production b. Inventory turns i. What is formula A. Based on cost of goods? B. Will it annualize most recent data to trend? C. Obsolescence D. Vendor Performance E. Finished Goods Delivery F. Profit G. Other (specify)					

46

Module Material Requirements Planning

Responsibility ______________
Compl Due Date ______________
Approved By ______________
Rev# ________ Date ________

(con't) Implementation Checklist	Action Req'd	Start Date	Actual Start	Compl Date	Actual Compl
23. How and when will MRP effectiveness be measured?					
a. Decreased PO commitment					
b. Increased productivity					
c. Increased customer service					
d. Decreased cost					
e. Decreased inventory					
f. Decreased scrap rate					
g. Improved bottom line					
h. Other (specify)					
24. How many copies of each report will be run?					
25. What is plan to convert MRP output to short interval schedule or kanban?					
26. How will finite schedule outputs be reconciled to the MRP due date?					
27. How will J-I-T drop zones be utilized with MRP schedule?					
28. What rescheduling criteria will be used by planners?					
29. How will rate-based schedules be reconciled to MRP work order schedules?					

47

D.6.5.10 Cost Accounting

Module Cost Accounting

Responsibility ______________
Compl Due Date ______________
Approved By ______________
Rev# ________ Date ________

Implementation Checklist	Action Req'd	Start Date	Actual Start	Compl Date	Actual Compl
1. What costing system will be used:					
a. Actual					
b. Standard					
c. Average					
d. Flexible budgeting					
e. Activity based accounting					
2. Will it be inclusive of:					
a. Material					
b. Labor					
c. Burden					
d. Outside Vendor					
e. Semi-variable					
3. Have all control numbers been assigned to all controlled documents, such as:					
a. Material Requisition					
b. Move to Stock					
c. Scrap Notice					
d. Receiver					
e. Interdepartmental Material Transfer					
f. Inventory Adjustment					
g. Vendor Corrective Action Notice					
h. Returned Goods Document					
i. Shippers					
j. Inter-stockroom transfers, etc.					

48

Module Cost Accounting

Responsibility ____________________
Compl Due Date ____________________
Approved By ____________________
Rev#__________Date __________

(con't) Implementation Checklist	Action Req'd	Start Date	Actual Start	Compl Date	Actual Compl
4. Are controlled documents accounted for and reconciled each month?					
5. If on a standard cost system: a. When and how will variances be reported on: i. Material and Overhead A. Purchased B. Manufactured C. Assembled ii. Labor and Overhead b. What method will be used to ensure that variances are accurate? Corrective Action?					
6. What data will be reported to the system? a. Standards b. Actual i. Material ii. Labor					
7. What reports will be needed? a. Inventory Valuation b. WIP Valuation c. Purchase Price Variance d. Labor Usage Variance e. Labor Rate Variance f. Absorption Variance g. Costed Transaction Register h. Cash Flow Analysis i. Complete Work Order Variances j. Job Cost Report					

49

Module Cost Accounting

Responsibility ____________________
Compl Due Date ____________________
Approved By ____________________
Rev#__________Date __________

(con't) Implementation Checklist	Action Req'd	Start Date	Actual Start	Compl Date	Actual Compl
8. How often will WIP Inventory be taken: a. Weekly? b. Monthly? c. Quarterly? d. Semi-Annually? e. Annually?					
9. Are physical inventory procedures in place? a. Are boxes marked with item number, unit of measure, and quantity? b. Have briefing and training session been held? c. Is an instruction sheet given to each counter and auditor? d. Are scales, calculators, scratch pads, etc. available for counters and auditors?					
10. Are cycle inventory procedures in place? a. Has a tentative date been set with external auditors whereby the cycle inventory will replace the annual physical inventory?					
11. Are costing methods accurate in light of just-in-time and CIM strategies?					
12. What is measurement methodology to track continuous improvement plans?					
13. What is plan to implement "projecting future variances" approach to managerial cost assistance?					

50

D.6.5.11 Outside Processing

Module Outside Processing

Responsibility ____________
Compl Due Date ____________
Approved By ____________
Rev# ________ Date ________

Implementation Checklist	Action Req'd	Start Date	Actual Start	Compl Date	Actual Compl
1. Does outside processing (also called farm out and subcontracting) occur within your company? 2. Will outside processing be tracked by: a. Purchase Order? i. Normal PO? ii. Special PO? b. Work Order (shop order)? c. Both? 3. What reports will be used to identify the location of the material? 4. If controlled by Purchase Order: a. Is there a formalized procedure written? b. Is there a document to authorize sending outside (shipper)? i. If yes, does it have a document number? ii. Will the document be used to decrease inventory? iii. If not, what will be used to decrease authorize shipment? (Explain) c. How will the cost of the processing be recorded on the PO? i. When will the unit cost be recorded on the PO? A. At point of origin of PO? B. After material is received, and job estimated by outside processor? (Vendor) C. Upon material receipt back from vendor? D. Upon receiving invoice from vendor? E. Combination of above (explain) F. Other (specify)					

51

Module Outside Processing

Responsibility ____________
Compl Due Date ____________
Approved By ____________
Rev# ________ Date ________

(con't) Implementation Checklist	Action Req'd	Start Date	Actual Start	Compl Date	Actual Compl
ii. How will the scrap be recorded on PO? A. Who is responsible B. How will it be audited? • By whom d. How will the delivery date (due date) be determined? i. When will it be put on the PO? ii. Who will ensure that the date is provided? A. If the date is not known at the origin of PO, what dating will be used: **Note:** If the following are given: "will advise," "Unknown," "ASAP," etc., then the computer must assume not date, which will default to past due, or many years in the future. iii. If higher level assemblies are rescheduled, what mechanism is there to update this due date? e. What mechanism will be used to receive material back in house? i. Normal Receipt ii. Special Receipt A. Is procedure written? B. Has cost accounting approved the procedure? C. How will it be controlled? D. Who is responsible?					

52

Module Outside Processing

Responsibility ____________
Compl Due Date ____________
Approved By ____________
Rev# ________ Date ________

(con't) Implementation Checklist	Action Req'd	Start Date	Actual Start	Compl Date	Actual Compl
f. After passing receiving inspection, will the material proceed to:					
i. Stock Room					
A. Has procedure been written?					
ii. Applicable work center					
A. Has procedure been written					
iii. Further outside processing					
A. How will it be controlled?					
B. How will further processing be tracked?					
C. How will "value added" be accounted for?					
D. Is the process acceptable to your external auditing firm (CPA)?					
E. Has procedure been written?					
g. Will the material be received under the same item number as sent out?					
i. How will unprocessed parts be differentiated from processed parts within the system?					
ii. How does routing reflect outside processing?					
iii. How will you distinguish preprocessed versus postprocessed material:					
A. In stock bin					
B. In receiving inspection					
C. At dock					
D. In staging area					
E. On inventory records					
F. For planning purposes					
G. For product costing					

53

Module Outside Processing

Responsibility ____________
Compl Due Date ____________
Approved By ____________
Rev# ________ Date ________

(con't) Implementation Checklist	Action Req'd	Start Date	Actual Start	Compl Date	Actual Compl
Note: Recommend assigning unique item numbers—if production control is to track outside processing, it is necessary to assign a unique item number.					
5. If controlled by Work Order:					
a. Is there a formalized procedure written?					
b. Is there a document to authorize sending outside? (Shipping)					
i. If yes, does it have a document number?					
ii. Will the document be used to decrease inventory?					
iii. If not, what will be used to authorize shipment (explain)?					
c. How will the "value added" (cost to process) be recorded on Work Order?					
i. When will the processing cost be recorded on a WO?					
A. At point of originating shipping document?					
B. After material is received and job estimated by outside processor?					
C. Upon material receipt back from vendor?					
D. Combination of the above? (Explain)					
E. Other (specify)					
d. How will the delivery date (due date) be determined?					
i. When will it be put on the WO?					
ii. Who will ensure that the date is provided?					

54

Module Outside Processing

Responsibility ______________
Compl Due Date ______________
Approved By ______________
Rev# ______________ Date ______________

(con't) Implementation Checklist	Action Req'd	Start Date	Actual Start	Compl Date	Actual Compl
A. If the date is not known at the origin of the WO, what dating will be used? **Note:** If the following are given: "will advise," "unknown," "ASAP," etc., then the computer must assume no date, which will default to past due or, many years into the future. iii. If higher level assemblies are rescheduled, what mechanism will be used to receive material back in house? e. What Mechanism will be used to receiver material back in house? i. Normal Receipt ii. Special Receipt A. Is a procedure written? B. Has cost accounting approved the procedure? C. How will it be controlled D. Who is responsible? f. After passing receiving inspection will material proceed to: i. Stock Room A. Has procedure been written? ii. Applicable Work Center A. Has procedure been written? iii. Further Outside Processing A. How will it be controlled? B. How will further processing be tracked? C. How will "value added" be accounted for?					

55

Module Outside Processing

Responsibility ______________
Compl Due Date ______________
Approved By ______________
Rev# ______________ Date ______________

(con't) Implementation Checklist	Action Req'd	Start Date	Actual Start	Compl Date	Actual Compl
Further Outside Processing D. Who is responsible for the flow E. How will variances be handled? F. Is the process acceptable to your external auditing firm? G. Has procedure been written? g. Will the material be received under the same item number as sent out? i. How will unprocessed parts be differentiated from the processed parts within the system? ii. How does routing affect outside processing? iii. How will you distinguish preprocessed versus postprocessed material? A. In stock bin B. In receiving inspection C. At dock D. In staging area E. On inventory records F. For planning purposes G. For product costing **Note:** Recommend assigning unique item numbers—if production control is to track outside processing, it is necessary to assign a unique number. 6. Is a kit list required? a. Is more than on component supplied by you needed for the outside process? i. Has a Bill of Material been defined?					

56

Module Outside Processing

Responsibility ______________
Compl Due Date ______________
Approved By ______________
Rev# ________ Date ________

(con't) Implementation Checklist	Action Req'd	Start Date	Actual Start	Compl Date	Actual Compl
7. If outside processing is done only when excessive load in-house a. Who will authorize going outside instead of in-house i. When will authorization be given? b. How will it be planned? i. If planned, how will it be rescheduled? c. How will standard cost be carried? i. How will variances be accounted for?					
8. Will you plan capacity for outside processing?					
9. How will scrap and shrinkage be accounted for? a. How will vendor be charged back for excess? b. How will scrap/shrinkage be communicated to you by the vendor? How Frequently? c. What objectives/incentives are in place with the vendor to reduce loss due to scrap/shrinkage?					

57

D.6.5.12 Purchase Requisition

Module Purchase Requisition

Responsibility ______________
Compl Due Date ______________
Approved By ______________
Rev# ________ Date ________

Implementation Checklist	Action Req'd	Start Date	Actual Start	Compl Date	Actual Compl
1. Will Purchase Requisitions be used?					
2. Will the same form be used for: a. Production materials b. Supplies c. Engineering samples d. Maintenance, Repair, and Office Supplies e. Other (specify)					
3. Is Purchase Requisition numerically controlled? a. By whom b. How is it reconciled? c. Who is responsible					
4. Who will be trained? a. Inventory Control b. Engineering i. Product ii. Design iii. Industrial iv. Manufacturing v. Quality c. Production Control d. User Groups (List)					
5. Will account numbers be required information? a. If yes, who will insure that documents are not processed without account number? b. Who will verify account numbers? c. How often will chart of accounts be updated?					

58

Module Purchase Requisition

Responsibility ____________
Compl Due Date ____________
Approved By ____________
Rev# ________ Date ________

(con't) Implementation Checklist	Action Req'd	Start Date	Actual Start	Compl Date	Actual Compl
6. Will there be an authorized approval listing? a. How will listing be updated? b. Who is responsible? c. Is there a procedure to be followed if the approved name is not on the list? 7. What mechanism is in place to ensure that Purchase Requisitions are not duplicated? a. Purchasing b. Inventory Control c. User Departments d. Maintenance e. Other (specify) 8. Is there a procedure on: a. Processing the Purchase Requisition for Production Material? b. Processing the Purchase Requisition for Maintenance, Repair, and Office Supplies? c. Processing the Purchase Requisition for local purchase? d. Processing the Purchase Requisition for Supplies? e. Other (specify) 9. On Order Point Controlled Material, can the "Below Order Point" report be used as a Purchase Requisition?					

59

Module Purchase Requisition

Responsibility ____________
Compl Due Date ____________
Approved By ____________
Rev# ________ Date ________

(con't) Implementation Checklist	Action Req'd	Start Date	Actual Start	Compl Date	Actual Compl
10. What information is required on the Purchase Requisition? a. Item Number b. Quantity c. Department d. Estimated Cost e. Description f. Unit of Measure g. Date h. Initiator i. Approval j. Other (specify) 11. What mechanism will the buyer use to establish a priority on purchase requisitions awaiting ordering? a. Pecking order of organization? b. ASAP? c. Determining easiest items? d. Place order only if quotations are all in? e. Due date? i. Will this require determining critical points of lead time, processing time, etc.? 12. How will buyer communicate trouble areas back to initiator? a. Is a procedure written?					

60

D.6.5.13 Vendor Supplied Material

Module Vendor Supplied Material

Responsibility ____________
Compl Due Date ____________
Approved By ____________
Rev# ________ Date ________

Implementation Checklist	Action Req'd	Start Date	Actual Start	Compl Date	Actual Compl
1. Is there ever an occasion to supply vendors with material?					
2. Is material ever drop-shipped from one vendor (Vendor 1) to another vendor (Vendor 2)?					
a. If yes:					
i. How will notification of receipt be communicated?					
ii. How is material inspected?					
iii. What or who authorizes the payment of invoice?					
A. How is receipt at Vendor 2 verified?					
iv. How will scrap be controlled?					
v. Is a procedure formalized?					
vi. How does Vendor 1 and Vendor 2 know about arrangements?					
vii. How will the integrity of the system be audited?					
A. Who is responsible?					
viii. How will computer keep track of accountability?					
3. Is material ever sent from stores to a vendor for processing?					
a. How will it be tracked?					
b. What document will be used to relieve from inventory?					
i. Shipper?					

61

Module Vendor Supplied Material

Responsibility ____________
Compl Due Date ____________
Approved By ____________
Rev# ________ Date ________

(con't) Implementation Checklist	Action Req'd	Start Date	Actual Start	Compl Date	Actual Compl
4. Is material ever sent from a work center to a vendor for processing? (see outside processing guide)					
5. Material should not be sent from receiving dock to vendor-is it?					
6. Has procedure been formalized?					
7. What other aspects not covered above need to be resolved? (specify)					

62

D.6.5.14 Tooling Control

Module Tooling Control

Responsibility __________
Compl Due Date __________
Approved By __________
Rev# __________ Date __________

Implementation Checklist	Action Req'd	Start Date	Actual Start	Compl Date	Actual Compl
1. Will tooling be controlled?					
2. What tools will be controlled? a. Jigs, fixtures, drills, etc.? List. b. Extrusion/injection molding tools? c. Others (specify)					
3. How will tooling requirements be forecasted? a. As a result of MRP? b. Reorder point? c. As they break? d. Other (specify)					
4. Will tooling impact: a. Capacity requirements? b. Router c. Work Order releasing and kitting?					
5. Scheduling Tools: a. Is capacity of tool known? b. Is location of tool known? c. Is maintenance history retained? d. Is tool producing quality product at full capacity? i. If not, what is percentage? e. Has material handling, special equipment (cranes, etc.), and move time been considered?					

63

Module Tooling Control

Responsibility __________
Compl Due Date __________
Approved By __________
Rev# __________ Date __________

(con't) Implementation Checklist	Action Req'd	Start Date	Actual Start	Compl Date	Actual Compl
5. Scheduling Tools: f. Is there a formal inspection and maintenance schedule? g. Is the tool storage area being given a pull/return tool schedule? i. Via kitting mechanism? ii. Via production schedule? iii. Other (specify) h. How is tool usage prioritized? i. How is availability of tooling checked prior to release of the shop order?					
6. Modification, qualification, and purchase of new tooling: a. Does a schedule exist? b. Are procedures formalized? c. Is approval list maintained and up-to-date? d. Is capitalization mechanism included? e. Is quality assurance interface defined? f. Who coordinates this tooling activity? i. Manufacturing Engineering ii. Marketing iii. Purchasing iv. Production Control v. Other (specify)					
7. Is there a tool location list?					
8. How often are tools inventoried? a. Is there a procedure established?					

64

Module Tooling Control

Responsibility ____________
Compl Due Date ____________
Approved By ____________
Rev# ________ Date ________

(con't) Implementation Checklist	Action Req'd	Start Date	Actual Start	Compl Date	Actual Compl
9. How are tool issues authorized? a. Production Control initiated requisition? b. Shop traveler? c. Kit list d. Router operation? e. Shop order? f. Special tool request? 10. Will tool issue requests be initiated in sufficient lead time so as to prep tool, adjust tool, etc.? 11. What shop paper will specify tooling required? a. Route Sheet? b. Work Order? c. Other (specify) 12. How is tool obsolescence controlled? a. Who authorizes it? b. Who determines if old tool can be modified to current configuration? c. Can salvage value be obtained? 13. How is history being retained on tools? a. Who is responsible? b. How long will it be maintained? c. What data will be retained? d. How will data be reported?					

65

Module Tooling Control

Responsibility ____________
Compl Due Date ____________
Approved By ____________
Rev# ________ Date ________

(con't) Implementation Checklist	Action Req'd	Start Date	Actual Start	Compl Date	Actual Compl
14. If tool is causing quality problems, how will the tool be recalled? a. Is a procedure established? b. How will product produced with faulty tool be identified? i. Reworked? ii. Is there a procedure? 15. How are tools, which either make more than one unique item or less than a unique item, controlled for order quantity? a. For example, what procedure is in place to avoid: the left hand side of a housing has 5,000 each on-hand, 100,000 on-order, and the right hand side has 10,000 on-hand, zero on-order. i. How are complementary items handled? ii. How does it interface with the capacity? iii. Schedule one part early to obtain capacity 16. How is mix of products requiring tooling controlled? 17. Alternate Source Qualification: a. How are back-up tools and vendors qualified? b. What criteria are used for back-up qualification?					

66

Module Tooling Control

Responsibility ____________
Compl Due Date ____________
Approved By ____________
Rev# ________ Date ________

(con't) Implementation Checklist	Action Req'd	Start Date	Actual Start	Compl Date	Actual Compl
18. Control of tooling at vendors: a. What document is used to send tool to vendor? i. Shipper? ii. Purchase Order? iii. Work Order? iv. Router? v. Other (specify) b. Is tool due back after completion of the job? i. How is due in status tracked ii. Will tool arrive with material? iii. How will tooling be received, inspected, and accepted? iv. Is a procedure written? **Note:** See Outside Processing Checklist for further detail on controlling tooling at vendors. 19. First Articles: a. What scheduling technique is used? b. If first articles are outside the specifications, is a functional test performed? i. Who authorizes acceptance? 20. Tool Storage: a. Are tools stored close to the shop floor?					

67

Module Tooling Control

Responsibility ____________
Compl Due Date ____________
Approved By ____________
Rev# ________ Date ________

(con't) Implementation Checklist	Action Req'd	Start Date	Actual Start	Compl Date	Actual Compl
21. Tooling Maintenance: a. Does a formal preventative maintenance schedule exist? b. Is tooling downtime scheduled? c. If outside maintenance is required, how is process controlled? 22. Has tool been described using basic noun as first entry in the description? 23 How will tool control benefits be measured? a. Reduction of tool inventory b. Reduction of tool shortage c. Reduction of shop order delays d Planned capacity for: i. Tool Handling ii. Tool inspection iii. Tool maintenance e. Reduction in rework due to properly maintained tools f. Increased tool utilization through proper planning					

68

D.6.5.15 Product Change/New Product

Module Product Change/New Product

Responsibility ____________
Compl Due Date ____________
Approved By ____________
Rev# ________ Date ________

Implementation Checklist	Action Req'd	Start Date	Actual Start	Compl Date	Actual Compl
1. Will product change and new product release be controlled by the same procedure? (Also known as Engineering Change Order (ECO), Engineering Change Notice [ECN], etc.)					
2. How will Purchased Finished Goods changes be handled?					
3. What is the scope of the Product Change Procedure?					
a. Does it include process changes to router?					
b. Does it include machinery modification?					
c. Etc.					
4. Who is responsible for the following tasks:					
a. Laboratory testing					
b. Cost study					
c. Product and process evaluation					
d. Field trial					
e. Tooling and capital equipment authorization					
f. Release design preproduction					
g. Develop implementation plan					
h. Issue item tested specification					
i. Requisition tooling					
j. Order tooling					
k. Test and approve first articles					
l. Request sample dimensional inspection					
m. Inspect and approve quality for production run					
n. Prepare a report on mold inspection					

69

Module Product Change/New Product

Responsibility ____________
Compl Due Date ____________
Approved By ____________
Rev# ________ Date ________

(con't) Implementation Checklist	Action Req'd	Start Date	Actual Start	Compl Date	Actual Compl
4. Who is responsible for the following tasks:					
o. Prepare a report on functional test					
p. Issue approved product authorization					
q. Enter part number on master file					
r. Initiate Bill of Material change					
s. Schedule production run					
5. Has one function been assigned the responsibility to coordinate all product change activities from beginning to end?					
a. If yes, who is responsible?					
b. If no, how will the coordinating function occur?					
i. Who will be held responsible for delays, excess costs, obsolescence failure to implement, etc.?					
6. Has an approval cycle been determined?					
a. What is approval cycle?					
b. What is the approval routing					
c. Does approval occur in ascending or descending management level sequence?					
7. Can a hot product change be implemented with only verbal authorization?					
a. If yes, what is the procedure?					
b. Who is responsible for the ramifications if problems occur?					

70

Module Product Change/New Product

Responsibility ______________
Compl Due Date ______________
Approved By ______________
Rev# ______________ Date ______________

(con't) Implementation Checklist	Action Req'd	Start Date	Actual Start	Compl Date	Actual Compl
8. Have authorized dispositions been established? a. If disposition reflects immediate implementation, have obsolete parts been costed and are all parties in the approval cycle aware of the costs? b. What method will be used to initiate proper documentation on obsolete parts? i. Who will ensure that proper action is taken on such parts?					
9. What method will be used to identify when manufacturing has physically switched over?					
10. If first batch or lot number is needed in order to make field announcements of change, what method will be used to track and convey the implementation to marketing? a. Who is responsible for this activity?					
11. What is plan to incorporate concurrent engineering methodology needed to: a. Release BoM the way it will be manufactured b. Incorporate simplification changes c. Reduce time to market lead time d. Incorporate process oriented operations when appropriate (replacing batch orientation) e. Reduce setup, move, and queue					

71

D.6.5.16 Example of a Module Checklist

Below is an example of a completed form, with values filled in appropriately

Example

Module Work Order

Responsibility Bob Jamison
Compl Due Date 7/23
Approved By Larry Giles
Rev# ______________ Date 6/14

Implementation Checklist	Action Req'd	Start Date	Actual Start	Compl Date	Actual Compl
1. Will work orders be retained on the system? a. Will WO file be purged prior to loading?	Yes ea. mo.	4/20	5/1	5/30	
2. What is the source for the WO number	PC	4/20		5/1	
3. Who will be responsible for assigning WO numbers? a. Will cancelled and closed WOs be reconciled each month?	PC	4/20		5/20	
4. How will accuracy be measured? a. Who will be responsible?	PC/FIN	5/1		6/1	
5. Will the shop floor be purged at the end of each month? (Will all WOs with zero work and behind schedule be reclaimed back into production control at the end of each month?)	PC/FIN	5/1		5/31	
6. Will splits be permitted? a. How will they be controlled?	PC/FIN	6/1		6/30	
7. Will WOs be assigned to: a. Fab b. Subassembly c. Assembly d. Final Assembly	 Yes Yes Yes Yes				

72

Module Work Order

Responsibility Bob Jamison
Compl Due Date 7/23
Approved By Larry Giles
Rev#______________Date 6/14

(con't) Implementation Checklist	Action Req'd	Start Date	Actual Start	Compl Date	Actual Compl
8. Lot Splitting:					
a. Determine number of splits/month	x	6/27	7/1	7/5	7/10
b. Flowchart splitting process	x	6/27	7/5	7/10	7/15
c. Determine impact on QC	x	6/27	7/5	7/12	7/15
9. Etc.					

73

D.6.5.17 "Blank" Module Checklist Form

Module

Responsibility ______________
Compl Due Date ______________
Approved By ______________
Rev#______________Date ______________

Implementation Checklist	Action Req'd	Start Date	Actual Start	Compl Date	Actual Compl

74

D.7 Sample Questions Asked When Reviewing Operational Documents for Procedural Impact

Note: It is helpful to review this section concurrent with Section D.6.5.8 (Forms)

Example 1: Returned Goods Inventory

When considering the establishment of this procedure, it may appear awesome in Toto; however, if each individual function is considered independently of the whole initially, and viewing the interactivity with other functions, the procedure seems probable. The points to consider are as follows:

1. Prior to receipt at dock
   a. What document will be used?
   b. How will the document be put into the system?
   c. Who is responsible for putting the document into the system?
   d. Who will make decision if a discrepancy exists?
   e. Will the goods have a part number? Be assigned a part number when notification of due? Or be handled similar to supply items?
   f. Should there be one universal purchase order (PO) number with multiple lines for each return? Or should the document number be PO number? Or assign an authentic PO number and prefix with a special letter or character?
   g. What will cost field of the PO contain?
   h. Will returned spares be handled the same as returned systems? Will warranty returns be handled the same as returns due to upgrades?
2. Receipts at dock
   a. What specification sheet will be used to compare what was due-in versus what was received?
   b. If the item received was different than the item due-in,
      i. Will a change purchase order be written?
      ii. Will there be a return goods variance?
   c. How will the material be marked to identify as "used" versus "new" material at dock? Movable stanchions? Locked room? Placards?
   d. Will a review board (similar to Material Review Board [MRB]) review the condition of material and recommend action?
3. Receipt into inspection
   a. Will the inspection be at the dock? If not, what document will authorize movement of goods? How will the computer be notified of movement?
   b. Will there be rejections? How will individual item rejections be handled? Total system rejections?
   c. What will authorize disposition of damaged returns? Who will authorize? Will parts be cannibalized? Where will canned parts go? How will they be cosseted?
   d. Where will the returned goods go after inspection? How will they be kept separate from "new" assemblies if they enter production areas?
4. Scheduling
   a. What priority will returned goods be given?
   b. Will factory orders be created? How will progress be tracked?
   c. Will there be a route sheet created for returns? Will it be "general" or "specific" in nature?
5. Costing
   a. Will returned goods have a devalued standard cost (i.e., 40% of "new" goods standard) and flow similar to standard costing? If yes, where will variances occur? What account(s) will the variances affect?
   b. How will requisitioned store items be handled? Where will reconditioned returns be made?
   c. How will labor be accounted for? Variances?

D.8 Questions for Review When Formalizing a Cost Accounting System

Example 2: Questions to Be Addressed When Establishing a Cost Accounting System

1. What are objectives? (One or more of below)
   a. Provide information, which may be used to evaluate profitability?
   b. Provide standard costs review and revise costs periodically?
   c. Inclusive of material, labor, and overhead?
   d. Report variances from standard
      i. To aid in correcting problem areas?
      ii. To determine cost of goods for future pricing?
      iii. To adjust standards?
      iv. Absorption reporting techniques?
      v. Method of paralleling accounting systems when considering cost of goods manufactured?
2. Material cost accounting
   a. Technique of reporting purchase and receipt of material
      i. What are variances called?
      ii. When are variances recorded?
         A. At dock?
         B. At stock?
      iii. Is freight included? Is there a freight standard?
   b. Withdrawals into WIP
      i. How is kitting handled? Dollars? Units? Both?
      ii. Miscellaneous withdrawals—How do you handle?
         A. Inclusive into account charging for comparison to standard?
      iii. How is material usage variance handled?
   c. When are variances accumulated?
      i. How are they accumulated?
      ii. Do they pass from work center to work center?
      iii. How do you account for variances when WIP inventory is taken?
      iv. How are your supplier invoices handled?
   d. How do you handle material returns?
      i. What do you do if value has been added?
   e. What are internal and external auditor's rules and regulations?
3. Labor cost accounting
   a. What technique is used for labor reporting?
      i. Router
      ii. Work Order?
         A. How will the above be maintained?
         B. Who is responsible for maintenance?
         C. How will accuracy be tested?
      iii. How will it be reconciled to payroll?
   b. WIP reconciliation
      i. How often will reconciliation be done?
         A. Weekly?
         B. Monthly?
         C. Semi-annually?
         D. Annually?
      ii. Which accounts will variances be reported to?
      iii. How is overhead handled?
         A. Variances handled?

- iv. Will a WIP inventory be taken monthly?
  - A. If yes, how will total of standard last-level labor costs be transferred and into what account?
  - B. If no, how will WIP be valued?
  - C. How will it be audited?
  - D. Who will be responsible for the audit?
  - E. How will variances be flagged?
  - F. When will variances be identified?
  - G. How will labor performance reporting be handled?
  - H. How do you handle labor usage variance?

4. Inventory
   - a. How will physical inventory procedure handle unreported losses?
   - b. What are account classifications?
5. General steps to follow
   - a. Variable and fixed expenses—How do they relate to production volumes?
   - b. What role does chart of accounts play?
   - c. Work order completion reporting control
     - i. What will be control points?
     - ii. Who is responsible for accuracy?
   - d. Material control and document control?
     - i. Who is responsible?
     - ii. What technique will be used for reconciliation?
   - e. Updating standards
     - i. Who is responsible?
     - ii. When and who will audit?
   - f. Procedures for journal entries?
   - g. Reporting procedure monitor and audit?
   - h. Management control
     - i. Planning
     - ii. Budgeting
     - iii. Accounting
     - iv. Reporting

Representative example "T" accounts used in a cost accounting system are given as follows:
Work in process

Actual labor + Overhead	
WIP1	SR1
WIP2	
WIP3	SR2

(WIP1) A. Actual labor charged to work order

Labor variance	
	Payroll – actual charges to work order

(WIP2) B. Overhead = Actual labor × overhead rate

Overhead variance

	Actual overhead expense – applied overhead

(WIP3) C. Materials (purchased, outside processing, etc.)

Purchase price variance

	Actual cost – standard cost

Stockroom raw materials, assemblies at Standard (material, labor, and overhead)

SR1	
SR2	COS1

(SR1) A. Standard labor cost on completed work order

Labor performance variance

	Actual – last-level labor cost on work order

(SR2) B. Overhead = Standard labor × overhead rate

Overhead performance variance

	Labor performance variance × overhead rate

Cost of sales (COS)

COS1	

(COS1) A. Material, labor, and overhead at standard totaled (rolled up)

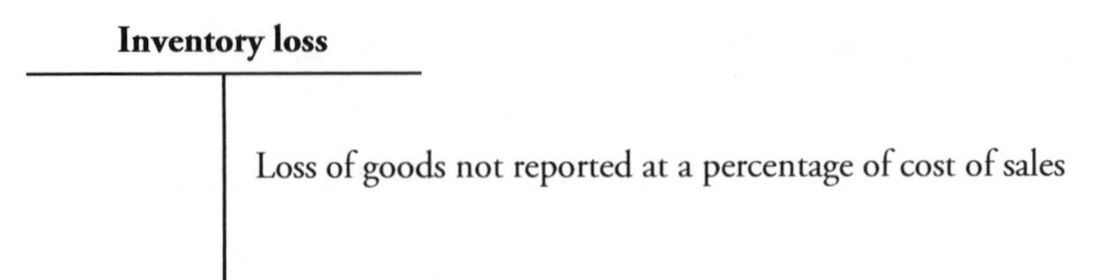

1. Actual materials used (material usage)

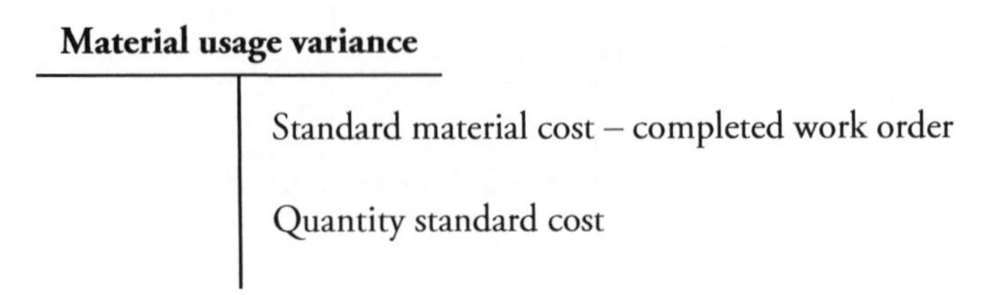

2. Actual labor used (labor usage)

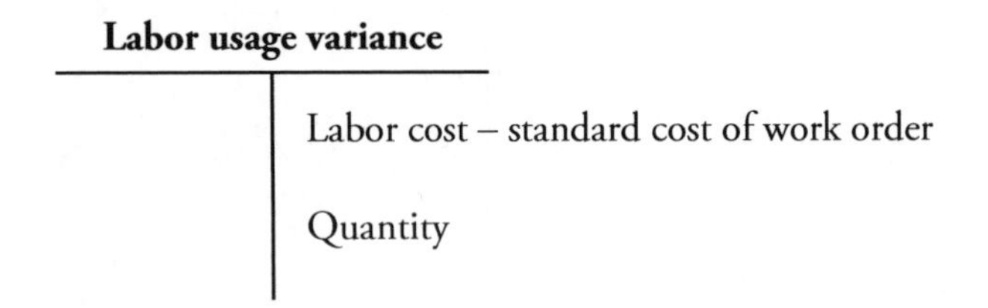

3. Actual overhead (overhead usage)

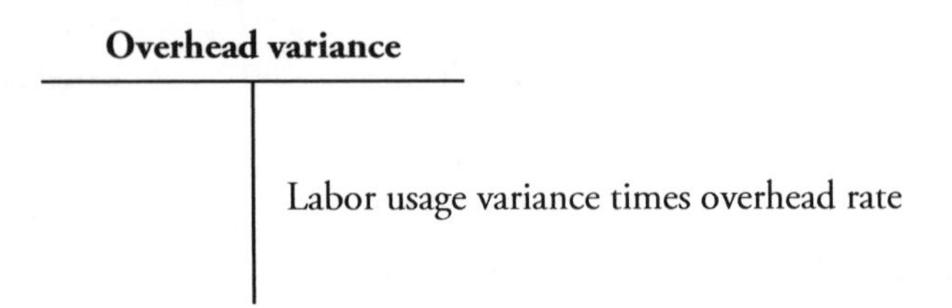

4. Expense

Expense chart of accounts

Material, labor, and overhead for inventory withdrawn for nonproduction-related activities

Appendix E (Chapter 7)

A few representative project management snippets, which show selected best practice excerpts representing project organization, tracking, and reporting, are discussed. The purpose for inclusion is to give the project core team a starter kit to assist their visioning and project planning effort.

There is a substantial project management body of knowledge available on the Internet to augment this starter kit readily available to the project core team.

E.1 Project Core Team Members' Roles and Responsibility Matrix (RACI)

Legend:
R = Responsible
A = Accountable
C = Consulted
I = Informed

Responsibility	Project Sponsor	Leadership Team	Project Manager	Advisory Board	Change Control	Config Mgmt	SW Engr/Platform Engr/DBA	Systems Engineer	Project Controller	Systems Champion	Subject Matter Expert	Project Methodologist	QA	Functional Manager	Stakeholder	Communications Manager	Subcontract Manager	Risk Manager	Issues Facilitator	Documentation Facilitator
1. A working knowledge of the project health process, IT book of business and procedures, and portfolio management	A	R	R					R	R					R	R					
2. Recommend project team direction with regard to changes	C	R	R	C	C	C	C	C	C	C	C	C	C	C	A	C	C	C	C	C
3. Actively promote and communicate project status to executive management	A	R	R	C										R	R					
4. Act as the primary interface between the technical staff, users, and project manager								A												
5. Administer the overall poject change process	R		R		A									R	R					
6. Approve any changes in schedule, budget, or scope	A	R	R	C	C	C	C	C	C	C	C	C	C	R	R	C	C	C	C	C
7. Assist in the creation of the cost accounts and appropriate budgeting process			R					A												
8. Collaborate with extended team on schedules, issue/risk identification and resolution, change management, configuration management, status processes, and communications			A					R						R	R					
9. Collaborate with project manager to determine/define project priorities	A																			
10. Collaborate closely with the IT, functional managers, system engineer, and project manager to ensure alignment of objectives and commitments	A	R	R	C					R					R	R			R		
11. Collaborate with extended team on similar or concurrent product life cycle management projects	C	C	A	C	C	C	C	R	C	C	C	C	C	R	R	C	C	C	C	C
12. Communicate key business priorities and project milestones of similar product life cycle management projects to the project team	C	C	R	C										R	R					
13. Communicate project status to leadership team, Advisory Board, and sponsors	C	C	A	C										R	R					
14. Complete earned value training			R						A											
15. Complete IT project management training	C	C	A	C	C	C	C	C	C	C	C	C	C	C	C	C	C	C	C	C
16. Comprehensive knowledge and experience in product life cycle management, product data management, applicable technical disciplines, procedures, standards, and related desk instructions			R					A						R	R					
17. Comprehensive knowledge of systems engineering and CMMI processes								A												
18. Concur with quality indicators reported in program management review	R		A											R	R					
19. Follow-up with project manager to ensure open items are resolved on schedule and escalate noncompliances in a timely manner													A							
20. Coordinate and oversee defect reviews			R										A	R	R					
21. Coordinate and oversee QA reviews													A							
22. Coordinate project use of project portfolio database			R						A					R	R					
23. Collaborate on flow-down communications			R													A				
24. Collaborate and lead schedule meetings and assist in resolving schedule/resource conflicts			A																	
25. Develop an overall project schedule to evaluate interdependencies and impacts between varous levels of resources areas conflicts			A					R						R	R					
26. Establish contacts within, serve as technical liaison between all support departments, and facilitate communications and information sharing			R					A						R	R					
27. Evaluate and assess impacts derived from new schedules and request for change			A											R	R					
28. Extensive project management experience			A																	
29. Help ensure appropriate resources are committed to successfully execute the project	A	R	R	C	C	C	C	C	C	C	C	C	C	R	R	C	C	C	C	C
30. Help ensure committed resources successfully execute on schedule	A	R	R	C	C	C	C	C	C	C	C	C	C	R	R	C	C	C	C	C
31. Help ensure that changes are input into appropriate database and communicated to all interested parties			A					R						R	R					

Legend:
R = Responsible
A = Accountable
C = Consulted
I = Informed

Responsibility	Project Sponsor	Leadership Team	Project Manager	Advisory Board	Change Control	Config Mgmt	SW Engr/Platform Engr/DBA	Systems Engineer	Project Controller	Systems Champion	Subject Matter Expert	Project Methodologist	QA	Functional Manager	Stakeholder	Communications Manager	Subcontract Manager	Risk Manager	Issues Facilitator	Documentation Facilitator
32. Help ensure executive alignment in support of project strategies	A	R	R	C										R	R					
33. Help ensure issues are captured and communicated and resolved			R																A	
34. Help ensure IT performance is consistent with customer requirements and service-level agreement	A		R																	
35. Help ensure IT performance exceeds customer expectations regarding cost, quality, schedule, and risk	A	R	R	R					R				R		R			R		
36. Help ensure project management effectively executes project charter	R	A	C	C	C	C	C	C	C	C	C	C	C	C	C	C	C	C	C	C
37. Help ensure QA reviews are defined, scheduled, and performed on schedule			R										A							
38. Help ensure risks are logged and statused regularly			R					A										R		
39. Help ensure that stated business requirements are met and business results achieved	A	R	R	C	C	C	C	C	C	C	C	C	C	R	R	C	C	C	C	C
40. Help ensure team is using guidelines and templates developed for the project			R					A												
41. Help ensure all processes and documentation meet the architecture and CMMI standards where applicable								A					R							
42. Help ensure that all verification and validation testing can be traced to customer requirements, the decomposed functional requirements, and allocated component requirements; and that the test results are properly documented								A												
43. Help ensure that CMMI and architecture requirements management process are properly adhered			R					A					R							
44. Help ensure that risk reduction procedures within the design process are followed and properly documented			R					A					R					R		
45. Lead process tailoring and template development process								A												
46. Help ensure that all changes are documented, analyzed for impact, and tracked to completion			A					R						R	R					
47. Manage, coordinate, and communicate changes to the documentation																				A
48. Manage, coordinate, and communicate changes to the configuration baselines and participate on the change control board					I	R		A												
49. Initiate/participate in quick response decision sessions as required	I	I	A	I	C	C	C	C	C	C	C	C	C	C	C	C	C	C	C	C
50. Initiate/participate in quick response adjudication sessions as required	A	R		C										R	R					
51. Meet with project team to provide direction as required			A																	
52. Meet as required	C	C	A	C	C	C	C	C	C	C	C	C	C	R	R	C	C	C	C	C
53. Meet weekly	C	C	A	C	C	C	C	C	C	C	C	C	C	C	C	C	C	C	C	C
54. Collaborates and mitigates variances to scope, scheduling, resources, risks, and issues management			A											R	R			R		
55. Meet with project team regularly to discuss project scheduling, scope, issues, and ad hoc status			A											R	R					
56. Meet with technical and user team to discuss project scheduling, scope, issues, and ad hoc status			A																	
57. Meet with user team regularly to manage expectations and update as needed			A																	
58. Mentor other staff in various systems engineering and technical disciplines		R						A												
59. Must be available for requirements gathering and definition and end product testing								A		R		R		R	R					I
60. Must complete "all up" process training and QA reviewer training with a minimum acceptable quality score													A							
61. Must have a good working understanding of MS project and the earned value template			A					R	R											
62. Must have agreement with project management for both assignment and changes to resource commitments	R													A	R					
63. Must understand architecture policies and procedures as they relate to project management and scheduling								A						R	R					
64. Participate in development of and concur with the QA section of the project plan, participate in peer reviews, and provide process guidance and coaching to project management and technical team		R											A		R					
65. Participate in project risk management activities		R	R					R						R	R			A		

Legend:
R = Responsible
A = Accountable
C = Consulted
I = Informed

Responsibility	Project Sponsor	Leadership Team	Project Manager	Advisory Board	Change Control	Config Mgmt	SW Engr/Platform Engr/DBA	Systems Engineer	Project Controller	Systems Champion	Subject Matter Expert	Project Methodologist	QA	Functional Manager	Stakeholder	Communications Manager	Subcontract Manager	Risk Manager	Issues Facilitator	Documentation Facilitator
66. Participate in project status preparation			A					R						R	R					
67. Participate in any rebaseling or replanning activities	R		A					R	R					R	R			C		
68. Perform all user assigned tasks throughout the project including the planning and execution of cutover to new/modified systems											R	R		A	R					
69. Periodically manage, coordinate, and communicate high and medium priority issue items and corrective active plans throughout the project life cycle to the project manager										R									A	
70. Periodically manage, coordinate, and communicate high and medium priority risk items and mitigation plans throughout the project life cycle to the project manager										R								A		
71. Plan, establish, and maintain the integrity of the issue management activities using issue identification, issue assessment, issue tracking, and issue closure processes			R											R					A	
72. Plan, establish, and maintain the integrity of the risk management activities using risk identification, risk analysis, risk prioritization, risk mitigation planning, risk tracking, and risk monitoring and closure processes			R					R						R	R			A		
73. Plan, establish, and maintain the integrity of the work products using configuration identification, configuration control, configuration status accounting, and configuration audits			R		R	A					R	R		R	R					
74. Plan, establish, and maintain the user and systems documentation								R						R	R					A
75. Primary IT interface into the internal customer business element	A																			
76. Provide a quality gate and single focal point for external communication			R										R			A				
77. Provide an escalation pathway for project management	R	A	R	R										R	R					
78. Provide analysis and/or insight into project feasibility and technical risk identification								A										R		
79. Provide communication of intitial customer requirements to IT and sponsor integrated IT teams in support of delivering results to defined requirements	C	C		C										R	A					
80. Provide consultation, support, and guidance in the use of effective processes, methods, and tools								A		R	R			R						
81. Provide customer satisfaction results to project team monthly	C	C	A	C										R	R					
82. Provide executive guidance and direction to project management, as needed	R	A	C	R	C	C	C	C	C	C	C	C	C	C	C	C	C	C	C	C
83. Provide funding for solution	R	R	R	R					A					R	R					
84. Provide guidance for project prioritization	A	R	R	R										R	R					
85. Provide input into project planning activities		R	A	R		R		R						R	R			R		
86. Provide leadership and direction for nonrecurring and recurring investments	A								R											
87. Provide adjudication and mediation for any deadlocked issues between IT and functional teams	A	R	R	R										R	R					
88. Provide overall guidance to project management	A	R		R	C	C	C	C	C	C	C	C	C	R	R	C	C	R	C	C
89. Provide overall technical guidance to team members and lead the efforts for prototyping and developing conceptual models			R					A						R	R					
90. Provide support for operational readiness, to include such items as new/updated customer policies and procedures	A	R	R											R	R					
91. Proxy for decisions for area of responsibility	A	R												R	R					
92. Responsible for acceptance of QA processes	C	C	A	C									R	R	R					
93. Responsible for acceptance of risk mitigation plans	R	R	R	R				A						R	R					
94. Responsible for all conversion, interface, and reports of the system including applicable software support tools, architecture standards, hardware, and infrastructure			R				A	R						R						
95. Responsible for all technical elements of the system including baseline software, configuration and customization, reports, interfaces, and applicable hardware and infrastructure			R				R	A			R			R						
96. Responsible for collecting hours across the project and ensuring that correct resources are charging to the correct charge numbers			R				R		A					R						

Legend:
R = Responsible
A = Accountable
C = Consulted
I = Informed

Responsibility	Project Sponsor	Leadership Team	Project Manager	Advisory Board	Change Control	Config Mgmt	SW Engr/Platform Engr/DBA	Systems Engineer	Project Controller	Systems Champion	Subject Matter Expert	Project Methodologist	QA	Functional Manager	Stakeholder	Communications Manager	Subcontract Manager	Risk Manager	Issues Facilitator	Documentation Facilitator
97. Responsible for communicating technical infrastructure requirements to support the system including computer hardware, operating systems, peripherals, networks, and database							R	A						R						
98. Responsible for contractor PO management									R								A			
99. Responsible for defining the format, content, frequency, and participants in project communications														R	R	A				
100. Responsible for detailed schedule development/progress			A											R	R					
101. Responsible for development of technical SoW			R				R	A						R			R			
102. Responsible for escalation resolution	A	R	R	R										R	R					
103. Responsible for filling out the user satisfaction survey monthly	R		A												R					
104. Responsible to help ensure that issues and risks are properly communicated, escalated, mitigated, and managed			A						R					R	R			R		
105. Responsible for identification, communication, and management of technical issues/risks			R				R	A						R	R			R		
106. Responsible for identifying, mitigating, and managing project risks		R	R					A						R	R					
107. Responsible to ensure that development processes are in compliance with architectural standards							R	A												
108. Responsible to ensure earned value reporting is done on time			R						A					R	R					
109. Responsible to ensure the information is provided to the technical team in a timely manner			A					R												
110. Responsible to ensure the overall quality of software product							R	A												
111. Responsible to ensure overall requirements are properly met	R		R				R	A		R	R			R	R					
112. Responsible to ensure the information provided to the technical and user teams is valid and current			A				R	R						R						
113. Responsible to ensure workplans meet the project schedule requirements	R		R				R		R	R				A	R					
114. Responsible for management of scope, schedule, budget, and quality of project deliverables	R	R	A	R				R				R		R	R					
115. Responsible for negotiating price on enterprise software and support needs																	A			
116. Responsible for overall IT participation	A		R	R										R	R					
117. Responsible for overall project execution	C	R	A	R					I			I		R						
118. Responsible for overall project performance and success	R	R	A	R	C	C	C	C	C	C	C	C	C	R	R	C	C	C	C	C
119. Responsible for providing guidance on approved products acceptable software solutions								A												
120. Responsible for providing test scenarios and criteria										A	R			R						
121. Responsible for publishing results of the program management review results to project health monthly			A						R					R	R					
122. Responsible for publishing the project health worksheet			A						R					R	R					
123. Responsible for recommendations on overall project scope, budget, cost, and schedule	A	R	R	R					R					R	R					
124. Responsible for executing the user acceptance testing process			R				R	R		R	A	R		C	C					
125. Responsible for status reporting to executive leadership and functional management	A		R	R																
126. Responsible for status reporting to project manager weekly									R					A	R			R	R	R

Legend:
R = Responsible
A = Accountable
C = Consulted
I = Informed

Responsibility	Project Sponsor	Leadership Team	Project Manager	Advisory Board	Change Control	Config Mgmt	SW Engr/Platform Engr/DBA	Systems Engineer	Project Controller	Systems Champion	Subject Matter Expert	Project Methodologist	QA	Functional Manager	Stakeholder	Communications Manager	Subcontract Manager	Risk Manager	Issues Facilitator	Documentation Facilitator
127. Responsible for the management of committed resources	A	R		R					R					R	R					
128. Responsible for the capabilities, skills inventory, performance of those resources including training, commitment of availability, timeliness of completion of deliverables, and quality of resource product delivery	A	R					R							R	R					
129. Responsible for the day-to-day leadership of the overall project			A												R	R				
130. Responsible for the day-to-day leadership of the project technical resources			R				R			C	C			A	R					
131. Responsible for the functional financial elements of the project									A											
132. Responsible for the IT financial elements of the project									A											
133. Responsible for the monthly budget variance reports			R						A					R	R					
134. Responsible for the monthly earned value hours/dollars/graph on functional and IT results			R						A					R	R					
135. Responsible for overall quality of the "AS IS" and TO BE project documentation			R					A		R	R									
136. Responsible for the customer requirements and traceability document							R	A		R	R									R
137. Responsible for the technical and process requirements document			R					A						R	R					R
138. Responsible for reporting chargeable hours	R	R	R	R	R	R	R	R	R	R	R	R	R	R	R	R	R	R	R	R
139. Responsible to review project schedule, issues, and status	C	C	A	C	C	C	C	C	C	C	C	C	C	R	R	C	C	C	C	C
140. Review and approve all formal project documentation, project plan, and schedule			A							R	R			R	R					R
141. Review and evaluate performance metrics	A		R						R					R	R					
142. Review and approve requirements documentation								A		R	R			R	R					R
143. Review weekly QA process metrics with project management			R										A							
144. Review/approve expenditures for capital and expense computer equipment	A								C					R	R					
145. Review/approve expenditures for direct, indirect, labor, and nonlabor expenses	A	R	R	I					R					R	R					
146. Review/approve other related schedules/workplans to ensure stated project milestones meet project requirements			A				R	R		R	R			R	R					
147. Serve as a focal point for all project change requests and change-related metrics			A		R									R	R					
148. Sets direction and priorities for system champions and super users			R											A	R					
149. Supports project manager in allocating budget to project entities and contract POs			R						A											
150. Supports project manager in reclassifying budget to proper project labor charge numbers			R						A											
151. Supports project manager in generating budget versus actual expenditures as needed			R						A					R						
152. Track project actions and trigger reminders as needed			A											R	R					
153. Use and report earned value to support the project (cost and schedule) requirements, contribute to lessons learned activities, and archive appropriate documents			R						A											R
154. Collaborate with customer to manage changing project priorities	C	A												R	R					
155. Work with software engineering, platform engineering, network engineering, and DBAs to provide overall technical leadership to ensure integrity and validity within the architecture to conform to customer requirements								A	R					R	R					

E.2 Sample High-Level Project Schedule

There is a need to track enterprise resource planning (ERP) deliverable progress. This is an example of a high-level (or milestone) software segment of the ERP implementation briefing chart. It may be expanded to include nonsoftware-related deliverables as well.

In addition to this high-level briefing tool, the project core team will likely use MicroSoft (MS) Project to detail track all the tasks essential for the ERP implementation. The detailed task includes start/complete dates, responsible resource, dependencies, and the percentage complete.

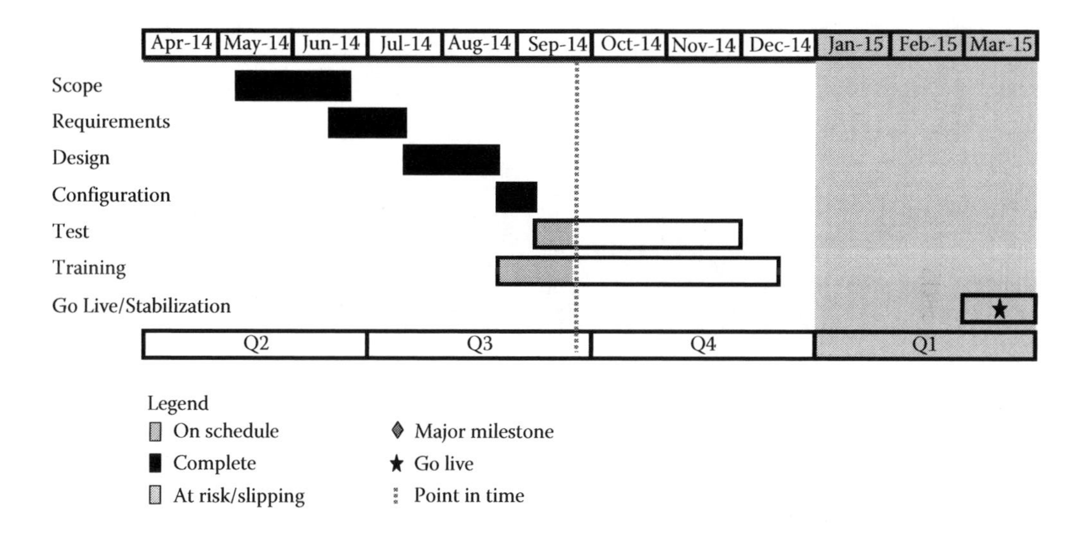

E.3 Sample Requirements Tracking

E.3.1 Completed by Phase

As discussed in Chapter 2 and reinforced in Chapter 7, documenting the ERP requirements is a cornerstone critical success factor of an ERP project. Therefore, requirements completion tracking is essential to help ensure that the project proceeds according to schedule. This briefing chart example shows software requirements (%) completed by software project life cycle phase (see legend) and would be used by the software engineering team to track progress by resource team deployed on the project.

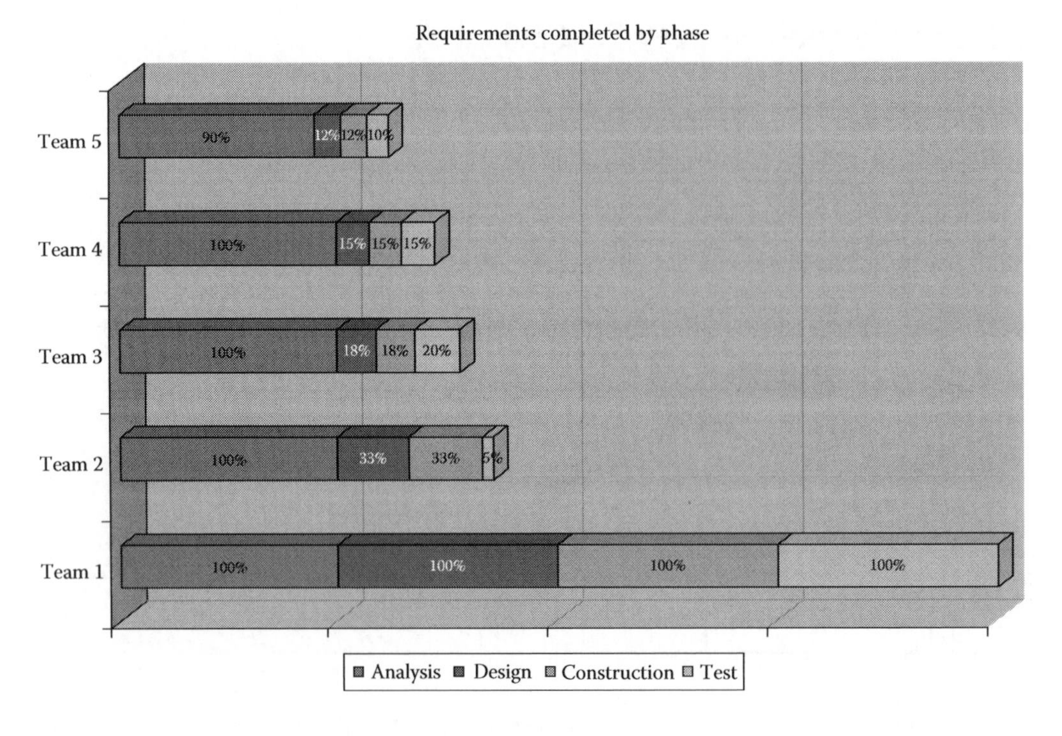

E.3.2 Completed by Team

This briefing chart example shows software customization requirements (%) completed by software project life cycle phase (see legend) and would be used by the software engineering team to track progress by resource group deployed on the project.

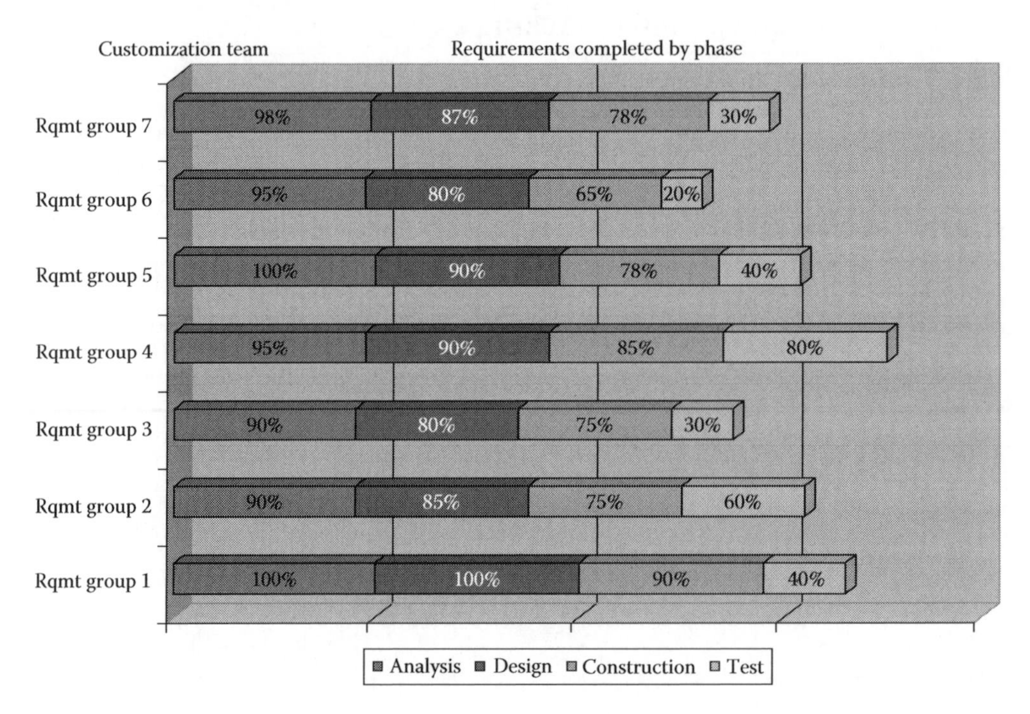

E.4 Sample Integrated Data Environment Report Diagram

The integrated data environment is a set of nested core capabilities and a subset of an ERP system, which may be used as a functionality starter kit for a new business unit. It is also transferable and may be adapted for new product development core functionality.

This may be a helpful tool to assist the first-time ERP user project team trigger what a "core" (or minimum) capability might include in their ERP effort. Attempting to "bite off more than you can chew" is a potential derailment event. This example shows what a core capability might include, how each core application relates to other core applications, and a phased approach to deployment.

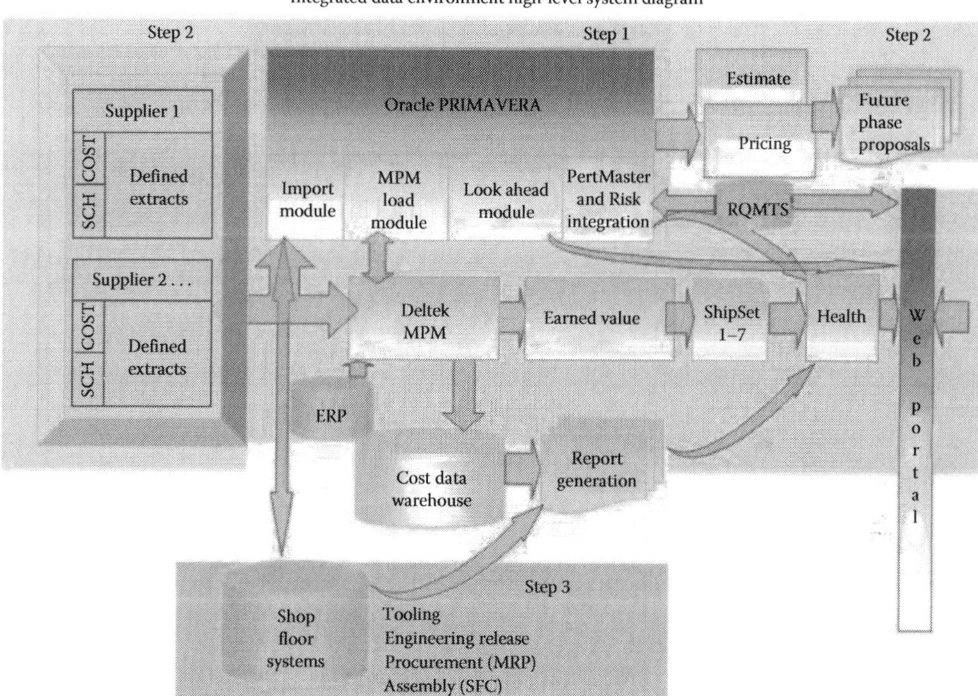

E.5 Sample Data Mapping Form

This will be a helpful tool if you are converting data from a legacy system to an ERP system for the first time. Disparate legacy data typically requires "conditioning" to help ensure that field sizes (From/To) are compatible (don't try to put a 20-character legacy field into a target 15-numeric field, it won't work). This tools helps map conversion data so that the final conversion pass has good data entering the new ERP application successfully. First-time users typically *dry run* conversion passes into a copy of the new ERP production environment using various iterations until the data is pristine in the target ERP.

Data mapping (module name) detail

1. Description:

2. Requirement(s) traceability:

3. Legacy data selection (record level)

4. Table layout

	Target table				Legacy source table					
#	Field	Description	Type	Len	Table	Field	Description	Type	Len	Mapping rules
1										
2										
3										
4										
5										
6										
7										
8										
9										
10										
11										
12										
13										
14										
15										
16										
17										
18										
19										
20										
21										
22										
23										
24										
25										

E.6 Sample Milestone Progress Report

This milestone progress briefing chart example shows milestones ahead or behind schedule across a tracking time horizon. The project manager would include a chart like this for an executive audience briefing to display milestone progress.

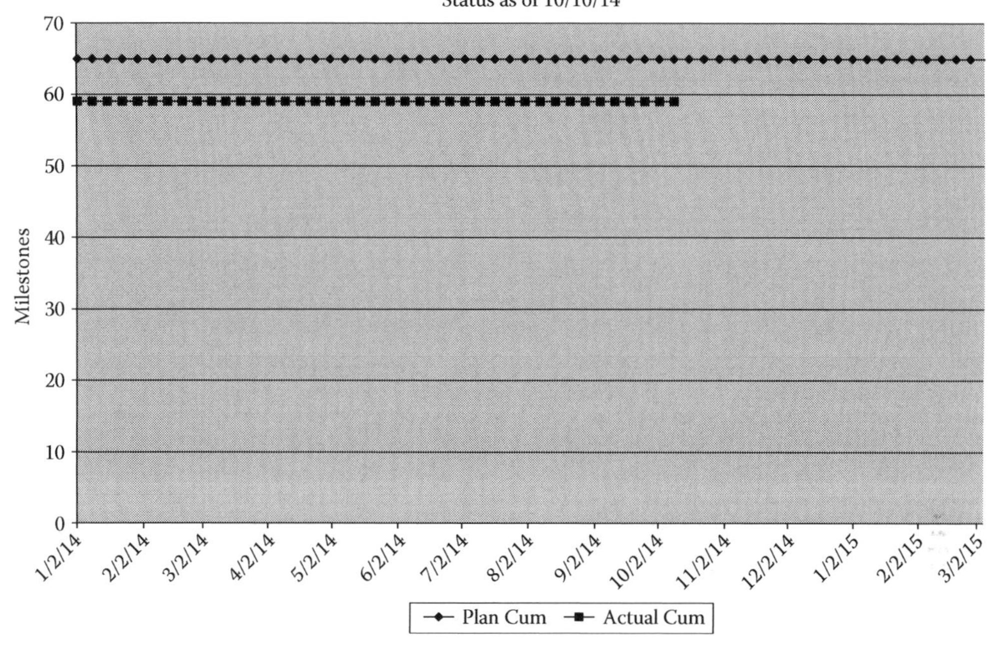
Milestones ahead/behind count
Status as of 10/10/14
Milestones
70
60
50
40
30
20
10
0
1/2/14
2/2/14
3/2/14
4/2/14
5/2/14
6/2/14
7/2/14
8/2/14
9/2/14
10/2/14
11/2/14
12/2/14
1/2/15
2/2/15
3/2/15
Plan Cum
Actual Cum

Appendix F (Chapter 9)

F.1 Overarching Goal of Project Success*

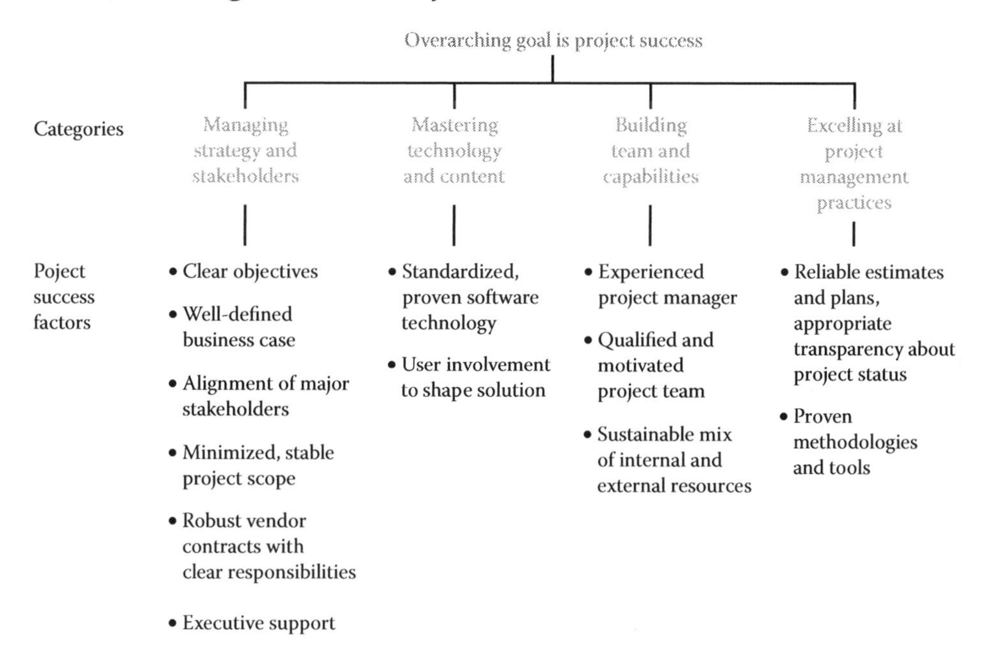

* Michael Bloch, Sven Blumberg, and Jürgen Laartz, *Delivering Large-Scale IT Projects on Time, on Budget, and on Value*, McKinsey & Company, New York, October 2012.

Index

Note: Locators followed by "*f*" and "*t*" denote figures and tables in the text

F

G

H

I

J

K

L

M

N

O

P

Q

R